Communications and Control Engineering

Springer
*London
Berlin
Heidelberg
New York
Barcelona
Budapest
Hong Kong
Milan
Paris
Santa Clara
Singapore
Tokyo*

Published titles include:

Sampled-Data Control Systems
J. Ackermann

Interactive System Identification
T. Bohlin

The Riccatti Equation
S. Bittanti, A.J. Laub and J.C. Willems (Eds)

Analysis and Design of Stream Ciphers
R.A. Rueppel

Sliding Modes in Control Optimization
V.I. Utkin

Fundamentals of Robotics
M. Vukobratović

Parametrizations in Control, Estimation and Filtering Problems: Accuracy Aspects
M. Gevers and G. Li

Parallel Algorithms for Optimal Control of Large Scale Linear Systems
Zoran Gajić and Xuemin Shen

Loop Transfer Recovery: Analysis and Design
A. Saberi, B.M. Chen and P. Sannuti

Markov Chains and Stochastic Stability
S.P. Meyn and R.L. Tweedie

Robust Control: Systems with Uncertain Physical Parameters
J. Ackermann in co-operation with A. Bartlett, D. Kaesbauer, W. Sienel and R. Steinhauser

Optimization and Dynamical Systems
U. Helmke and J.B. Moore

Optimal Sampled-Data Control Systems
Tongwen Chen and Bruce Francis

Nonlinear Control Systems (3rd edition)
Alberto Isidori

The ZODIAC

Theory of Robot Control

Carlos Canudas de Wit, Bruno Siciliano and Georges Bastin (Eds)

With 31 Figures

Springer

Professor Carlos Canudas de Wit, PhD
Laboratoire d'Automatique de Grenoble
École Nationale Supérieure d'Ingénieurs Electriciens de Grenoble
Rue de la Houille Blanche, Domaine Universitaire
38402 Saint-Martin-d'Hères, France

Professor Bruno Siciliano, PhD
Dipartimento di Informatica e Sistemistica
Università degli Studi di Napoli Federico II
Via Claudio 21, 80125 Napoli, Italy

Professor Georges Bastin, PhD
Centre d'Ingénierie des Systèmes, d'Automatique et de Mécanique Appliquée
Université Catholique de Louvain
4 Avenue G. Lemaître, 1348 Louvain-la-Neuve, Belgium

Series Editors
B.W. Dickinson • A. Fettweis • J.L. Massey • J.W. Modestino
E.D. Sontag • M. Thoma

ISBN 3-540-76054-7 Springer-Verlag Berlin Heidelberg New York

British Library Cataloguing in Publication Data
Theory of robot control. - (Communications and control engineering series)
  1.Robots - Control systems
  I.Canudas de Wit, Carlos II.Siciliano, Bruno III.Bastin, G. (Georges)
  629.8'92
ISBN 3540760547

Library of Congress Cataloging-in-Publication Data
A catalog record for this book is available from the Library of Congress

Apart from any fair dealing for the purposes of research or private study, or criticism or review, as permitted under the Copyright, Designs and Patents Act 1988, this publication may only be reproduced, stored or transmitted, in any form or by any means, with the prior permission in writing of the publishers, or in the case of reprographic reproduction in accordance with the terms of licences issued by the Copyright Licensing Agency. Enquiries concerning reproduction outside those terms should be sent to the publishers.

© Springer-Verlag London Limited 1996
Printed in Great Britain
2nd printing with corrections, 1997

The use of registered names, trademarks, etc. in this publication does not imply, even in the absence of a specific statement, that such names are exempt from the relevant laws and regulations and therefore free for general use.

The publisher makes no representation, express or implied, with regard to the accuracy of the information contained in this book and cannot accept any legal responsibility or liability for any errors or omissions that may be made.

Typesetting: Camera ready by editors
Printed and bound at the Athenæum Press Ltd, Gateshead
69/3830-54321 Printed on acid-free paper

The ZODIAC are:

Georges Bastin
Université Catholique de Louvain, Belgium

Bernard Brogliato
Laboratoire d'Automatique de Grenoble, France

Guy Campion
Université Catholique de Louvain, Belgium

Carlos Canudas de Wit
Laboratoire d'Automatique de Grenoble, France

Brigitte d'Andréa-Novel
École Nationale Supérieure des Mines de Paris, France

Alessandro De Luca
Università degli Studi di Roma "La Sapienza", Italy

Wisama Khalil
École Centrale de Nantes, France

Rogelio Lozano
Université de Technologie de Compiègne, France

Roméo Ortega
Université de Technologie de Compiègne, France

Claude Samson
INRIA Centre de Sophia-Antipolis, France

Bruno Siciliano
Università degli Studi di Napoli Federico II, Italy

Patrizio Tomei
Università degli Studi di Roma "Tor Vergata", Italy

Contents

Preface	xiii
Synopsis	xv

I Rigid manipulators   1

1 Modelling and identification   3
- 1.1 Kinematic modelling 4
  - 1.1.1 Direct kinematics 5
  - 1.1.2 Inverse kinematics 12
  - 1.1.3 Differential kinematics 16
- 1.2 Dynamic modelling 21
  - 1.2.1 Lagrange formulation 21
  - 1.2.2 Newton-Euler formulation 26
  - 1.2.3 Model computation 29
- 1.3 Identification of kinematic parameters 38
  - 1.3.1 Model for identification 38
  - 1.3.2 Kinematic calibration 41
  - 1.3.3 Parameter identifiability 42
- 1.4 Identification of dynamic parameters 44
  - 1.4.1 Use of dynamic model 45
  - 1.4.2 Use of energy model 47
- 1.5 Further reading 48
- References 49

2 Joint space control   59
- 2.1 Dynamic model properties 61
- 2.2 Regulation 63
  - 2.2.1 PD control 63

		2.2.2	PID control	68

        2.2.2   PID control   68
        2.2.3   PD control with gravity compensation   71
  2.3   Tracking control 72
        2.3.1   Inverse dynamics control   73
        2.3.2   Lyapunov-based control   74
        2.3.3   Passivity-based control   78
  2.4   Robust control   80
        2.4.1   Constant bounded disturbance: integral action . . .   81
        2.4.2   Model parameter uncertainty: robust control   84
  2.5   Adaptive control   95
        2.5.1   Adaptive gravity compensation   95
        2.5.2   Adaptive inverse dynamics control   98
        2.5.3   Adaptive passivity-based control   101
  2.6   Further reading   103
  References 108

3 Task space control   115
  3.1   Kinematic control   116
        3.1.1   Differential kinematics inversion   116
        3.1.2   Inverse kinematics algorithms   124
        3.1.3   Extension to acceleration resolution   129
  3.2   Direct task space control   131
        3.2.1   Regulation   131
        3.2.2   Tracking control   133
  3.3   Further reading   134
  References 135

4 Motion and force control   141
  4.1   Impedance control   142
        4.1.1   Task space dynamic model   143
        4.1.2   Inverse dynamics control   145
        4.1.3   PD control   147
  4.2   Parallel control 150
        4.2.1   Inverse dynamics control   151
        4.2.2   PID control   153
  4.3   Hybrid force/motion control   156
        4.3.1   Constrained dynamics   157
        4.3.2   Inverse dynamics control   161
        4.3.3   Hybrid task specification and control   166
  4.4   Further reading   168
  References 170

II Flexible manipulators　　　　　　　　　　　　　　　　177

5 Elastic joints　　　　　　　　　　　　　　　　　　　　　179
　5.1　Modelling 181
　　　5.1.1　Dynamic model properties 184
　　　5.1.2　Reduced models 186
　　　5.1.3　Singularly perturbed model 187
　5.2　Regulation 189
　　　5.2.1　Single link 189
　　　5.2.2　PD control using only motor variables ... 191
　5.3　Tracking control 195
　　　5.3.1　Static state feedback 196
　　　5.3.2　Two-time scale control 199
　　　5.3.3　Dynamic state feedback 202
　　　5.3.4　Nonlinear regulation 209
　5.4　Further reading 211
　References 213

6 Flexible links　　　　　　　　　　　　　　　　　　　　　219
　6.1　Modelling of a single-link arm 221
　　　6.1.1　Euler-Bernoulli beam equations 221
　　　6.1.2　Constrained and unconstrained modal analysis ... 224
　　　6.1.3　Finite-dimensional models 228
　6.2　Modelling of multilink manipulators 231
　　　6.2.1　Direct kinematics 231
　　　6.2.2　Lagrangian dynamics 233
　　　6.2.3　Dynamic model properties 235
　6.3　Regulation 237
　　　6.3.1　Joint PD control 237
　　　6.3.2　Vibration damping control 240
　6.4　Joint tracking control 242
　　　6.4.1　Inversion control 242
　　　6.4.2　Two-time scale control 246
　6.5　End-effector tracking control 248
　　　6.5.1　Frequency domain inversion 250
　　　6.5.2　Nonlinear regulation 252
　6.6　Further reading 254
　References 256

III Mobile robots — 263

7 Modelling and structural properties — 265
- 7.1 Robot description — 266
  - 7.1.1 Conventional wheels — 267
  - 7.1.2 Swedish wheel — 269
- 7.2 Restrictions on robot mobility — 269
- 7.3 Three-wheel robots — 276
  - 7.3.1 Type (3,0) robot with Swedish wheels — 277
  - 7.3.2 Type (3,0) robot with castor wheels — 277
  - 7.3.3 Type (2,0) robot — 278
  - 7.3.4 Type (2,1) robot — 279
  - 7.3.5 Type (1,1) robot — 280
  - 7.3.6 Type (1,2) robot — 281
- 7.4 Posture kinematic model — 282
  - 7.4.1 Generic models of wheeled robots — 283
  - 7.4.2 Mobility, steerability and manoeuvrability — 286
  - 7.4.3 Irreducibility — 287
  - 7.4.4 Controllability and stabilizability — 288
- 7.5 Configuration kinematic model — 290
- 7.6 Configuration dynamic model — 294
  - 7.6.1 Model derivation — 294
  - 7.6.2 Actuator configuration — 296
- 7.7 Posture dynamic model — 300
- 7.8 Further reading — 302
- References — 303

8 Feedback linearization — 307
- 8.1 Feedback control problems — 308
  - 8.1.1 Posture tracking — 308
  - 8.1.2 Point tracking — 308
  - 8.1.3 Velocity and torque control — 309
- 8.2 Static state feedback — 310
  - 8.2.1 Omnidirectional robots — 310
  - 8.2.2 Restricted mobility robots — 312
- 8.3 Dynamic state feedback — 318
  - 8.3.1 Dynamic extension algorithm — 319
  - 8.3.2 Differential flatness — 321
  - 8.3.3 Avoiding singularities — 322
  - 8.3.4 Solving the posture tracking problem — 325
  - 8.3.5 Avoiding singularities for Type (2,0) robots — 326

CONTENTS   xi

	8.4	Further reading	328
	References	328
9	**Nonlinear feedback control**	**331**
	9.1	Unicycle robot	331
		9.1.1 Model transformations	332
		9.1.2 Linear approximation	333
		9.1.3 Smooth state feedback stabilization	334
	9.2	Posture tracking	335
		9.2.1 Linear feedback control	336
		9.2.2 Nonlinear feedback control	337
	9.3	Path following	339
		9.3.1 Linear feedback control	341
		9.3.2 Nonlinear feedback control	341
	9.4	Posture stabilization	343
		9.4.1 Smooth time-varying control	344
		9.4.2 Piecewise continuous control	349
		9.4.3 Time-varying piecewise continuous control	354
	9.5	Further reading	356
	References	358
A	**Control background**	**363**
	A.1	Lyapunov theory	363
		A.1.1 Autonomous systems	363
		A.1.2 Nonautonomous systems	367
		A.1.3 Practical stability	371
	A.2	Singular perturbation theory	371
	A.3	Differential geometry theory	374
		A.3.1 Normal form	375
		A.3.2 Feedback linearization	376
		A.3.3 Stabilization of feedback linearizable systems	377
	A.4	Input–output theory	378
		A.4.1 Function spaces and operators	379
		A.4.2 Passivity	380
		A.4.3 Robot manipulators as passive systems	383
		A.4.4 Kalman-Yakubovich-Popov lemma	384
	A.5	Further reading	386
	References	386
Index		**389**

Preface

The advent of new high-speed microprocessor technology together with the need for high-performance robots created substantial and realistic place for control theory in the field of robotics. Since the beginning of the 80's, robotics and control theory have greatly benefited from a mutual fertilization. On one hand, robot models (inherently highly nonlinear) have been used as good case studies for exemplifying general concepts of analysis and design of advanced control theory; on the other hand, robot manipulator performance has been improved by using new control algorithms. Furthermore, many interesting robotics problems, e.g., in mobile robots, have brought new control theory research lines and given rise to the development of new controllers (time-varying and nonlinear). Robots in control are more than a simple case study. They represent a natural source of inspiration and a great pedagogical tool for research and teaching in control theory.

Several advanced control algorithms have been developed for different types of robots (rigid, flexible and mobile), based either on existing control techniques, e.g., feedback linearization and adaptive control, or on new control techniques that have been developed on purpose. Most of those results, although widely spread, are nowadays rather dispersed in different journals and conference proceedings.

The purpose of this book is to collect some of the most fundamental and current results on theory of robot control in a unified framework, by editing, improving and completing previous works in the area. The text is addressed to graduate students as well as to researchers in the field.

This book has originated from the lecture notes prepared for the European Summer School on Theory of Robot Control, held from September 7 to 11, 1992, at the Laboratory of Automatic Control of Grenoble (LAG) of ENSIEG, with the financial support of CNRS and MEN. As the scientific coordinator of the school, Carlos Canudas de Wit would like to thank the other eleven contributors for their motivation during the project, and for making a considerable effort in putting the material together. Thanks to all the attendees (more than one hundred) coming from different countries who made the school a true success. A note of thanks goes also to J.M. Dion, Director of LAG, for providing us with a site for the school and for supporting the realization of the book.

Although the book is the outcome of a joint work, individual contributions can be attributed as follows: Chapter 1 (W. Khalil and B. Siciliano), Chapter 2 (B. Brogliato and C. Canudas de Wit), Chapter 3 (B. Siciliano), Chapter 4 (A. De Luca and B. Siciliano), Chapter 5 (A. De Luca and P. Tomei), Chapter 6 (A. De Luca and B. Siciliano), Chapter 7 (G. Bastin, G. Campion and B. d'Andréa-Novel), Chapter 8 (G. Bastin, G. Campion and B. d'Andréa-Novel), Chapter 9 (C. Canudas de Wit and C. Samson), and Appendix A (R. Lozano and R. Ortega).

The editing of the book was realized by the two of us with the precious help of Georges Bastin. During this time, Bruno Siciliano took on the laborious task of unifying and shaping the presentation of the material, as well as of coordinating the whole group. A note of thanks goes also to most of the other nine authors for providing further revisions of their own and others' contributions during the critical stage of the editorial process.

About the ZODIAC nickname, this was created to gather all the authors into a single entity, so as to reflect the truly cooperative spirit of the project. The ZODIAC describes the well-known twelve-star constellation in Greek mythology. About the number, it agrees with the twelve authors; about the stars ...

Europe, March 1996 *Carlos Canudas de Wit and Bruno Siciliano*

Synopsis

This book is divided in three major parts. Part I deals with modelling and control of rigid robot manipulators. Part II is concerned with modelling and control of flexible robot manipulators. Part III is focused on modelling and control of mobile robots. An Appendix contains a review of some basic definitions and mathematical background in control theory that are preliminary to the study of the book contents.

The sequence of parts, besides being quite rational from a pedagogical viewpoint, reflects the gradual development of theory of robot control in the last fifteen years. In this respect, most of the results on rigid robot manipulators in Part I are now well assessed, with the exception of constrained motion control. On the other hand, in the area of flexible manipulators in Part II not all the problems have been solved yet and there is certainly need for new theoretical work, especially with regard to the case of flexible links. Finally, the results reported in Part III are meant to present the most updated findings on control of mobile robots, which appears to be a very challenging area for future investigation.

The parts are organized into chapters, and each chapter is complemented with a further reading section which provides the reader with a detailed guide through the numerous end-of-chapter references.

Part I: Rigid manipulators

Chapter 1 is devoted to modelling and identification of rigid robot manipulators. Kinematic and dynamic models are derived. Methods for identifying both the kinematic and the dynamic parameters are illustrated.

Chapter 2 presents the properties of the dynamic model useful for control design. It describes the most standard control algorithms in the joint space, such as PID regulators and inverse dynamics control, and then it illustrates more advanced control schemes, such as robust control and adaptive control. Particular attention is paid to stability analysis.

Chapter 3 addresses the control problem in the task space. Kinematic control is introduced which is based on an inverse kinematics algorithm to generate the reference inputs to some joint space controller. Alternatively, direct task space control is discussed in order to extend some of the previous joint space control schemes.

Chapter 4 describes control algorithms for tasks requiring end-effector contact with environment, where both motion and force control are of concern. Impedance control, parallel control and hybrid force/motion control schemes are presented, and the influence of the environment is discussed.

Part II: Flexible manipulators

Chapter 5 considers robot manipulators with elastic joints. Dynamic models are derived and their properties are pointed out. Linear control algorithms are analyzed for the regulation problem, and nonlinear control algorithms are introduced for the tracking problem.

Chapter 6 covers robot manipulators with flexible links. Approximate finite-dimensional approximate models are derived from exact infinite-dimensional dynamic models. Both linear and nonlinear control algorithms are presented to solve the regulation and the tracking problem, respectively, with special concern to the choice of the system output.

Part III: Mobile robots

Chapter 7 presents a general formalism for modelling of wheeled mobile robots. It includes both two-degree-of-freedom robots equipped with classical nondeformable wheels and omnidirectional robots equipped with the so-called Swedish wheels in a unified description. Various structural properties of these models are discussed.

Chapter 8 addresses feedback linearization of mobile robots, with special emphasis to the tracking control problem. Both static state feedback and dynamic state feedback are presented. Intrinsic limitations of this approach are underlined when dealing with the stabilization problem.

Chapter 9 presents several nonlinear feedback control strategies for both the regulation and tracking problems. The limitations of smooth time-invariant state feedback control are first emphasized. Then, the need for developing more efficient schemes satisfying the control requirements is shown. Two types of such control laws are analyzed; namely, smooth time-varying state feedback control and nonsmooth state feedback control.

Part I
Rigid manipulators

Chapter 1

Modelling and identification

From a mechanical viewpoint, a robotic system is in general constituted by a locomotion apparatus (legs, wheels) to move in the environment and by a manipulation apparatus to operate on the objects present in the environment. It is then important to distinguish between the class of *mobile robots* and the class of *robot manipulators*.

The mechanical structure of a robot manipulator consists of a sequence of links connected by means of joints. Links and joints are usually made as rigid as possible so as to achieve high precision in robot positioning. The presence of elasticity at the joint transmission or the use of lightweight materials for the links poses a number of interesting issues which lead to separating the study of *flexible robot manipulators* from that of *rigid robot manipulators*.

Completion of a generic task requires the execution of a specific motion prescribed to the manipulator end effector. The motion can be either unconstrained, if there is no physical interaction between the end effector and the environment, or constrained if contact forces arise between the end effector and the environment.

The correct execution of the end-effector motion is entrusted to the control system which shall provide the joint actuators of the manipulator with the commands consistent with the desired motion trajectory. Control of end-effector motion demands an accurate analysis of the characteristics of the mechanical structure, actuators, and sensors. The goal of this analysis is the derivation of mathematical models of robot components. Modelling a robot manipulator is therefore a necessary premise to finding motion control strategies.

In order to characterize the mechanical structure of a robot manipulator, it is opportune to consider the following two subjects.

- *Kinematic modelling* of a manipulator concerns the description of the manipulator motion with respect to a fixed reference frame by ignoring the forces and moments that cause motion of the structure. The formulation of the kinematics relationship allows studying both direct kinematics and inverse kinematics. The former consists of determining a systematic, general method to describe the end-effector motion as a function of the joint motion. The latter consists of transforming the desired motion naturally prescribed to the end effector in the workspace into the corresponding joint motion.

- *Dynamic modelling* of a manipulator concerns the derivation of the equations of motion of the manipulator as a function of the forces and moments acting on it. The availability of the dynamic model is very useful for mechanical design of the structure, choice of actuators, determination of control strategies, and computer simulation of manipulator motion. It is worth emphasizing that kinematics of a manipulator represents the basis of a systematic, general derivation of its dynamics.

Both kinematic and dynamic modelling rely on an accurate knowledge of a number of constant *parameters* characterizing the mechanical structure, such as link lengths and angles, link masses and inertial properties. It is therefore necessary to dispose of *identification* techniques aimed at providing best estimates of kinematic and dynamic parameters.

The material of this chapter is organized as follows. First, *kinematic modelling* of rigid manipulators is considered in terms of *direct kinematics* and *inverse kinematics*, and the Jacobian is introduced as the fundamental tool to describe *differential kinematics*. Then, *dynamic modelling* of rigid manipulators is presented by using both the *Lagrange formulation* and the *Newton-Euler formulation*, and *model computation* aspects are discussed in detail. The problem of *identification* of both *kinematic parameters* and *dynamic parameters* is treated in the last part of the chapter.

1.1 Kinematic modelling

In the study of robot manipulator kinematics, it is customary to distinguish between kinematics and differential kinematics. The former is concerned

1.1. KINEMATIC MODELLING 5

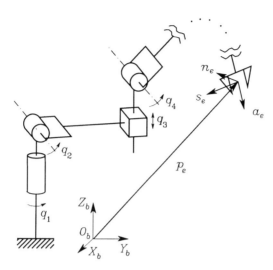

Figure 1.1: Schematic of an open-chain robot manipulator with base frame and end-effector frame.

with the study of the mapping between positions, while the latter is concerned with the study of the mapping between velocities. Let us start by considering direct kinematics of a robot manipulator.

1.1.1 Direct kinematics

A robot manipulator consists of a kinematic chain of $n+1$ *links* connected by means of n *joints*. Joints can essentially be of two types: *revolute* and *prismatic*; complex joints can be decomposed into these simple joints. Revolute joints are usually preferred to prismatic joints in view of their compactness and reliability. One end of the chain is connected to the base link, whereas an *end effector* is connected to the other end. The basic structure of a manipulator is the open kinematic chain which occurs when there is only one sequence of links connecting the two ends of the chain. Alternatively, a manipulator contains a closed kinematic chain when a sequence of links forms a loop. In Fig. 1.1, an open-chain robot manipulator is illustrated with conventional representation of revolute and prismatic joints.

Direct kinematics of a manipulator consists of determining the mapping between the joint variables and the end-effector position and orientation with respect to some reference frame. From classical rigid body mechanics, the direct kinematics equation can be expressed in terms of the (4 × 4) homogeneous transformation matrix

$$^bT_e(q) = \begin{pmatrix} ^bR_e(q) & ^bp_e(q) \\ 0 \quad 0 \quad 0 & 1 \end{pmatrix}, \qquad (1.1)$$

where q is the $(n \times 1)$ vector of joint variables, ^bp_e is the (3×1) vector of end-effector position and $^bR_e = (\,^bn_e \quad ^bs_e \quad ^ba_e\,)$ is the (3×3) rotation matrix of the end-effector frame e with respect to the base frame b (Fig. 1.1); the superscript preceding the quantity denotes the frame in which that is expressed. Notice that the matrix ^bR_e is orthogonal, and its columns ^bn_e, ^bs_e, ^ba_e are the unit vectors of the end-effector frame axes X_e, Y_e, Z_e referred to frame b.

Denavit-Hartenberg notation

An effective procedure for computing the direct kinematics function for a general robot manipulator is based on the so-called modified *Denavit-Hartenberg* notation. According to this notation, a coordinate frame is attached to each link of the chain and the overall transformation matrix from link 0 to link n is derived by composition of transformations between consecutive frames. With reference to Fig. 1.2, let joint i connect link $i-1$ to link i, where the links are assumed to be rigid; frame i is attached to link i and can be defined as follows.

- Choose axis Z_i aligned with the axis of joint i.
- Choose axis X_i along the common normal to axes Z_i and Z_{i+1} with direction from joint i to joint $i+1$.
- Choose axis Y_i so as to complete a right-handed frame.

Once the link frames have been established, the position and orientation of frame i with respect to frame $i-1$ are completely specified by the following *kinematic parameters*:

α_i angle between Z_{i-1} and Z_i about X_{i-1} measured counter-clockwise,

ℓ_i distance between Z_{i-1} and Z_i along X_{i-1},

ϑ_i angle between X_{i-1} and X_i about Z_i measured counter-clockwise,

d_i distance between X_{i-1} and X_i along Z_i.

1.1. KINEMATIC MODELLING

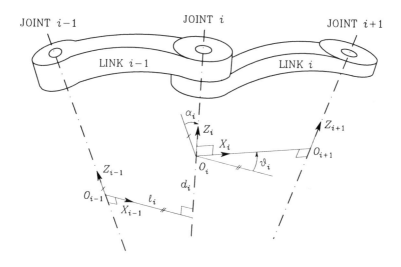

Figure 1.2: Kinematic parameters with modified Denavit-Hartenberg notation.

Let Rot(K, δ) (Trans(K, δ)) denote the homogeneous transformation matrix expressing the rotation (translation) about (along) axis K by an angle (distance) δ. Then, the coordinate transformation of frame i with respect to frame $i-1$ can be expressed in terms of the above four parameters by the matrix

$$
\begin{aligned}
{}^{i-1}T_i &= \mathrm{Rot}(X, \alpha_i)\mathrm{Trans}(X, \ell_i)\mathrm{Rot}(Z, \vartheta_i)\mathrm{Trans}(Z, d_i) \\
&= \begin{pmatrix}
\cos\vartheta_i & -\sin\vartheta_i & 0 & \ell_i \\
\cos\alpha_i \sin\vartheta_i & \cos\alpha_i \cos\vartheta_i & -\sin\alpha_i & -d_i \sin\alpha_i \\
\sin\alpha_i \sin\vartheta_i & \sin\alpha_i \cos\vartheta_i & \cos\alpha_i & d_i \cos\alpha_i \\
0 & 0 & 0 & 1
\end{pmatrix} \\
&= \begin{pmatrix}
{}^{i-1}R_i & {}^{i-1}p_i \\
0 \quad 0 \quad 0 & 1
\end{pmatrix}
\end{aligned}
\qquad (1.2)
$$

where ${}^{i-1}R_i$ is the (3×3) matrix defining the orientation of frame i with respect to frame $i-1$, and ${}^{i-1}p_i$ is the (3×1) vector defining the origin of frame i with respect to frame $i-1$.

Dually, the transformation matrix defining frame $i-1$ with respect to frame i is given by

$$^iT_{i-1} = \text{Trans}(Z, -d_i)\text{Rot}(Z, -\vartheta_i)\text{Trans}(X, -\ell_i)\text{Rot}(X, -\alpha_i)$$

$$= \begin{pmatrix} & & & -\ell_i \cos\vartheta_i \\ & ^{i-1}R_i^T & & \ell_i \sin\vartheta_i \\ & & & -d_i \\ 0 & 0 & 0 & 1 \end{pmatrix}. \tag{1.3}$$

Two of the four parameters (ℓ_i and α_i) are always constant and depend only on the size and shape of link i. Of the remaining two parameters, only one is variable (*degree of freedom*) depending on the type of joint that connects link $i-1$ to link i. If q_i denotes joint i variable, then it is

$$q_i = \bar{\xi}_i \vartheta_i + \xi_i d_i \tag{1.4}$$

where $\bar{\xi}_i = 1 - \xi_i$, i.e.,

- $\xi_i = 0$ if joint i is *revolute* ($q_i = \vartheta_i$),

- $\xi_i = 1$ if joint i is *prismatic* ($q_i = d_i$).

In view of (1.4), the equation

$$\bar{q}_i = \xi_i \vartheta_i + \bar{\xi}_i d_i \tag{1.5}$$

gives the constant parameter at each joint to add to α_i and ℓ_i.

The above procedure does not yield a unique definition of frames 0 and n which can be chosen arbitrarily. Also, in all cases of nonuniqueness in the definition of the frames, it is convenient to make as many link parameters zero as possible, since this will simplify kinematics computation.

Remarks

- A simple choice to define frame 0 is to take it coincident with frame 1 when $q_1 = 0$; this makes $\alpha_1 = 0$ and $\ell_1 = 0$, and $\bar{q}_1 = 0$.

- A similar choice for frame n is to take X_n along X_{n-1} when $q_n = 0$; this makes $\bar{q}_n = 0$.

- If joint i is prismatic, the direction of Z_i is fixed while its location is arbitrary; it is convenient to locate Z_i either at the origin of frame $i-1$ ($\ell_i = 0$) or at the origin of frame $i+1$ ($\ell_{i+1} = 0$).

1.1. KINEMATIC MODELLING

- When the joint axes i and $i+1$ are parallel, it is convenient to locate X_i so as to achieve either $d_i = 0$ or $d_{i+1} = 0$ if either joint is revolute.

By composition of the individual link transformations, the coordinate transformation describing the position and orientation of frame n with respect to frame 0 is given by

$$^0T_n(q) = {}^0T_1(q_1){}^1T_2(q_2)\ldots{}^{n-1}T_n(q_n). \tag{1.6}$$

In order to derive the direct kinematics equation in the form of (1.1), two further *constant* transformations have to be introduced; namely, the transformation from frame b to frame 0 ($^bT_0$) and the transformation from frame n to frame e (^nT_e), i.e.,

$$^bT_e(q) = {}^bT_0\,{}^0T_n(q)\,{}^nT_e. \tag{1.7}$$

Subscripts and superscripts can be omitted when the relevant frames are clear from the context.

Direct kinematics of the anthropomorphic manipulator

As an example of open-chain robot manipulator, consider the *anthropomorphic manipulator*. With reference to the frames illustrated in Fig. 1.3, the Denavit-Hartenberg parameters are specified in Tab. 1.1.

i	α_i	ℓ_i	ϑ_i	d_i
1	0	0	q_1	0
2	$\pi/2$	0	q_2	0
3	0	ℓ_3	q_3	0
4	$-\pi/2$	0	q_4	d_4
5	$\pi/2$	0	q_5	0
6	$-\pi/2$	0	q_6	0

Table 1.1: Denavit-Hartenberg parameters of the anthropomorphic manipulator.

Computing the transformation matrices in (1.2) and composing them as in (1.6) gives

$$^0T_6 = \begin{pmatrix} {}^0n_6 & {}^0s_6 & {}^0a_6 & {}^0p_6 \\ 0 & 0 & 0 & 1 \end{pmatrix} \tag{1.8}$$

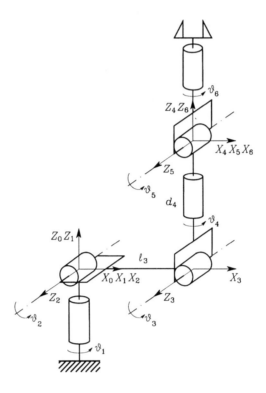

Figure 1.3: Anthropomorphic manipulator with frame assignment.

where

$$^0p_6 = \begin{pmatrix} c_1(c_2\ell_3 - s_{23}d_4) \\ s_1(c_2\ell_3 - s_{23}d_4) \\ s_2\ell_3 + c_{23}d_4 \end{pmatrix} \quad (1.9)$$

for the position, and

$$^0n_6 = \begin{pmatrix} c_1\big(c_{23}(c_4c_5c_6 - s_4s_6) - s_{23}s_5c_6\big) - s_1(s_4c_5c_6 + c_4s_6) \\ s_1\big(c_{23}(c_4c_5c_6 - s_4s_6) - s_{23}s_5c_6\big) + c_1(s_4c_5c_6 + c_4s_6) \\ s_{23}(c_4c_5c_6 - s_4s_6) + c_{23}s_5c_6 \end{pmatrix} \quad (1.10)$$

$$^0s_6 = \begin{pmatrix} c_1\big(-c_{23}(c_4c_5s_6 + s_4c_6) + s_{23}s_5s_6\big) + s_1(s_4c_5s_6 - c_4c_6) \\ s_1\big(-c_{23}(c_4c_5s_6 + s_4c_6) + s_{23}s_5s_6\big) - c_1(s_4c_5s_6 - c_4c_6) \\ -s_{23}(c_4c_5s_6 + s_4c_6) - c_{23}s_5s_6 \end{pmatrix} \quad (1.11)$$

1.1. KINEMATIC MODELLING   11

$$^0a_6 = \begin{pmatrix} -c_1(c_{23}c_4s_5 + s_{23}c_5) + s_1s_4s_5 \\ -s_1(c_{23}c_4s_5 + s_{23}c_5) - c_1s_4s_5 \\ -s_{23}c_4s_5 + c_{23}c_5 \end{pmatrix} \qquad (1.12)$$

for the orientation, where $c_i = \cos\vartheta_i$, $s_i = \sin\vartheta_i$, $c_{23} = \cos(\vartheta_2 + \vartheta_3)$ and $s_{23} = \sin(\vartheta_2 + \vartheta_3)$.

Joint space and task space

If a task has to be assigned to the end effector, it is necessary to specify both end-effector position and orientation. This is easy for the position p_e. Instead, specifying the orientation through the unit vector triple (n_e, s_e, a_e) is difficult, since their nine components must be guaranteed to satisfy the orthonormality constraint imposed by the relation $R_e^T R_e = I$.

The problem of describing end-effector orientation admits a natural solution if a minimal representation is adopted to describe the rotation of the end-effector frame with respect to the base frame, e.g., Euler or RPY angles. This allows introducing an $(m \times 1)$ vector as

$$x = \begin{pmatrix} p_e \\ \phi_e \end{pmatrix}, \qquad (1.13)$$

where p_e describes the end-effector position and ϕ_e its orientation. Notice that, from a rigorous mathematical viewpoint, it is not correct to consider the quantity ϕ_e as a vector since the commutative property does not hold.

This representation of position and orientation allows the description of the end-effector task in terms of a number of inherently independent parameters. The vector x is defined in the space in which the manipulator task is specified; hence, this space is typically called *task space*. The dimension of the task space is at most $m = 6$, since 3 coordinates specify position and 3 angles specify orientation. Nevertheless, depending on the geometry of the task, a reduced number of task space variables may be specified; for instance, for a planar manipulator it is $m = 3$, since two coordinates specify position and one angle specifies orientation.

On the other hand, the *joint space* (configuration space) denotes the space in which the $(n \times 1)$ vector of joint variables q is defined. Accounting for the dependence of position and orientation from the joint variables, we can write the direct kinematics equation in a form other than (1.2), i.e.,

$$x = k(q). \qquad (1.14)$$

It is worth noticing that the explicit dependence of the function $k(q)$ from the joint variables for the orientation components is not available except for simple cases. In fact, on the most general assumption of a six-dimensional task space ($m = 6$), the computation of the three components of the function $\phi_e(q)$ cannot be performed in closed-form but goes through the computation of the elements of the rotation matrix.

The notion of joint space and task space naturally allows introducing the concept of *kinematic redundancy*. This occurs when the dimension of the task space is smaller than the dimension of the joint space ($m < n$). Redundancy is, anyhow, a concept *relative* to the task assigned to the manipulator; a manipulator can be redundant with respect to a task and nonredundant with respect to another, depending on the number of task space variables of interest.

For instance, a three-degree-of-freedom planar manipulator becomes redundant if end-effector orientation is of no concern ($m = 2$, $n = 3$). Yet, the typical example of redundant manipulator is constituted by the human arm that has seven degrees of freedom: three in the shoulder, one in the elbow and three in the wrist, without considering the degrees of freedom in the fingers ($m = 6$, $n = 7$).

1.1.2 Inverse kinematics

The direct kinematics equation, either in the form (1.2) or in the form (1.14), establishes the functional relationship between the joint variables and the end-effector position and orientation. *Inverse kinematics* concerns the determination of the joint variables q corresponding to a given end-effector position p_e and orientation R_e. The solution to this problem is of fundamental importance in order to translate the specified motion, naturally assigned in the task space, into the equivalent joint space motion that allows execution of the desired task.

With regard to the direct kinematics equation (1.2), the end-effector position and rotation matrix are uniquely computed, once the joint variables are known. In general, this cannot be said for eq. (1.14) too, since the Euler or RPY angles are not uniquely defined. On the other hand, the inverse kinematics problem is much more complex for the following reasons.

- The equations to solve are in general nonlinear equations for which it is not always possible to find closed-form solutions.

- Multiple solutions may exist.

- Infinite solutions may exist, e.g., in the case of a kinematically redundant manipulator.

- There might not be admissible solutions, in view of the manipulator kinematic structure.

1.1. KINEMATIC MODELLING

For what concerns existence of solutions, this is guaranteed if the given end-effector position and orientation belong to the manipulator workspace.

On the other hand, the problem of multiple solutions depends not only on the number of degrees of freedom but also on the Denavit-Hartenberg parameters; in general, the greater is the number of nonnull parameters, the greater is the number of admissible solutions. For a 6-degree-of-freedom manipulator without mechanical joint limits, there are in general up to 16 admissible solutions. This occurrence demands some criterion to choose among admissible solutions.

The computation of closed-form solutions requires either algebraic intuition to find out those significant equations containing the unknowns or geometric intuition to find out those significant points on the structure with respect to which it is convenient to express position and orientation. On the other hand, in all those cases when there are no —or it is difficult to find— closed-form solutions, it might be appropriate to resort to *numerical solution* techniques; these clearly have the advantage to be applicable to any kinematic structure, but in general they do not allow computation of all admissible solutions.

Most of the existing manipulators are kinematically simple, since they are typically formed by an arm (three or more degrees of freedom) which provides mobility and by a wrist which provides dexterity (three degrees of freedom). This choice is partly motivated by the difficulty to find solutions to the inverse kinematics problem in the general case. In particular, a *six-degree-of-freedom* manipulator has closed-form inverse kinematics solutions if three consecutive revolute joint axes intersect at a common point. This situation occurs when a manipulator has a so-called *spherical wrist* which is characterized by

$$\ell_5 = d_5 = \ell_6 = 0 \qquad \xi_4 = \xi_5 = \xi_6 = 0, \qquad (1.15)$$

with $\sin \alpha_5 \neq 0$ and $\sin \alpha_6 \neq 0$ so as to avoid parallel axes (degenerate manipulator). In that case, it is possible to articulate the inverse kinematics problem into two subproblems, since the solution for the *position* is *decoupled* from that for the *orientation*.

In the case of a three-degree-of-freedom arm, for given end-effector position ^0p_e and orientation ^0R_e, the inverse kinematics can be solved according to the following steps:

- compute the wrist position $^0p_4$ from ^0p_e;
- solve inverse kinematics for (q_1, q_2, q_3);
- compute $^0R_3(q_1, q_2, q_3)$;

- compute $^3R_6(q_4, q_5, q_6) = {}^3R_0\,{}^0R_e\,{}^eR_6$;
- solve inverse kinematics for (q_4, q_5, q_6).

Therefore, on the basis of this kinematic decoupling, it is possible to solve the inverse kinematics for the arm separately from the inverse kinematics for the spherical wrist.

Inverse kinematics of the anthropomorphic manipulator

Consider the anthropomorphic manipulator in Fig. 1.3, whose direct kinematics was given in (1.8). It is desired to find the vector of joint variables q corresponding to given end-effector position ^0p_e and orientation ^0R_e; without loss of generality, assume that $^0p_e = {}^0p_6$ and $^6R_e = I$.

Observing that $^0p_6 = {}^0p_4$, the first three joint variables can be solved from (1.9) which can be rewritten as

$$\begin{pmatrix} p_x \\ p_y \\ p_z \end{pmatrix} = \begin{pmatrix} c_1(c_2\ell_3 - s_{23}d_4) \\ s_1(c_2\ell_3 - s_{23}d_4) \\ s_2\ell_3 + c_{23}d_4 \end{pmatrix}. \qquad (1.16)$$

From the first two components of (1.16), it is

$$q_1 = \text{Atan2}(p_y, p_x) \qquad (1.17)$$

where Atan2 is the arctangent function of two arguments which allows the correct determination of an angle in a range of 2π. Notice that another solution is

$$q_1 = \pi + \text{Atan2}(p_y, p_x). \qquad (1.18)$$

The second joint variable can be found by squaring and summing the first two components of (1.16), i.e.,

$$p_x^2 + p_y^2 = (c_2\ell_3 - s_{23}d_4)^2; \qquad (1.19)$$

then, squaring the third component and summing it to (1.19) leads to the solution

$$q_3 = \text{Atan2}(s_3, c_3) \qquad (1.20)$$

where

$$s_3 = \frac{\ell_3^2 + d_4^2 - p_x^2 - p_y^2 - p_z^2}{2\ell_3 d_4} \qquad c_3 = \pm\sqrt{1 - s_3^2}.$$

1.1. KINEMATIC MODELLING

Substituting q_3 in (1.19), taking the square root thereof and combining the result with the third component of (1.16) leads to a system of equations in the unknowns s_2 and c_2; its solution can be found as

$$s_2 = \frac{(\ell_3 - s_3 d_4) p_z - c_3 d_4 \sqrt{p_x^2 + p_y^2}}{p_x^2 + p_y^2 + p_z^2}$$

$$c_2 = \frac{(\ell_3 - s_3 d_4) \sqrt{p_x^2 + p_y^2} + c_3 d_4 p_z}{p_x^2 + p_y^2 + p_z^2},$$

and thus the second joint variable is

$$q_2 = \mathrm{Atan2}(s_2, c_2). \tag{1.21}$$

Notice that four admissible solutions are obtained according to the values of q_1, q_2, q_3; namely, shoulder-right/elbow-up, shoulder-left/elbow-up, shoulder-right/elbow-down, shoulder-left/elbow-down.

In order to solve for the three joint variables of the wrist, let us proceed as follows. Given the matrix

$$^0 R_6 = \begin{pmatrix} n_x & s_x & a_x \\ n_y & s_y & a_y \\ n_z & s_z & a_z \end{pmatrix}, \tag{1.22}$$

the matrix $^0 R_3$ can be computed from the first three joint variables via (1.2), and thus the following equation is to be considered

$$\begin{pmatrix} ^3 n_x & ^3 s_x & ^3 a_x \\ ^3 n_y & ^3 s_y & ^3 a_y \\ ^3 n_z & ^3 s_z & ^3 a_z \end{pmatrix} = \begin{pmatrix} c_4 c_5 c_6 - s_4 s_6 & -c_4 c_5 s_6 - s_4 c_6 & -c_4 s_5 \\ s_5 c_6 & -s_5 s_6 & c_5 \\ -s_4 c_5 c_6 - c_4 s_6 & s_4 c_5 s_6 - c_4 c_6 & s_4 s_5 \end{pmatrix}. \tag{1.23}$$

The elements of the matrix on the right-hand side of (1.23) have been obtained by computing $^3 R_6$ via (1.2), whereas the elements of the matrix on the left-hand side of (1.23) can be computed as $^3 R_0\,^0 R_6$ with $^0 R_6$ as in (1.22), i.e.,

$$\begin{aligned} ^3 n_x &= c_{23}(c_1 n_x + s_1 n_y) + s_{23} n_z \\ ^3 n_y &= -s_{23}(c_1 n_x + s_1 n_y) + s_{23} n_z \\ ^3 n_z &= s_1 n_x - c_1 n_y; \end{aligned} \tag{1.24}$$

the other elements $(^3 s_x, {}^3 s_y, {}^3 s_z)$ and $(^3 a_x, {}^3 a_y, {}^3 a_z)$ can be computed from (1.24) by replacing (n_x, n_y, n_z) with (s_x, s_y, s_z) and (a_x, a_y, a_z), respectively.

At this point, inspecting (1.23) reveals that from the elements $[1,3]$ and $[3,3]$, q_4 can be computed as

$$q_4 = \text{Atan2}(^3a_z, -^3a_x). \qquad (1.25)$$

Then, q_5 can be computed by squaring and summing the elements $[1,3]$ and $[3,3]$, and from the element $[2,3]$ as

$$q_5 = \text{Atan2}(\sqrt{(^3a_x)^2 + (^3a_z)^2}, {}^3a_y). \qquad (1.26)$$

Finally, q_6 can be computed from the elements $[2,1]$ and $[2,2]$ as

$$q_6 = \text{Atan2}(-^3s_y, {}^3n_y). \qquad (1.27)$$

It is worth noticing that another set of solutions is given by the triplet

$$q_4 = \text{Atan2}(-^3a_z, {}^3a_x) \qquad (1.28)$$
$$q_5 = \text{Atan2}(-\sqrt{(^3a_x)^2 + (^3a_z)^2}, {}^3a_y) \qquad (1.29)$$
$$q_6 = \text{Atan2}(^3s_y, -^3n_y). \qquad (1.30)$$

Notice that both sets of solutions degenerate when $^3a_x = {}^3a_z = 0$; in this case, q_4 is arbitrary and simpler expressions can be found for q_5 and q_6.

In conclusion, 4 admissible solutions have been found for the arm and 2 admissible solutions have been found for the wrist, resulting in a total of 8 admissible inverse kinematics solutions for the anthropomorphic manipulator with a spherical wrist.

1.1.3 Differential kinematics

The mapping between the $(n \times 1)$ vector of joint velocities $\dot{q}$ and the (6×1) vector of end-effector velocities v is established by the *differential kinematics* equation

$$v = \begin{pmatrix} \dot{p} \\ \omega \end{pmatrix} = J(q)\dot{q}, \qquad (1.31)$$

where $\dot{p}$ is the (3×1) vector of linear velocity, ω is the (3×1) vector of angular velocity, and $J(q)$ is the $(6 \times n)$ *Jacobian* matrix. The computation of this matrix usually follows a geometric procedure that is based on computing the contributions of each joint velocity to the linear and angular end-effector velocities. Hence, $J(q)$ can be termed as the *geometric Jacobian* of the manipulator.

1.1. KINEMATIC MODELLING

Geometric Jacobian

The velocity contributions of each joint to the linear and angular velocities of link n gives the following relationship

$$\begin{pmatrix} \dot{p}_n \\ \omega_n \end{pmatrix} = \begin{pmatrix} \xi_1 a_1 + \bar{\xi}_1(a_1 \times p_{1n}) & \cdots & \xi_n a_n + \bar{\xi}_n(a_n \times p_{nn}) \\ \bar{\xi}_1 a_1 & \cdots & \bar{\xi}_n a_n \end{pmatrix} \begin{pmatrix} \dot{q}_1 \\ \vdots \\ \dot{q}_n \end{pmatrix}$$

$$= J_n(q)\dot{q} \qquad (1.32)$$

where a_k is the unit vector of axis Z_k and p_{kn} denotes the vector from the origin of frame k to the origin of frame n. Notice that J_n is a function of q through the vectors a_k and p_{kn} which can be computed on the basis of direct kinematics.

The geometric Jacobian can be computed with respect to any frame i; in that case, the k-th column of ${}^i J_n$ is given by

$${}^i j_{nk} = \begin{pmatrix} \xi_k{}^i a_k + \bar{\xi}_k{}^i R_k S({}^k a_k){}^k p_n \\ \bar{\xi}_k{}^i a_k \end{pmatrix} \qquad (1.33)$$

where ${}^k p_n = {}^k p_{kn}$ and $S(\cdot)$ is the skew-symmetric matrix operator performing the vector product, i.e., $S(a)p = a \times p$. In view of the expression of ${}^k a_k = (0\ 0\ 1)$, eq. (1.33) can be rewritten as

$${}^i j_{nk} = \begin{pmatrix} \xi_k{}^i a_k + \bar{\xi}_k(-{}^k p_{ny}{}^i n_k + {}^k p_{nx}{}^i s_k) \\ \bar{\xi}_k{}^i a_k \end{pmatrix} \qquad (1.34)$$

where ${}^k p_{nx}$ and ${}^k p_{ny}$ are the x and y components of ${}^k p_n$.

Remarks

- The transformation of the Jacobian from frame i to a different frame l can be obtained as

$${}^l J_n = \begin{pmatrix} {}^l R_i & 0 \\ 0 & {}^l R_i \end{pmatrix} {}^i J_n. \qquad (1.35)$$

- The Jacobian relating the end-effector velocity to the joint velocities can be computed either by using (1.32) and replacing p_{kn} with p_{ke}, or by using the relationship

$${}^i J_e = \begin{pmatrix} I & -S({}^i p_{ne}) \\ 0 & I \end{pmatrix} {}^i J_n. \qquad (1.36)$$

A Jacobian ^iJ_n can be decomposed as the product of three matrices, where the first two are full-rank, while the third one has the same rank as ^iJ_n but contains simpler elements to compute. To achieve this, the Jacobian of link n can be expressed as a function of a generic Jacobian

$$J_{n,h} = \begin{pmatrix} \xi_1 a_1 + \bar{\xi}_1(a_1 \times p_{1h}) & \cdots & \xi_n a_n + \bar{\xi}_n(a_n \times p_{nh}) \\ \bar{\xi}_1 a_1 & \cdots & \bar{\xi}_n a_n \end{pmatrix} \quad (1.37)$$

giving the velocity of a frame fixed to link n attached instantaneously to frame h. Then J_n can be computed via (1.36) as

$$J_n = \begin{pmatrix} I & -S(p_{hn}) \\ 0 & I \end{pmatrix} J_{n,h} \quad (1.38)$$

which can be expressed with respect to frame i, giving

$$^iJ_n = \begin{pmatrix} I & -S(^iR_h{}^hp_n) \\ 0 & I \end{pmatrix} {}^iJ_{n,h}. \quad (1.39)$$

Combining (1.35) with (1.39) yields the result that the matrix ^lJ_n can be computed as the product of three matrices

$$^lJ_n = \begin{pmatrix} ^lR_i & 0 \\ 0 & ^lR_i \end{pmatrix} \begin{pmatrix} I & -S(^iR_h{}^hp_n) \\ 0 & I \end{pmatrix} {}^iJ_{n,h}, \quad (1.40)$$

where remarkably the first two matrices are full-rank. In general, the values of h and i leading to the Jacobian $^iJ_{n,h}$ of simplest expression are given by

$$i = \text{int}(n/2) \qquad h = \text{int}(n/2) + 1.$$

Hence, for a manipulator with 6 degrees of freedom, the matrix $^3J_{6,4}$ is expected to have the simplest expression; if the wrist is spherical ($p_{46} = 0$), then the second matrix in (1.40) is identity and $^3J_{6,4} = {}^3J_6$.

As an example, the geometric Jacobian for the anthropomorphic manipulator in Fig. 1.3 can be computed on the basis of the matrix

$$^3J_6 = \begin{pmatrix} 0 & l_3 s_3 - d_4 & -d_4 & 0 & 0 & 0 \\ 0 & l_3 c_3 & 0 & 0 & 0 & 0 \\ -l_3 c_2 + d_4 s_{23} & 0 & 0 & 0 & 0 & 0 \\ s_{23} & 0 & 0 & 0 & s_4 & -c_4 s_5 \\ c_{23} & 0 & 0 & 1 & 0 & c_5 \\ 0 & 1 & 1 & 0 & c_4 & s_4 s_5 \end{pmatrix}. \quad (1.41)$$

1.1. KINEMATIC MODELLING

Analytical Jacobian

If the end-effector position and orientation are specified in terms of a minimum number of parameters in the task space as in (1.14), it is possible to compute the Jacobian matrix by direct differentiation of the direct kinematics equation, i.e.,

$$\dot{x} = \begin{pmatrix} \dot{p}_e \\ \dot{\phi}_e \end{pmatrix} = J_a(q)\dot{q}, \tag{1.42}$$

where the matrix $J_a(q) = \partial k/\partial q$ is termed *analytical* Jacobian.

The relationship between the analytical Jacobian and the geometric Jacobian is expressed as

$$J = \begin{pmatrix} I & 0 \\ 0 & T(\phi_e) \end{pmatrix} = T_a(\phi_e)J_a, \tag{1.43}$$

where $T(\phi_e)$ is a transformation matrix that depends on the particular set of parameters used to represent end-effector orientation.

We can easily recognize that the two Jacobians are in general different; note, however, that the two coincide for the positioning part. Concerning their use, the geometric Jacobian is adopted when physical quantities are of interest while the analytical Jacobian is adopted when task space quantities are of interest. It is always possible to pass from one Jacobian to the other, except when the transformation matrix is singular; the orientations at which the determinant of $T(\phi_e)$ vanishes are called *representation singularities* of ϕ_e.

Singularities

The differential kinematics equation (1.31) defines a linear mapping between the vector of joint velocities $\dot{q}$ and the vector of end-effector velocities v. The Jacobian is in general a function of the arm configuration q; those configurations at which J is rank-deficient are called *kinematic singularities*. In the following, rank deficiencies of T (representation singularities) are not considered since those are related to the particular representation of orientation chosen and not to the geometric characteristics of the manipulator; then, the analysis is based on the geometric Jacobian without loss of generality.

The simplest means to find singularities is to compute the determinant of the Jacobian matrix. For instance, for the above Jacobian in (1.41) it is

$$\det(^3 J_6) = \ell_3 d_4 c_3 s_5 (d_4 s_{23} - \ell_3 c_2) \tag{1.44}$$

leading to three types of singularities ($\ell_3, d_4 \neq 0$). These are the *elbow singularity*

$$c_3 = 0$$

occurring when link 2 and link 3 are aligned; the *shoulder singularity*

$$d_4 s_{23} - \ell_3 c_2 = 0$$

occurring when origin of frame 4 is along axis Z_0; and the *wrist singularity*

$$s_5 = 0$$

occurring when axes Z_4 and Z_6 are aligned. Notice that the elbow singularity is not troublesome since it occurs at the boundary of the manipulator workspace ($q_3 = \pm\pi/2$). The shoulder singularity is characterized in the task space and thus it can be avoided when planning an end-effector trajectory. Instead, the wrist singularity is characterized in the joint space ($q_5 = 0, \pi$), and thus it is difficult to predict when planning an end-effector trajectory.

An effective tool to analyze the linear mapping from the joint velocity space into the task velocity space defined by (1.31) is offered by the *singular value decomposition* (SVD) of the Jacobian matrix; this is given by

$$J = U\Sigma V^T = \sum_{i=1}^{m} \sigma_i u_i v_i^T, \qquad (1.45)$$

where U is the $(m \times m)$ matrix of the output singular vectors u_i, V is the $(n \times n)$ matrix of the input singular vectors v_i, and $\Sigma = (\, S \quad 0 \,)$ is the $(m \times n)$ matrix whose $(m \times m)$ diagonal submatrix S contains the singular values σ_i of the matrix J. If r denotes the rank of J, the following properties hold:

- $\sigma_1 \geq \sigma_2 \geq \ldots \geq \sigma_r > \sigma_{r+1} = \ldots = \sigma_m = 0$,
- $\mathcal{R}(J) = \text{span}\{u_1, \ldots, u_r\}$,
- $\mathcal{N}(J) = \text{span}\{v_{r+1}, \ldots, v_n\}$.

The null space $\mathcal{N}(J)$ is the set of joint velocities that yield null task velocities at the current configuration; these joint velocities are termed *null space joint velocities*. A base of $\mathcal{N}(J)$ is given by the $(n - r)$ last input singular vectors, which represent independent linear combinations of the joint velocities. Hence, one effect of a singularity is to increase the dimension of $\mathcal{N}(J)$ by introducing a linear combination of joint velocities that produce a null task velocity.

The range space $\mathcal{R}(J)$ is the set of task velocities that can be obtained as a result of all possible joint velocities; these task velocities are termed *feasible space task velocities*. A base of $\mathcal{R}(J)$ is given by the first r output singular vectors, which represent independent linear combinations of the single components of task velocities. Accordingly, another effect of a singularity is to decrease the dimension of $\mathcal{R}(J)$ by eliminating a linear combination of task velocities from the space of feasible velocities.

The singular value decomposition (1.45) shows that the i-th singular value of J can be viewed as a gain factor relating the joint velocity along the v_i direction to the task velocity along the u_i direction. When a singularity is approached, the r-th singular value tends to zero and the task velocity produced by a fixed joint velocity along v_r is decreased proportionally to σ_r. At the singular configuration, the joint velocity along v_r is in the null space and the task velocity along u_r becomes infeasible.

In the general case, the joint velocity has components in any v_i direction, and the resulting task velocity can be obtained as a combination of the single components along each output singular vector direction.

1.2 Dynamic modelling

Dynamic modelling of a robot manipulator consists of finding the mapping between the forces exerted on the structures and the joint positions, velocities and accelerations. Two formulations are mainly used to derive the *dynamic model*; namely, the Lagrange formulation and the Newton-Euler formulation, the former being simpler and more systematic and the latter being more efficient from a computational viewpoint.

1.2.1 Lagrange formulation

Since the joint variables q_i constitute a set of *generalized coordinates* of the system, the dynamic model can be derived by the Lagrange's equations

$$\frac{d}{dt}\frac{\partial L}{\partial \dot{q}_i} - \frac{\partial L}{\partial q_i} = \tau_i \qquad i = 1,\ldots,n \qquad (1.46)$$

where

$$L = T - U \qquad (1.47)$$

is the Lagrangian expressed as the difference between *kinetic energy* and *potential energy*, and τ_i is the generalized force at joint i; a torque for a revolute joint and a force for a prismatic joint, respectively. Typically the generalized forces are shortly referred to as torques, since most joints of a manipulator are revolute.

The kinetic energy is a quadratic form of the joint velocities, i.e.,

$$T = \frac{1}{2}\dot{q}^T H(q)\dot{q} \tag{1.48}$$

where the $(n \times n)$ matrix $H(q)$ is the *inertia matrix* of the robot manipulator which is symmetric and positive definite. Substituting (1.48) in (1.47) and taking the derivatives needed by (1.46) leads to the *equations of motion*

$$H(q)\ddot{q} + C(q,\dot{q})\dot{q} + g(q) = \tau \tag{1.49}$$

where τ is the $(n \times 1)$ vector of joint torques, $g(q)$ is the $(n \times 1)$ vector of gravity forces with

$$g_i(q) = \frac{\partial U}{\partial q_i}, \tag{1.50}$$

and

$$C(q,\dot{q})\dot{q} = \dot{H}(q)\dot{q} - \frac{1}{2}\left(\frac{\partial}{\partial q}(\dot{q}^T H(q)\dot{q})\right)^T \tag{1.51}$$

is the $(n \times 1)$ vector of Coriolis and centrifugal forces. This term is quadratic in the joint velocities, and thus we can write its generic element as

$$C_{ij} = \sum_{k=1}^{n} c_{ijk}\dot{q}_j\dot{q}_k. \tag{1.52}$$

There exist several factorizations of the term $C(q,\dot{q})\dot{q}$ according to the choice of the elements C_{ij} of the matrix C satisfying (1.52). The choice

$$c_{ijk} = \frac{1}{2}\left(\frac{\partial H_{ij}}{\partial q_k} + \frac{\partial H_{ik}}{\partial q_j} - \frac{\partial H_{jk}}{\partial q_i}\right), \tag{1.53}$$

where the c_{ijk}'s are termed *Christoffel symbols of the first type* makes the matrix $\dot{H} - 2C$ *skew-symmetric*; this property will be very useful for control design purposes.

The dynamic model in the form (1.49) is represented by a set of n second-order coupled and nonlinear differential equations relating the joint positions, velocities and accelerations to the joint torques. The elements of H, C and g are a function of the kinematic and dynamic parameters of the links, as can be obtained by expressing the kinetic energy and potential energy in terms of the joint positions and velocities.

The kinetic energy of the manipulator is given by the sum of the contributions of each link, i.e.,

$$T = \sum_{i=1}^{n} T_i \tag{1.54}$$

1.2. DYNAMIC MODELLING

where T_i denotes the kinetic energy of link i. This can be computed by referring all quantities to link frame i as

$$T_i = \frac{1}{2}\left({}^i\omega_i^T {}^iI_i {}^i\omega_i + m_i {}^i\dot{p}_i^T {}^i\dot{p}_i + 2m_i {}^ir_i^T({}^i\dot{p}_i \times {}^i\omega_i)\right). \tag{1.55}$$

In (1.55), m_i is the *mass* of link i, ${}^iI_i$ is the *inertia tensor* of link i with respect to the origin of frame i

$$
{}^iI_i = \begin{pmatrix} I_{ixx} & I_{ixy} & I_{ixz} \\ I_{ixy} & I_{iyy} & I_{iyz} \\ I_{ixz} & I_{iyz} & I_{izz} \end{pmatrix}, \tag{1.56}
$$

and $m_i {}^ir_i$ is the *first moment of inertia* with respect to the origin of frame i

$$m_i {}^ir_i = \begin{pmatrix} m_i r_{ix} & m_i r_{iy} & m_i r_{iz} \end{pmatrix}^T, \tag{1.57}$$

being $r_i = p_{ci} - p_i$, where p_{ci} is the position vector of the center of mass of link i and p_i is the position vector of the origin of link frame i.

In view of the previous geometric Jacobian computation, we can compute the linear and angular velocities of link i as

$$\begin{pmatrix} {}^i\dot{p}_i \\ {}^i\omega_i \end{pmatrix} = \begin{pmatrix} {}^ij_{i1} & \cdots & {}^ij_{ii} & 0 & \cdots & 0 \end{pmatrix} \begin{pmatrix} \dot{q}_1 \\ \vdots \\ \dot{q}_i \\ \dot{q}_{i+1} \\ \vdots \\ \dot{q}_n \end{pmatrix} \tag{1.58}$$

where, via (1.34), it is

$$
{}^ij_{ik} = \begin{pmatrix} \xi_k {}^ia_k + \bar{\xi}_k(-{}^kp_{iy}{}^in_k + {}^kp_{ix}{}^is_k) \\ \bar{\xi}_k {}^ia_k \end{pmatrix}. \tag{1.59}
$$

Proceeding as for the kinetic energy, the potential energy is given by

$$U = \sum_{i=1}^n U_i \tag{1.60}$$

where U_i denotes the potential energy of link i

$$U_i = -m_i {}^0\hat{g}^{T\,0}p_{ci} = -{}^0\hat{g}^T(m_i {}^0p_i + {}^0R_i m_i {}^ir_i), \tag{1.61}$$

being $\hat{g}$ the vector of gravity acceleration.

It is worth noticing that both the kinetic energy and the potential energy of link i are *linear* with respect to the set of 10 constant *dynamic parameters*, namely the link mass, the six elements of the link inertia tensor, and the three components of the link first moment of inertia. In view of (1.46), the property of linearity in the dynamic parameters holds also for the dynamic model (1.49). Also, notice that the torque at joint i is a function only of the dynamic parameters of links i to n.

Regarding the joint torque, each joint is driven by an actuator (direct drive or gear drive); in general, the following torque contributions appear

$$\tau_i = u_i - \tau_{mi} - \tau_{fi} + \tau_{ei} \tag{1.62}$$

where u_i is the actual driving torque at the joint, τ_{mi} denotes the inertia torque due to the rotor of motor i, τ_{fi} is the torque due to joint friction, and τ_{ei} is the torque caused by external force and moment applied to the end effector.

In detail, the torque due to motor inertia is

$$\tau_{mi} = I_{mi}\ddot{q}_i \tag{1.63}$$

where I_{mi} is the moment of inertia of the rotor and gyroscopic effects have been neglected; accounting for the torque in (1.63) is equivalent to augmenting the elements H_{ii} of the inertia matrix by the term I_{mi}.

Joint friction is difficult to model accurately; an approximate model of the friction torque is

$$\tau_{fi} = F_{si}\mathrm{sgn}(\dot{q}_i) + F_{vi}\dot{q}_i \tag{1.64}$$

where F_{si} and F_{vi} respectively denote the static and viscous friction coefficients.

Finally, to compute the torque due to an external end-effector force γ and moment μ, we can resort to the principle of virtual work. Since a rigid manipulator is a system with holonomic and time-independent constraints, its configurations depend only on the joint coordinates and not on time; this implies that virtual displacements coincide with elementary displacements. Hence, equating the elementary works performed by the two force systems (end-effector and joint) for the manipulator *at the equilibrium* gives

$$\tau_e^T dq = \gamma^T dp + \mu^T \omega dt \tag{1.65}$$

where dq is the elementary joint displacement, and dp (ωdt) is the elementary end-effector linear (angular) displacement. In view of (1.31), eq. (1.65) gives

$$\tau_e = J^T f \tag{1.66}$$

1.2. DYNAMIC MODELLING

where $f = (\gamma^T \ \mu^T)^T$.

By incorporating the motor inertia and the two friction coefficients into the set of dynamic parameters, we can write the dynamic model as

$$u = \Phi(q, \dot{q}, \ddot{q})\pi \tag{1.67}$$

where u is the $(n \times 1)$ vector of driving torques, π is a $(13n \times 1)$ vector of dynamic parameters

$$\pi = (\pi_1 \ \ldots \ \pi_n)^T \tag{1.68}$$

with

$$\pi_i = (I_{ixx} \ \ I_{ixy} \ \ I_{ixz} \ \ I_{iyy} \ \ I_{iyz} \ \ I_{izz} \tag{1.69}$$
$$m_i r_{ix} \ \ m_i r_{iy} \ \ m_i r_{iz} \ \ m_i \ \ I_{mi} \ \ F_{si} \ \ F_{vi})^T,$$

and $\Phi(\cdot)$ is an $(n \times 13n)$ matrix which is usually termed *regressor* of the dynamic model.

Dynamic model of the anthropomorphic manipulator

As an example of dynamic model computation, consider the first three links of the anthropomorphic manipulator in Fig. 1.3. Derivation of the complete dynamic model would be tedious and error prone. For manipulators with more than 3 degrees of freedom, it is convenient to resort to symbolic software packages to derive the dynamic model.

Link angular and linear velocities can be computed as in (1.58), giving

$$^0\omega_0 = 0$$
$$^1\omega_1 = (0 \ \ 0 \ \ 1)^T$$
$$^2\omega_2 = (s_2 \dot{q}_1 \ \ c_2 \dot{q}_1 \ \ \dot{q}_2)^T$$
$$^3\omega_3 = (s_{23} \dot{q}_1 \ \ c_{23} \dot{q}_1 \ \ \dot{q}_2 + \dot{q}_3)^T$$

and

$$^0\dot{p}_0 = 0$$
$$^1\dot{p}_1 = 0$$
$$^2\dot{p}_2 = 0$$
$$^3\dot{p}_3 = (\ell_3 s_3 \dot{q}_2 \ \ \ell_3 c_3 \dot{q}_2 \ \ -\ell_3 c_2 \dot{q}_1)^T,$$

respectively. It follows that the inertia matrix is

$$H = \begin{pmatrix} H_{11} & H_{12} & H_{13} \\ H_{12} & H_{22} & H_{23} \\ H_{13} & H_{23} & H_{33} \end{pmatrix} \tag{1.70}$$

where

$$H_{11} = I_{m1} + I_{1zz} + s_2^2 I_{2xx} + 2s_2 c_2 I_{2xy} + c_2^2 I_{2yy}$$
$$\quad + s_{23}^2 I_{3xx} + 2s_{23} c_{23} I_{3xy} + c_{23}^2 I_{3yy} + 2\ell_3 c_2 c_{23} m_3 r_{3x}$$
$$\quad - 2\ell_3 c_2 s_{23} m_3 r_{3y} + \ell_3^2 c_2^2 m_3$$
$$H_{12} = s_2 I_{2xz} + c_2 I_{2yz} + s_{23} I_{3xz} + c_{23} I_{3yz} - \ell_3 s_2 m_3 r_{3z}$$
$$H_{13} = s_{23} I_{3xz} + c_{23} I_{3yz}$$
$$H_{22} = I_{m2} + I_{2zz} + 2\ell_3 c_3 m_3 r_{3x} - 2\ell_3 s_3 m_3 r_{3y} + \ell_3^2 m_3 + I_{3zz}$$
$$H_{23} = I_{3zz} + \ell_3 c_3 m_3 r_{3x} - \ell_3 s_3 m_3 r_{3y}$$
$$H_{33} = I_{m3} + I_{3zz},$$

from which the elements of the matrix C can be computed as in (1.52) and (1.53). Furthermore, assuming that

$$^0\hat{g} = (0 \quad 0 \quad \hat{g}_0)^T,$$

the vector of gravity forces is

$$g = (g_1 \quad g_2 \quad g_3)^T \tag{1.71}$$

where

$$g_1 = 0$$
$$g_2 = -\hat{g}_0(c_2 m_2 r_{2x} - s_2 m_2 r_{2y} + c_{23} m_3 r_{3x} - s_{23} m_3 r_{3y} + \ell_3 c_2 m_3)$$
$$g_3 = -\hat{g}_0(c_{23} m_3 r_{3x} - s_{23} m_3 r_{3y}).$$

1.2.2 Newton-Euler formulation

The Newton-Euler formulation allows computing the dynamic model of a rigid manipulator without deriving the explicit expressions of the terms H, C and g. The equations of motion are obtained as the result of two recursive computations; namely, a forward recursion to compute link velocities and accelerations from link 0 to link n, and a backward recursion to compute link forces and moments (and thus joint torques) from link n to link 0.

The linear velocity of the origin of link frame i is given by

$$\dot{p}_i = \dot{p}_{i-1} + \omega_{i-1} \times p_{i-1,i} + \xi_i \dot{q}_i a_i, \tag{1.72}$$

whereas the angular velocity of link i is given by

$$\omega_i = \omega_{i-1} + \bar{\xi}_i \dot{q}_i a_i. \tag{1.73}$$

1.2. DYNAMIC MODELLING

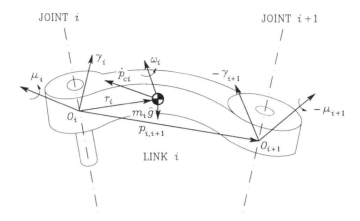

Figure 1.4: Characterization of link i for Newton-Euler formulation.

Notice that eqs. (1.72) and (1.73), when referred to frame i, are in general more efficient than (1.58) for computing link velocity.

Differentiation of (1.72) and (1.73) respectively gives

$$\ddot{p}_i = \ddot{p}_{i-1} + \dot{\omega}_{i-1} \times p_{i-1,i} + \omega_{i-1} \times (\omega_{i-1} \times p_{i-1,i}) + \xi_i(\ddot{q}_i a_i + 2\omega_{i-1} \times \dot{q}_i a_i) \tag{1.74}$$

and

$$\dot{\omega}_i = \dot{\omega}_{i-1} + \bar{\xi}_i(\ddot{q}_i a_i + \omega_{i-1} \times \dot{q}_i a_i). \tag{1.75}$$

Furthermore, the acceleration of the center of mass of link i is given by

$$\ddot{p}_{ci} = \ddot{p}_i + \dot{\omega}_i \times r_i + \omega_i \times (\omega_i \times r_i). \tag{1.76}$$

With reference to Fig. 1.4, the *Newton equation* gives a balance of forces acting on link i in the form of

$$\gamma_i = m_i \ddot{p}_{ci} - m_i \hat{g} + \gamma_{i+1} \tag{1.77}$$

where γ_i denotes the force exerted from link $i-1$ on link i at the origin of link frame i. Substituting (1.76) in (1.77) gives

$$\gamma_i = m_i \ddot{p}_i + \dot{\omega}_i \times m_i r_i + \omega_i \times (\omega_i \times m_i r_i) + \gamma_{i+1}. \tag{1.78}$$

The effect of $m_i \hat{g}$ will be introduced automatically by taking $\ddot{p}_0 = -\hat{g}$.

The *Euler equation* gives a balance of moments acting on link i (referred to the center of mass) in the form of

$$\mu_i = \hat{I}_i \dot{\omega}_i + \omega_i \times (\hat{I}_i \omega_i) + r_i \times \gamma_i + \mu_{i+1} + p_{ci,i+1} \times \gamma_{i+1}, \tag{1.79}$$

where $\hat{I}_i$ is the inertia tensor of link i with respect to its center of mass. Applying Steiner's theorem, the inertia tensor with respect to the origin of link frame i is given by

$$I_i = \hat{I}_i + m_i S^T(r_i) S(r_i), \tag{1.80}$$

and thus (1.79), via (1.76), can be rewritten as

$$\mu_i = I_i \dot{\omega}_i + \omega_i \times (I_i \omega_i) + m_i r_i \times \ddot{p}_i + \mu_{i+1} + p_{i,i+1} \times \gamma_{i+1}. \tag{1.81}$$

The joint driving force (torque) u_i can be obtained by projecting γ_i (μ_i) along axis Z_i, i.e.,

$$u_i = (\xi_i \gamma_i + \bar{\xi}_i \mu_i)^T a_i + I_{mi} \ddot{q}_i + F_{si} \mathrm{sgn}(\dot{q}_i) + F_{vi} \dot{q}_i \tag{1.82}$$

where the contributions of motor inertia and joint friction have been suitably added.

Recursive algorithm

On the basis of the above equations, it is possible to construct the following algorithm where it is worth referring link velocities, accelerations, forces and moments to the current frame i.

The initial conditions for a manipulator with fixed link 0 are

$$^0\omega_0 = 0 \qquad ^0\dot{\omega}_0 = 0 \qquad ^0\ddot{p}_0 = -{}^0\hat{g}$$

so as to incorporate gravity acceleration in the acceleration of link 0. The forward recursive equations for $i = 1, \ldots, n$ are:

$$^i\omega_i = {}^iR_{i-1}{}^{i-1}\omega_{i-1} + \bar{\xi}_i \dot{q}_i {}^i a_i \tag{1.83}$$

$$^i\dot{\omega}_i = {}^iR_{i-1}{}^{i-1}\dot{\omega}_{i-1} + \bar{\xi}_i (\ddot{q}_i {}^i a_i + {}^iR_{i-1}{}^{i-1}\omega_{i-1} \times \dot{q}_i {}^i a_i) \tag{1.84}$$

$$^i\ddot{p}_i = {}^iR_{i-1} \big({}^{i-1}\ddot{p}_{i-1} + {}^{i-1}\dot{\omega}_{i-1} \times {}^{i-1}p_i \tag{1.85}$$
$$+ {}^{i-1}\omega_{i-1} \times ({}^{i-1}\omega_{i-1} \times {}^{i-1}p_i)\big)$$
$$+ \xi_i (\ddot{q}_i {}^i a_i + 2 {}^iR_{i-1}{}^{i-1}\omega_{i-1} \times \dot{q}_i {}^i a_i)$$

where ${}^i a_i = (0 \ \ 0 \ \ 1)^T$.

For given terminal conditions ${}^{n+1}\gamma_{n+1}$ and ${}^{n+1}\mu_{n+1}$, the backward recursive equations for $i = n, \ldots, 1$ are:

$$^i\gamma_i = m_i {}^i\ddot{p}_i + {}^i\dot{\omega}_i \times m_i {}^i r_i + {}^i\omega_i \times ({}^i\omega_i \times m_i {}^i r_i) + {}^iR_{i+1}{}^{i+1}\gamma_{i+1} \tag{1.86}$$

$$^i\mu_i = {}^iI_i {}^i\dot{\omega}_i + {}^i\omega_i \times ({}^iI_i {}^i\omega_i) + m_i {}^i r_i \times {}^i\ddot{p}_i \tag{1.87}$$
$$+ {}^iR_{i+1}{}^{i+1}\mu_{i+1} + {}^ip_{i+1} \times {}^iR_{i+1}{}^{i+1}\gamma_{i+1}.$$

1.2. DYNAMIC MODELLING

Finally, the generalized force at joint i is computed as

$$u_i = (\xi_i{}^i \gamma_i + \bar{\xi}_i{}^i \mu_i)^T{}^i a_i + I_{mi} \ddot{q}_i + F_{si} \text{sgn}(\dot{q}_i) + F_{vi} \dot{q}_i. \tag{1.88}$$

It is worth noticing that both (1.86) and (1.87) are linear with respect to the set of dynamic parameters introduced in (1.69), and such a linearity is obviously preserved by (1.88).

1.2.3 Model computation

Direct dynamics and inverse dynamics

Computing the dynamic model of a robot manipulator is of interest both for simulation and control. For the former problem, we are concerned with the so-called *direct dynamics* of the manipulator, i.e., from a given set of joint torques u, compute the resulting joint positions, velocities and accelerations. For the latter problem, we are concerned with the so-called *inverse dynamics* of the manipulator, i.e., from a given set of joint positions, velocities and accelerations, compute the resulting joint torques.

The computational load of direct dynamics is expected to be larger than that of inverse dynamics, because the dynamic model naturally gives the mapping from the joint positions, velocities and accelerations to the joint torques. From the equations of motion in (1.49), via (1.62) and (1.65), the joint accelerations can be computed as

$$\ddot{q} = H^{-1}(q)\bigl(u - d(q, \dot{q})\bigr) \tag{1.89}$$

where $d(q, \dot{q}) = C\dot{q} + g + \tau_m + \tau_f - J^T f$. Setting (1.89) in state space form

$$\begin{pmatrix} \dot{q} \\ \ddot{q} \end{pmatrix} = \begin{pmatrix} \dot{q} \\ -H^{-1}(q)d(q,\dot{q}) \end{pmatrix} + \begin{pmatrix} 0 \\ H^{-1}(q)u \end{pmatrix} \tag{1.90}$$

gives a system of $2n$ nonlinear ordinary differential equations which can be integrated over time with known initial conditions to provide the state $q(t)$ and $\dot{q}(t)$.

The terms H and d can be computed with Lagrange formulation, but a more efficient procedure consists of using the above Newton-Euler algorithm based on (1.83)–(1.88). The vector $d(q, \dot{q})$ can be obtained by computing u with $\ddot{q} = 0$. Then, the column h_i of the matrix $H(q)$ can be computed as the torque u with $\hat{g} = 0$, $\dot{q} = 0$, $f = 0$, $\ddot{q}_i = 1$ and $\ddot{q}_j = 0$ for $j \neq i$. Iterating the procedure for $i = 1, \ldots, n$ leads to constructing the entire inertia matrix.

On the other hand, reducing the computational burden of inverse dynamics is crucial for model-based control algorithms which require on-line update of the terms in the dynamic model. Since several kinematic parameters may be unitary or null, redundant operations shall be avoided; also, intermediate variables can be introduced which appear multiple times in the computation. The result is a customized symbolic computation based on the recursive Newton-Euler algorithm which can lead to a savings by more than 50% in many practical cases.

Base dynamic parameters

In view of dynamic model computational load reduction, it is important to find the minimum set of dynamic parameters (*base dynamic parameters*) which are needed to compute the dynamic model. They can be deduced from the components of the vector π in (1.68) by eliminating those which have no effect on the dynamic model and by grouping those which always appear together. The determination of the base parameters is essential also for the identification of dynamic parameters treated in a following section, since they constitute the only identifiable parameters.

Since the kinetic energy and the potential energy are linear in the dynamic parameters, we can refer to the total energy (*Hamiltonian*) of the manipulator in the form

$$W = T + U = \sum_{j=1}^{p} w_j \pi_j \qquad (1.91)$$

where $w_j = \partial W / \partial \pi_j$ and $p = 13n$ if all 13 dynamic parameters are considered for each of the n links.

A parameter π_j has no effect on the dynamic model if w_j is a constant. A parameter π_j can be grouped with some other parameters $\pi_{j,1}, \ldots, \pi_{j,s}$ if

$$w_j = \sum_{k=1}^{s} \nu_{jk} w_{j,k} \qquad (1.92)$$

with ν_{jk} all constant. In that case, the parameter π_j shall be eliminated while the parameters $\pi_{j,1}, \ldots, \pi_{j,s}$ shall be changed into

$$\pi'_{j,k} = \pi_{j,k} + \nu_{jk} \pi_j. \qquad (1.93)$$

The use of (1.92) for manipulators with more than three degrees of freedom is lengthy and error prone, owing to the complicated expressions of w_j for the outer links. In the following, a set of general rules are established for symbolic grouping of dynamic parameters.

1.2. DYNAMIC MODELLING

Let the dynamic parameters of link i be given by the vector in (1.69) without considering the motor inertia nor the friction coefficients, i.e.,

$$\pi_i = \begin{pmatrix} I_{ixx} & I_{ixy} & I_{ixz} & I_{iyy} & I_{iyz} & I_{izz} & m_i r_{ix} & m_i r_{iy} & m_i r_{iz} & m_i \end{pmatrix}^T; \quad (1.94)$$

accordingly, the coefficients in (1.91) can be cast into a (10×1) vector

$$w_i = \begin{pmatrix} w_{ixx} & w_{ixy} & w_{ixz} & w_{iyy} & w_{iyz} & w_{izz} & w_{imx} & w_{imy} & w_{imz} & w_{im} \end{pmatrix}^T \quad (1.95)$$

where it can be shown that

$$w_{ixx} = \frac{1}{2}{}^i\omega_{ix}^2$$

$$w_{ixy} = {}^i\omega_{ix}{}^i\omega_{iy}$$

$$w_{ixz} = {}^i\omega_{ix}{}^i\omega_{iz}$$

$$w_{iyy} = \frac{1}{2}{}^i\omega_{iy}^2$$

$$w_{iyz} = {}^i\omega_{iy}{}^i\omega_{iz}$$

$$w_{izz} = \frac{1}{2}{}^i\omega_{iz}^2$$

$$w_{imx} = {}^i\omega_{iz}{}^i\dot{p}_{iy} - {}^i\omega_{iy}{}^i\dot{p}_{iz} - {}^0\hat{g}^{T0}n_i$$

$$w_{imy} = {}^i\omega_{ix}{}^i\dot{p}_{iz} - {}^i\omega_{iz}{}^i\dot{p}_{ix} - {}^0\hat{g}^{T0}s_i$$

$$w_{imz} = {}^i\omega_{iy}{}^i\dot{p}_{ix} - {}^i\omega_{ix}{}^i\dot{p}_{iy} - {}^0\hat{g}^{T0}a_i$$

$$w_{im} = \frac{1}{2}{}^i\dot{p}_i^T{}^i\dot{p}_i - {}^0\hat{g}^{T0}p_i.$$

By computing w_i as a function of w_{i-1}, general relations can be found which satisfy (1.92). Let ${}^{i-1}\Lambda_i$ be the (10×10) matrix expressing the transformation of dynamic parameters of link i into frame $i-1$, i.e.,

$$^{i-1}\pi_i = {}^{i-1}\Lambda_i \pi_i. \quad (1.96)$$

Let us distinguish between the two cases of a revolute and a prismatic joint. For a *revolute* joint, the following three relations always hold:

$$w_{ixx} + w_{iyy} = w_{i-1}^T ({}^{i-1}\lambda_{i,1} + {}^{i-1}\lambda_{i,4}) \quad (1.97)$$

$$w_{imz} = w_{i-1}^T {}^{i-1}\lambda_{i,9} \quad (1.98)$$

$$w_{im} = w_{i-1}^T {}^{i-1}\lambda_{i,10} \quad (1.99)$$

where ${}^{i-1}\lambda_{i,k}$ denotes the k-th column of matrix ${}^{i-1}\Lambda_i$ which is a function of the kinematic parameters of link i. By comparing (1.97)–(1.99) with (1.92), three parameters can be grouped with the other, the choice being not unique. By choosing to group the parameters I_{iyy}, $m_i r_{iz}$ and m_i, the sought relations can be found via (1.93), giving

$$I'_{ixx} = I_{ixx} - I_{iyy} \tag{1.100}$$

and

$$\pi'_{i-1} = \pi_{i-1} + I_{iyy}(^{i-1}\lambda_{i,1} + {}^{i-1}\lambda_{i,4}) + m_i r_{iz} {}^{i-1}\lambda_{i,9} + m_i {}^{i-1}\lambda_{i,10}. \tag{1.101}$$

In view of the structure of the columns of matrix ${}^{i-1}\Lambda_i$, the resulting grouped parameters via (1.100) and (1.101) are

$$\begin{aligned}
I'_{ixx} &= I_{ixx} - I_{iyy} \\
I'_{i-1xx} &= I_{i-1xx} + I_{iyy} + 2d_i m_i r_{iz} + d_i^2 m_i \\
I'_{i-1xy} &= I_{i-1xy} + \ell_i \sin\alpha_i m_i r_{iz} + \ell_i d_i \sin\alpha_i m_i \\
I'_{i-1xz} &= I_{i-1xz} - \ell_i \cos\alpha_i m_i r_{iz} - \ell_i d_i \cos\alpha_i m_i \\
I'_{i-1yy} &= I_{i-1yy} + \cos^2\alpha_i I_{iyy} + 2d_i \cos^2\alpha_i m_i r_{iz} \\
&\quad + (\ell_i^2 + d_i^2 \cos^2\alpha_i) m_i \\
I'_{i-1yz} &= I_{i-1yz} + \sin\alpha_i \cos\alpha_i I_{iyy} + 2d_i \sin\alpha_i \cos\alpha_i m_i r_{iz} \\
&\quad + d_i^2 \sin\alpha_i \cos\alpha_i m_i \\
I'_{i-1zz} &= I_{i-1zz} + \sin^2\alpha_i I_{iyy} + 2d_i \sin^2\alpha_i m_i r_{iz} \\
&\quad + (\ell_i^2 + d_i^2 \sin^2\alpha_i) m_i \\
m'_{i-1} r'_{i-1x} &= m_{i-1} r_{i-1x} + \ell_i m_i \\
m'_{i-1} r'_{i-1y} &= m_{i-1} r_{i-1y} - \sin\alpha_i m_i r_{iz} - d_i \sin\alpha_i m_i \\
m'_{i-1} r'_{i-1z} &= m_{i-1} r_{i-1z} + \cos\alpha_i m_i r_{iz} + d_i \cos\alpha_i m_i \\
m'_{i-1} &= m_{i-1} + m_i.
\end{aligned} \tag{1.102}$$

On the other hand, for a *prismatic* joint, the following six relations always hold:

$$w_{ixx} = w_{i-1}^T {}^{i-1}\lambda_{i,1} \tag{1.103}$$

$$w_{ixy} = w_{i-1}^T {}^{i-1}\lambda_{i,2} \tag{1.104}$$

$$w_{ixz} = w_{i-1}^T {}^{i-1}\lambda_{i,3} \tag{1.105}$$

$$w_{iyy} = w_{i-1}^T {}^{i-1}\lambda_{i,4} \tag{1.106}$$

$$w_{iyz} = w_{i-1}^T {}^{i-1}\lambda_{i,5} \tag{1.107}$$

$$w_{izz} = w_{i-1}^T {}^{i-1}\lambda_{i,6}. \tag{1.108}$$

1.2. DYNAMIC MODELLING

Therefore, six parameters can be grouped; choosing to group the six parameters of the inertia matrix of link i with those of link $i-1$ yields the relation

$$\pi'_{i-1} = \pi_{i-1} + I_{ixx}{}^{i-1}\lambda_{i,1} + I_{ixy}{}^{i-1}\lambda_{i,2} + I_{ixz}{}^{i-1}\lambda_{i,3} \quad (1.109)$$
$$+ I_{iyy}{}^{i-1}\lambda_{i,4} + I_{iyz}{}^{i-1}\lambda_{i,5} + I_{izz}{}^{i-1}\lambda_{i,6}.$$

In view of the structure of the columns of matrix ${}^{i-1}\Lambda_i$, the parameters of the inertia tensor ${}^iI_i$ can be grouped with the parameters of the inertia tensor ${}^{i-1}I_{i-1}$, and the grouped parameters are

$$^{i-1}I'_{i-1} = {}^{i-1}I_{i-1} + {}^{i-1}R_i{}^iI_i{}^iR_{i-1}$$

which gives

$$I'_{i-1xx} = I_{i-1xx} + \cos^2\vartheta_i I_{ixx} - 2\sin\vartheta_i\cos\vartheta_i I_{ixy} + \sin^2\vartheta_i I_{iyy}$$

$$I'_{i-1xy} = I_{i-1xy} + \sin\vartheta_i\cos\vartheta_i\cos\alpha_i I_{ixx}$$
$$+ (\cos^2\vartheta_i - \sin^2\vartheta_i)\cos\alpha_i I_{ixy} - \cos\vartheta_i\sin\alpha_i I_{ixz}$$
$$- \sin\vartheta_i\cos\vartheta_i\cos\alpha_i I_{iyy} + \sin\vartheta_i\sin\alpha_i I_{iyz}$$

$$I'_{i-1xz} = I_{i-1xz} + \sin\vartheta_i\cos\vartheta_i\sin\alpha_i I_{ixx}$$
$$+ (\cos^2\vartheta_i - \sin^2\vartheta_i)\sin\alpha_i I_{ixy} + \cos\vartheta_i\cos\alpha_i I_{ixz} \quad (1.110)$$
$$- \sin\vartheta_i\cos\vartheta_i\sin\alpha_i I_{iyy} - \sin\vartheta_i\cos\alpha_i I_{iyz}$$

$$I'_{i-1yy} = I_{i-1yy} + \sin^2\vartheta_i\cos^2\alpha_i I_{ixx}$$
$$+ 2\sin\vartheta_i\cos\vartheta_i\cos^2\alpha_i I_{ixy} - 2\sin\vartheta_i\sin\alpha_i\cos\alpha_i I_{ixz}$$
$$+ \cos^2\vartheta_i\cos^2\alpha_i I_{iyy} - 2\cos\vartheta_i\sin\alpha_i\cos\alpha_i I_{iyz} + \sin^2\alpha_i I_{izz}$$

$$I'_{i-1yz} = I_{i-1yz} + \sin^2\vartheta_i\sin\alpha_i\cos\alpha_i I_{ixx}$$
$$+ 2\sin\vartheta_i\cos\vartheta_i\sin\alpha_i\cos\alpha_i I_{ixy} + \sin\vartheta_i(\cos^2\alpha_i - \sin^2\alpha_i)I_{ixz}$$
$$+ \cos^2\vartheta_i\sin\alpha_i\cos\alpha_i I_{iyy} + \cos\vartheta_i(\cos^2\alpha_i - \sin^2\alpha_i)I_{iyz}$$
$$- \sin\alpha_i\cos\alpha_i I_{izz}$$

$$I'_{i-1zz} = I_{i-1zz} + \sin^2\vartheta_i\sin^2\alpha_i I_{ixx}$$
$$+ 2\sin\vartheta_i\cos\vartheta_i\sin^2\alpha_i I_{ixy} + 2\sin\vartheta_i\sin\alpha_i\cos\alpha_i I_{ixz}$$
$$+ \cos^2\vartheta_i\sin^2\alpha_i I_{iyy} + 2\cos\vartheta_i\sin\alpha_i\cos\alpha_i I_{iyz}$$
$$+ \cos^2\alpha_i I_{izz}.$$

Grouping or elimination of some parameters, other than the rules given above, may occur for particular geometries depending on the sequence of joints. Detection of such situations can be studied case by case using relations (1.92) and (1.93).

Additional grouping of inertial parameters will concern parameters $m_i r_{ix}$, $m_i r_{iy}$ and $m_i r_{iz}$ of the prismatic joints lying between joints r_1 and r_2, where r_1 is the first revolute joint and r_2 is the subsequent revolute joint whose axis is not parallel to joint r_1 axis. The following two cases are to be considered.

- The axis of prismatic joint i is not parallel to joint r_1 axis for $r_1 < i < r_2$. In this case, the coefficients w_{imx}, w_{imy} and w_{imz} satisfy the relation

$$^i a_{r_1 x} w_{imx} + {}^i a_{r_1 y} w_{imy} + {}^i a_{r_1 z} w_{imz} = \text{const} \qquad (1.111)$$

where $^i a_{r_1} = (\,^i a_{r_1 x} \quad {}^i a_{r_1 y} \quad {}^i a_{r_1 z}\,)^T$ is the unit vector of joint r_1 axis referred to frame i. Therefore, depending on the particular values of $^i a_{r_1 x}$, $^i a_{r_1 y}$ and $^i a_{r_1 z}$, one of the parameters $m_i r_{ix}$, $m_i r_{iy}$, $m_i r_{iz}$ can be grouped or eliminated.

- The axis of prismatic joint i is parallel to joint r_1 axis for $r_1 < i < r_2$. In this case, the following linear relation is obtained:

$$\begin{pmatrix} w_{imx} \\ w_{imy} \\ w_{imz} \end{pmatrix} = {}^i R_{i-1} \begin{pmatrix} w_{i-1mx} \\ w_{i-1my} \\ w_{i-1mz} \end{pmatrix} + \begin{pmatrix} 2\ell_i \cos\vartheta_i w_{kzz} \\ -2\ell_i \sin\vartheta_i w_{kzz} \\ 0 \end{pmatrix}, \qquad (1.112)$$

where k denotes the nearest revolute joint from i back to joint 1. Therefore, we can deduce that the parameter $m_i r_{iz}$ has no effect on the dynamic model and the parameters $m_i r_{ix}$ and $m_i r_{iy}$ can be grouped using the relations:

$$m'_{i-1} r_{i-1x} = m_{i-1} r_{i-1x} + \cos\vartheta_i m_i r_{ix} - \sin\vartheta_i m_i r_{iy} \qquad (1.113)$$
$$m'_{i-1} r_{i-1y} = m_{i-1} r_{i-1y} + \sin\vartheta_i \cos\alpha_i m_i r_{ix} + \cos\vartheta_i \cos\alpha_i m_i r_{iy}$$
$$m'_{i-1} r_{i-1z} = m_{i-1} r_{i-1z} + \sin\vartheta_i \sin\alpha_i m_i r_{ix} + \cos\vartheta_i \sin\alpha_i m_i r_{iy}$$
$$I'_{kzz} = I_{kzz} + 2\ell_i \cos\vartheta_i m_i r_{ix} - 2\ell_i \sin\vartheta_i m_i r_{iy}.$$

The following rules can be adopted to define all the parameters which can be grouped or eliminated, and the remaining parameters will constitute the set of base dynamic parameters of the dynamic model.

For $i = n, \ldots, 1$:

1. Use the general grouping relations (1.102) or (1.110) to eliminate I_{iyy}, $m_i r_{iz}$ and m_i if joint i is revolute, or I_{ixx}, I_{ixy}, I_{ixz}, I_{iyy}, I_{iyz} and I_{izz} if joint i is prismatic.

1.2. DYNAMIC MODELLING

2. If joint i is prismatic and a_i is parallel to a_{r_1} for $r_1 < i < r_2$, then eliminate $m_i r_{iz}$ and group $m_i r_{ix}$ and $m_i r_{iy}$ using (1.113).

3. If joint i is prismatic and a_i is not parallel to a_{r_1} for $r_1 < i < r_2$, then group or eliminate one of the parameters $m_i r_{ix}$, $m_i r_{iy}$, $m_i r_{iz}$ using the property (1.111).

4. If joint i is revolute and both a_i and a_{r_2} are parallel to a_{r_1} for $r_1 \leq i < r_2$, then eliminate I_{ixx}, I_{ixy}, I_{ixz} and I_{iyz}. Notice that I_{iyy} has also been eliminated using rule 1.

5. If joint i is revolute and a_i is parallel to a_{r_1}, a_k and $\hat{g}$ for $r_1 \leq i < r_2$ and $k < i$, then eliminate $m_i r_{ix}$ and $m_i r_{iy}$. Notice that $m_i r_{iz}$ has also been eliminated using rule 1.

6. If joint i is prismatic and $i < r_1$, then eliminate $m_i r_{ix}$, $m_i r_{iy}$ and $m_i r_{iz}$.

For a general robot manipulator with n revolute joints and $n > 2$, it can be found that the number of operations required by dynamic model computation using customized symbolic computation and the base dynamic parameters is less than $92n - 127$ multiplies and $81n - 117$ additions; this means that, for $n = 6$, less than 425 multiplies and 369 additions are required.

As an alternative to symbolic reduction of dynamic parameters, a numerical reduction can be achieved on the basis of the QR decomposition or the singular value decomposition of a matrix $\bar{\Phi}$ calculated from (1.67) for N random values of $q, \dot{q}, \ddot{q}$, i.e.,

$$\bar{\Phi} = \begin{pmatrix} \Phi(q_1, \dot{q}_1, \ddot{q}_1) \\ \vdots \\ \Phi(q_N, \dot{q}_N, \ddot{q}_N) \end{pmatrix} \quad (1.114)$$

such that the number of rows is (typically much) greater than the number of columns.

No matter how parameter reduction is accomplished, eq. (1.67) can be rewritten as

$$u = Y(q, \dot{q}, \ddot{q})\rho \quad (1.115)$$

where ρ is an $(r \times 1)$ vector of *base* dynamic parameters and $Y(\cdot)$ is an $(n \times r)$ matrix which is determined accordingly.

As an example of dynamic parameter reduction, consider the complete 6-degree-of-freedom anthropomorphic manipulator in Fig. 1.3. Using the general rules for $j = n, \ldots, 1$ gives the following sets of reductions:

$$I'_{6xx} = I_{6xx} - I_{6yy}$$
$$I'_{5xx} = I_{5xx} + I_{6yy}$$
$$I'_{5zz} = I_{5zz} + I_{6yy}$$
$$m'_5 r'_{5y} = m_5 r_{5y} + m_6 r_{6z}$$
$$m'_5 = m_5 + m_6$$

for link 6, and thus the base parameters are: I'_{6xx}, I_{6xy}, I_{6xz}, I_{6yz}, I_{6zz}, $m_6 r_{6x}$, $m_6 r_{6y}$;

$$I'_{5xx} = I_{5xx} + I_{6yy} - I_{5yy}$$
$$I'_{4xx} = I_{4xx} + I_{5yy}$$
$$I'_{4zz} = I_{4zz} + I_{5yy}$$
$$m'_4 r'_{4y} = m_4 r_{4y} - m_5 r_{5z}$$
$$m'_4 = m_4 + m_5 + m_6$$

for link 5, and thus the base parameters are: I'_{5xx}, I_{5xy}, I_{5xz}, I_{5yz}, I'_{5zz}, $m_5 r_{5x}$, $m_5 r_{5y}$;

$$I'_{4xx} = I_{4xx} + I_{5yy} - I_{4yy}$$
$$I'_{3xx} = I_{3xx} + I_{4yy} + 2d_4 m_4 r_{4z} + d_4^2(m_4 + m_5 + m_6)$$
$$I'_{3zz} = I_{3zz} + I_{4yy} + 2d_4 m_4 r_{4z} + d_4^2(m_4 + m_5 + m_6)$$
$$m'_3 r'_{3y} = m_3 r_{3y} + m_4 r_{4z} + d_4(m_4 + m_5 + m_6)$$
$$m'_3 = m_3 + m_4 + m_5 + m_6$$

for link 4, and thus the base parameters are: I'_{4xx}, I_{4xy}, I_{4xz}, I_{4yz}, I'_{4zz}, $m_4 r_{4x}$, $m'_4 r'_{4y}$;

$$I'_{3xx} = I_{3xx} - I_{3yy} + I_{4yy} + 2d_4 m_4 r_{4z} + d_4^2(m_4 + m_5 + m_6)$$
$$I'_{2xx} = I_{2xx} + I_{3yy}$$
$$I'_{2xz} = I_{2xz} - \ell_3 m_3 r_{3z}$$
$$I'_{2yy} = I_{2yy} + \ell_3^2(m_3 + m_4 + m_5 + m_6) + I_{3yy}$$
$$I'_{2zz} = I_{2zz} + \ell_3^2(m_3 + m_4 + m_5 + m_6)$$
$$m'_2 r'_{2x} = m_2 r_{2x} + \ell_3(m_3 + m_4 + m_5 + m_6)$$
$$m'_2 r'_{2z} = m_2 r_{2z} + m_3 r_{3z}$$
$$m'_2 = m_2 + m_3 + m_4 + m_5 + m_6$$

1.2. DYNAMIC MODELLING

for link 3, and thus the base parameters are: I'_{3xx}, I_{3xy}, I_{3xz}, I_{3yz}, I'_{3zz}, $m_3 r_{3x}$, $m'_3 r'_{3y}$;

$$I'_{2xx} = I_{2xx} - I_{2yy} - \ell_3^2 (m_3 + m_4 + m_5 + m_6)$$
$$I'_{1zz} = I_{1zz} + I_{2yy} + \ell_3^2 (m_3 + m_4 + m_5 + m_6) + I_{3yy}$$

for link 2, and thus the base parameters are: I'_{2xx}, I_{2xy}, I_{2xz}, I_{2yz}, I'_{2zz}, $m'_2 r'_{2x}$, $m_2 r_{2y}$; the parameters I_{1xx}, I_{1xy}, I_{1xz}, I_{1yy}, I_{1yz}, $m_1 r_{1x}$, $m_1 r_{1y}$, $m_1 r_{1z}$, m_1 have no effect on the dynamic model and the only parameter of link 1 is I'_{1zz}. In addition, concerning the motor inertias, it can be seen that I_{m1} can be grouped with I_{1zz} and I_{m2} with I_{2zz}.

The final result can be summarized as follows:

- The parameters having no effect on the model are: I_{1xx}, I_{1xy}, I_{1xz}, I_{1yy}, I_{1yz}, $m_1 r_{1x}$, $m_1 r_{1y}$, $m_1 r_{1z}$, m_1, $m_2 r_{2z}$, m_2.

- The parameters which are grouped are: I_{m1}, I_{2yy}, I_{m2}, I_{3yy}, $m_3 r_{3z}$, m_3, I_{4yy}, $m_4 r_{4z}$, m_4, I_{5yy}, $m_5 r_{5z}$, m_5, I_{6yy}, $m_6 r_{6z}$, m_6.

- The grouping relations are:

$$I'_{1zz} = I_{1zz} + I_{m1} + I_{2yy} + \ell_3^2(m_3 + m_4 + m_5 + m_6) + I_{3yy}$$
$$I'_{2xx} = I_{2xx} - I_{2yy} - \ell_3^2(m_3 + m_4 + m_5 + m_6)$$
$$I'_{2xz} = I_{2xz} - \ell_3 m_3 r_{3z}$$
$$I'_{2zz} = I_{2zz} + I_{m2} + \ell_3^2(m_3 + m_4 + m_5 + m_6)$$
$$m'_2 r'_{2x} = m_2 r_{2x} + \ell_3(m_3 + m_4 + m_5 + m_6)$$
$$I'_{3xx} = I_{3xx} - I_{3yy} + I_{4yy} + 2d_4 m_4 r_{4z} + d_4^2(m_4 + m_5 + m_6)$$
$$I'_{3zz} = I_{3zz} + I_{4yy} + 2d_4 m_4 r_{4z} + d_4^2(m_4 + m_5 + m_6)$$
$$m'_3 r'_{3y} = m_3 r_{3y} + m_4 r_{4z} + d_4(m_4 + m_5 + m_6)$$
$$I'_{4xx} = I_{4xx} + I_{5yy} - I_{4yy}$$
$$I'_{4zz} = I_{4zz} + I_{5yy}$$
$$m'_4 r'_{4y} = m_4 r_{4y} - m_5 r_{5z}$$
$$I'_{5xx} = I_{5xx} + I_{6yy} - I_{5yy}$$
$$I'_{5zz} = I_{5zz} + I_{6yy}$$
$$m'_5 r'_{5y} = m_5 r_{5y} + m_6 r_{6z}$$
$$I'_{6xx} = I_{6xx} - I_{6yy}.$$

It follows that the number of base dynamic parameters is 40. As a result of this reduction, it can be found that the number of operations required to compute the dynamic model with the recursive Newton-Euler algorithm is reduced from 294 multiplies and 283 additions to 253 multiplies and 238 additions.

1.3 Identification of kinematic parameters

Errors unavoidably occur when positioning the end effector of a robot manipulator. These are typically both of mechanical and of kinematic nature. The former may be due to joint friction, compliance, gear transmission and backlash. The latter may result not only from imprecise manufacturing of manipulator links and joints or position transducer offsets, but also from poor estimation of the *kinematic parameters* defining the location of the manipulator with respect to the base frame and the end-effector location with respect to the terminal link.

In the sequel, a method for *identification* of the kinematic parameters is presented which is aimed at correcting the nominal values of such parameters on the basis of suitable measurements. The identification is based on a model which gives the mapping between task space errors and kinematic parameter errors; this model is derived below.

1.3.1 Model for identification

The end-effector frame position and orientation can be computed with respect to the base frame as in (1.7). If the values of the kinematic parameters defining the transformation matrices are not accurately known, there will be a deviation between the real end-effector location and the location computed using the direct kinematic model.

The transformation between two consecutive link frames is defined by (1.2) in terms of four parameters for each link. Nevertheless, eq. (1.7) involves two additional transformations; namely, from the base frame b to link 0 frame, and from link n frame to the end-effector frame e. Let us determine the parameters that describe such transformations.

We can define the base frame b arbitrarily. In order to define frame 0 with respect to frame b, six parameters $(\gamma_b, b_b, \alpha_b, \ell_b, \vartheta_b, d_b)$ are needed, in general, as illustrated in Fig. 1.5 for the mutual position and orientation between frames $i-1$ and i. The transformation matrix is

$$^bT_0 = \text{Rot}(Z, \gamma_b)\text{Trans}(Z, b_b)\text{Rot}(X, \alpha_b)\text{Trans}(X, \ell_b)\text{Rot}(Z, \vartheta_b)\text{Trans}(Z, d_b). \tag{1.116}$$

The subsequent transformation from frame 0 to frame 1 is given by (1.2),

1.3. IDENTIFICATION OF KINEMATIC PARAMETERS

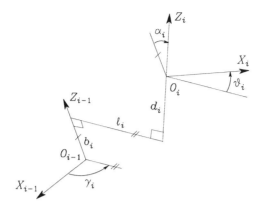

Figure 1.5: Definition of mutual position and orientation between two arbitrary frames.

i.e.,

$$^0T_1 = \text{Rot}(X, \alpha_1)\text{Trans}(X, \ell_1)\text{Rot}(Z, \vartheta_1)\text{Trans}(Z, d_1). \tag{1.117}$$

Combining (1.116) with (1.117) and observing that α_1 and ℓ_1 can be always taken equal to zero gives

$$\begin{aligned}^bT_1 = {}&\text{Rot}(X, \alpha_0)\text{Trans}(X, \ell_0)\text{Rot}(Z, \vartheta_0)\text{Trans}(Z, d_0) \cdot \\ &\text{Rot}(X, \alpha'_1)\text{Trans}(X, a'_1)\text{Rot}(Z, \vartheta'_1)\text{Trans}(Z, d'_1)\end{aligned} \tag{1.118}$$

where $\alpha_0 = 0$, $\ell_0 = 0$, $\vartheta_0 = \gamma_b$, $d_0 = b_b$ and $\alpha'_1 = \alpha_b$, $l'_1 = \ell_b$, $\vartheta'_1 = \vartheta_b + \vartheta_1$, $d'_1 = d_b + d_1$. As can be recognized from (1.118), the overall transformation from frame b to frame 1 can be formally defined in terms of two sets of four parameters, similarly to the case of two consecutive link frames.

Likewise, we can define the end-effector frame e arbitrarily, and then the transformation from frame n to frame e is

$$^nT_e = \text{Rot}(Z, \gamma_e)\text{Trans}(Z, b_e)\text{Rot}(X, \alpha_e)\text{Trans}(X, \ell_e)\text{Rot}(Z, \vartheta_e)\text{Trans}(Z, d_e) \tag{1.119}$$

where the six parameters $(\gamma_e, b_e, \alpha_e, \ell_e, \vartheta_e, d_e)$ are defined with reference to Fig. 1.5. Proceeding as above leads to expressing the overall transformation from frame $n-1$ to frame e as

$$\begin{aligned}^{n-1}T_e = {}&\text{Rot}(X, \alpha'_n)\text{Trans}(X, l'_n)\text{Rot}(Z, \vartheta'_n)\text{Trans}(Z, d'_n) \cdot \\ &\text{Rot}(X, \alpha_{n+1})\text{Trans}(X, \ell_{n+1})\text{Rot}(Z, \vartheta_{n+1})\text{Trans}(Z, d_{n+1})\end{aligned} \tag{1.120}$$

where $\alpha'_n = \alpha_n$, $\ell'_n = \ell_n$, $\vartheta'_n = \vartheta_n + \gamma_e$, $d'_n = d_n + b_e$ and $\alpha_{n+1} = \alpha_e$, $\ell_{n+1} = \ell_e$, $\vartheta_{n+1} = \vartheta_e$, $d_{n+1} = d_e$ are the two sets of four parameters defining the transformation.

As a consequence, the overall transformation from the base frame to the end-effector frame can be described by $n+2$ sets of 4 parameters.

A critical situation occurs when any two consecutive axes Z_{i-1} and Z_i are parallel, since a small deviation of orientation may cause a large error on d_i. In order to overcome such a drawback, it is worth introducing an additional parameter β_i representing a rotation about axis Y_{i-1}, so that eq. (1.2) is modified into

$$^{i-1}T_i = \text{Rot}(Y, \beta_i)\text{Rot}(X, \alpha_i)\text{Trans}(X, \ell_i)\text{Rot}(Z, \vartheta_i)\text{Trans}(Z, d_i). \quad (1.121)$$

The nominal value of β_i is zero. If Z_{i-1} is not parallel to Z_i, then the parameter β_i will not be identified.

Consider the direct kinematics equation (1.14) which can be rewritten as

$$x = k(\zeta) \quad (1.122)$$

where ζ is the $(z \times 1)$ vector of kinematic parameters, with at most $z = 4(n+2) - 2$; indeed, we can see that when β_i exists, then d_i will be cancelled also because $\alpha_0 = \ell_0 = 0$.

Let x_m be the measured location and x_n the nominal location that can be computed via (1.122) with the nominal values of the parameters ζ. The nominal values of the fixed parameters are set equal to the design data of the mechanical structure, whereas the nominal values of the joint variables are set equal to the data provided by the position transducers at the given manipulator configuration. The deviation $\Delta x = x_m - x_n$ gives a measure of accuracy at the given configuration. On the assumption of small deviations, at first approximation, it is possible to derive the following model for identification purposes:

$$\Delta x = \Psi(\zeta_n)\Delta\zeta \quad (1.123)$$

where $\Delta\zeta = \zeta_m - \zeta_n$ is the parameter error and Ψ is an $(m \times z)$ matrix which is usually termed *regressor* of the kinematic model. With reference to (1.13), the task space error can be partitioned as

$$\Delta x = \begin{pmatrix} \Delta p \\ \Delta \phi \end{pmatrix} \quad (1.124)$$

where

$$\Delta p = {}^b p_{em} - {}^b p_{en} \quad (1.125)$$

1.3. IDENTIFICATION OF KINEMATIC PARAMETERS   41

denotes the position error and

$$\Delta\phi = \frac{1}{2}(^b n_{en} \times {}^b n_{em} + {}^b s_{en} \times {}^b s_{em} + {}^b a_{en} \times {}^b a_{em}) \quad (1.126)$$

denotes the orientation error (see Chapter 3 for a formal derivation of this error).

The matrix Ψ plays the role of a Jacobian of the kind $\partial k/\partial \zeta$, and its columns can be computed by a procedure substantially analogous to that followed for the geometric Jacobian; to this purpose, notice that $\Delta\phi$ can be interpreted as an angular velocity for small orientation errors. Let then

$$^b T_i = \begin{pmatrix} ^b n_i & ^b s_i & ^b a_i & ^b p_i \\ 0 & 0 & 0 & 1 \end{pmatrix} \quad (1.127)$$

denote the transformation matrix from frame b to frame i. Without loss of generality, assume that $m = 6$ and that the measuring frame coincides with the base frame. Then, the columns of Ψ corresponding to the five types of kinematic parameters are:

$$\psi_{\beta_i} = \begin{pmatrix} ^b s_{i-1} \times {}^b p_{i-1,e} \\ ^b s_{i-1} \end{pmatrix} \quad (1.128)$$

$$\psi_{\alpha_i} = \begin{pmatrix} ^b n_{i-1} \times {}^b p_{i-1,e} \\ ^b n_{i-1} \end{pmatrix} \quad (1.129)$$

$$\psi_{\ell_i} = \begin{pmatrix} ^b n_{i-1} \\ 0 \end{pmatrix} \quad (1.130)$$

$$\psi_{\vartheta_i} = \begin{pmatrix} ^b a_i \times {}^b p_{i,e} \\ ^b a_i \end{pmatrix} \quad (1.131)$$

$$\psi_{d_i} = \begin{pmatrix} ^b a_i \\ 0 \end{pmatrix}. \quad (1.132)$$

1.3.2 Kinematic calibration

It is desired to compute $\Delta\zeta$ starting from the knowledge of ζ_n, x_n and the measurement of x_m. The end-effector location shall be measured with high precision for the effectiveness of the identification procedure. To this purpose, we may use a mechanical apparatus that allows constraining the end effector at given locations with a priori known precision. Alternatively, direct measurement systems of object position and orientation in the Cartesian space can be used which employ triangulation techniques. Notice, however, that an accurate measurement of orientation is difficult to obtain, and thus only the first three equations of (1.123) are typically taken into account.

Since (1.123) constitutes a system of m equations into z unknowns with $m < z$, a sufficient number of end-effector location measures has to be performed so as to obtain a system of at least z equations. Therefore, if measurements are made for a number of N locations, eq. (1.123) yields

$$\Delta \bar{x} = \begin{pmatrix} \Delta x_1 \\ \vdots \\ \Delta x_N \end{pmatrix} = \begin{pmatrix} \Psi_1 \\ \vdots \\ \Psi_N \end{pmatrix} \Delta \zeta = \bar{\Psi} \Delta \zeta. \tag{1.133}$$

As regards the nominal values of the parameters needed for the computation of the matrices Ψ_j, it should be observed that the kinematic parameters are all constant except the joint variables which depend on the manipulator configuration at location j.

In order to avoid ill-conditioning of matrix $\bar{\Psi}$, it is advisable to choose N so that $Nm \gg z$ and then solve (1.133) with a least-squares technique; in this case the solution is of the form

$$\Delta \zeta = (\bar{\Psi}^T \bar{\Psi})^{-1} \bar{\Psi}^T \Delta \bar{x} \tag{1.134}$$

where $(\bar{\Psi}^T \bar{\Psi})^{-1} \bar{\Psi}^T$ is the *left pseudoinverse* matrix of $\bar{\Psi}$. By computing $\bar{\Psi}$ with the nominal values of the parameters ζ_n, the first parameter *estimate* is given by

$$\zeta' = \zeta_n + \Delta \zeta. \tag{1.135}$$

This is a nonlinear parameter estimation problem and, as such, the procedure shall be iterated until $\Delta \zeta$ converges within a given threshold; this procedure is commonly known as *kinematic calibration*. At each iteration, the matrix $\bar{\Psi}$ is to be updated with the parameter estimates ζ' obtained via (1.135) at the previous iteration. In a similar manner, the deviation $\Delta \bar{x}$ is to be computed as the difference between the measured values for the N end-effector locations and the corresponding locations computed by the direct kinematics function with the values of the parameters at the previous iteration. As a result of the kinematic calibration procedure, more accurate estimates of the real manipulator geometric parameters as well as possible corrections to make on the joint transducers measurements are obtained.

1.3.3 Parameter identifiability

The above identification of kinematic parameters requires that the matrix $\bar{\Psi}$ in (1.134) is full-rank. If this is not the case, a loss of *identifiability* of some parameters occurs. This may be caused by two kinds of problems; namely, structural unidentifiability and selection of measuring locations.

Unidentifiability of some parameters constitutes a structural problem when some columns of the matrix $\bar{\Psi}$ are zero or they are linearly dependent,

1.3. IDENTIFICATION OF KINEMATIC PARAMETERS

whatever the number of measurements N used to construct the matrix $\bar{\Psi}$. In the former case, the corresponding parameters have to be eliminated from ζ and the corresponding null columns in $\bar{\Psi}$ have to be eliminated as well. In the latter case, it is necessary to eliminate the dependent columns from $\bar{\Psi}$ and the corresponding parameters from ζ; this gives the set of *base kinematic parameters* of the model for identification. The errors corresponding to the eliminated parameters can be added as a linear combination of the errors of the base parameters.

As a simple example, consider the case when Z_i and Z_{i-1} are parallel; in this case, the column $\bar{\psi}_{d_i}$ is equal to $\bar{\psi}_{d_{i-1}}$. Hence, $\bar{\psi}_{d_i}$ (or $\bar{\psi}_{d_{i-1}}$) can be eliminated from $\bar{\Psi}$, the parameter d_{i-1} (or d_i) can be eliminated from ζ, and the parameter error Δd_{i-1} (or Δd_i) will be equal to $\Delta d_{i-1} + \Delta d_i$.

In case of a rank-deficient matrix $\bar{\Psi}$, eq. (1.133) can be rewritten as

$$\Delta x = (\bar{\Psi}_1 \quad \bar{\Psi}_2) \begin{pmatrix} \Delta \zeta_1 \\ \Delta \zeta_2 \end{pmatrix} \quad (1.136)$$

where $\bar{\Psi}_1$ includes the r independent columns of $\bar{\Psi}$, $\bar{\Psi}_2$ includes the remaining $z - r$ dependent columns, and ζ_1, ζ_2 are determined accordingly. The columns of $\bar{\Psi}_2$ can be written as a linear combination of the columns of $\bar{\Psi}_1$ as

$$\bar{\Psi}_2 = \bar{\Psi}_1 V \quad (1.137)$$

where V is a suitable $(r \times (z - r))$ matrix with constant elements. Substituting (1.137) in (1.136) gives

$$\Delta \zeta = \bar{\Psi}_1 \Delta \zeta_r \quad (1.138)$$

where

$$\Delta \zeta_r = \Delta \zeta_1 + V \Delta \zeta_2. \quad (1.139)$$

In the identification process, eq. (1.138) shall be used instead of (1.134), and the solution will yield the correction on the base parameters; notice that the matrix V is not needed by the identification, the error components on the parameters ζ_2 are considered null, and the parameter error will be updated using only the parameters ζ_1.

It should be clear that the choice of the independent columns of $\bar{\Psi}$ is not unique. A good criterion is to select the columns corresponding to those parameters which appear explicitly in the symbolic kinematic model, such as:

- the joint variables (the corresponding errors represent offsets on the position transducer values),
- the distances ℓ_i and d_i whose nominal values are not equal to zero,

- the angles α_i and ϑ_i whose nominal values are not equal to $k\pi/2$ with k integer,

- the parameters defining link 0 with respect to the base frame,

- the parameters defining the end-effector frame with respect to link n frame.

Once such a selection is operated, it is sufficient to permute the columns of $\bar{\Psi}$ so that the independent columns come first as in (1.136).

As an alternative to symbolic reduction of the kinematic parameters, numerical techniques based either on SVD or on QR decomposition of the matrix $\bar{\Psi}$ can be used to determine the base kinematic parameters.

Ill-conditioning of the matrix $\bar{\Psi}$ in (1.134) may depend also on the number and type of *measuring locations*. As the measuring process is time consuming, the use of a limited number of good selected locations may be crucial. A good selection of the locations may be more important than its number to achieve good identification results.

An index to quantify the goodness of a location is given by the condition number of the matrix $\bar{\Psi}^T\bar{\Psi}$. Hence, the locations shall be chosen so as to minimize the condition number, and thus to obtain low sensitivity of the solution to noise and error modelling.

1.4 Identification of dynamic parameters

Simulation and control of a robot manipulator demands the knowledge of the values of the *dynamic parameters*. Computing such parameters from the design data of the mechanical structure is not simple. We may adopt CAD modelling techniques which allow computing the values of the dynamic parameters of the various components (links, actuators and transmissions) on the basis of their geometry and type of materials employed. Nevertheless, the estimates obtained by such techniques are inaccurate because of the simplification typically introduced by geometric modelling; moreover, complex dynamic effects, such as joint friction, cannot be taken into account.

A heuristic approach could be to dismantle the various components of the manipulator and perform a series of measurements to evaluate the dynamic parameters. Such a technique is not easy to implement and may be troublesome to measure the relevant quantities. In order to find accurate estimates of dynamic parameters, it is worth resorting to *identification* techniques which are illustrated below.

1.4.1 Use of dynamic model

The natural choice of a model for identification of dynamic parameters is the dynamic model of a manipulator, which has the nice property of linearity in the dynamic parameters. In particular, if a set of *base* parameters has been found, the model to consider is described by (1.115).

The goal of an identification method is to provide the best estimates of the parameter vector ρ from the measurements of joint torques u and of relevant quantities for the evaluation of the matrix $Y(\cdot)$, when suitable motion trajectories are imposed to the manipulator. The matrix Y can be obtained directly if a symbolic model of the manipulator (with Lagrange formulation) is available. Otherwise, its column y_i can be computed using the Newton-Euler algorithm based on (1.83)–(1.88) as the torque u with $\rho_i = 1$ and $\rho_j = 0$ for $j \neq i$.

On the assumption that the kinematic parameters in the matrix $Y(\cdot)$ are known with good accuracy, e.g., as a result of a kinematic calibration, measurements of joint positions q, velocities $\dot{q}$ and accelerations $\ddot{q}$ are required. Joint positions and velocities can be actually measured while numerical reconstruction of accelerations is needed; this can be performed on the basis of the position and velocity values recorded during the execution of the trajectories. The reconstructing filter does not work in real time and thus it can also be anti-causal, allowing an accurate reconstruction of the accelerations. As regards joint torques, in the unusual case of torque sensors at the joint, these can be measured directly. Otherwise, they can be evaluated from either wrist force measurements or current measurements in the case of electric actuators.

If measurements of joint torques, positions, velocities and accelerations have been obtained at given time instants $t_1, \ldots, t_N$ along a given trajectory, we may write

$$\bar{u} = \begin{pmatrix} u(t_1) \\ \vdots \\ u(t_N) \end{pmatrix} = \begin{pmatrix} Y(t_1) \\ \vdots \\ Y(t_N) \end{pmatrix} \rho = \bar{Y}\rho. \qquad (1.140)$$

The number of time instants sets the number of measurements to perform and shall be chosen as $Nn \geq r$ so as to obtain an overdetermined system of equations. Solving (1.140) by a least-squares technique leads to the solution in the form

$$\rho = (\bar{Y}^T \bar{Y})^{-1} \bar{Y}^T \bar{u} \qquad (1.141)$$

where $(\bar{Y}^T \bar{Y})^{-1} \bar{Y}^T$ is the *left pseudoinverse* matrix of $\bar{Y}$. The use of base dynamic parameters guarantees that the matrix $\bar{Y}$ is always full-rank. As pointed out above, the base parameters can be computed either symbolically or numerically.

It is worth observing that this method can be extended also to the identification of the dynamic parameters of an unknown payload at the end effector. In such a case, the payload can be regarded as a structural modification of the last link and we may proceed to identify the dynamic parameters of the modified link. To this purpose, if a force sensor is available at the manipulator wrist, it is possible to directly characterize the dynamic parameters of the payload starting from force sensor measurements.

A critical issue for identification of dynamic parameters is the type of trajectory imposed to the manipulator joints. The choice shall be oriented towards polynomial type trajectories which are sufficiently *exciting*, otherwise some parameters may become unidentifiable or very sensitive to noisy data. This can be achieved by minimizing the condition number of the *persistent excitation matrix* matrix $\bar{Y}^T\bar{Y}$ along the trajectory. On the other hand, such trajectories shall not excite any unmodelled dynamic effects such as joint elasticity or link flexibility that would naturally lead to obtaining unreliable estimates of the dynamic parameters to identify.

As pointed out previously, the torque at joint i is a function only of the dynamic parameters of links i to n. It follows that the matrix Y has a block upper-triangular structure so that

$$\begin{pmatrix} u_1 \\ \vdots \\ u_n \end{pmatrix} = \begin{pmatrix} y_{11}^T & \cdots & y_{1n}^T \\ \vdots & \ddots & \vdots \\ 0 & \cdots & y_{nn}^T \end{pmatrix} \begin{pmatrix} \rho_1 \\ \vdots \\ \rho_n \end{pmatrix} \qquad (1.142)$$

where y_{ij}^T is a row vector containing the coefficients of the base parameters of link j in the equation of motion for joint i. It follows that computation of parameter estimates could be simplified by resorting to a sequential procedure. Take the equation $u_n = y_{nn}^T \rho_n$ and solve it for ρ_n by specifying u_n and y_{nn}^T for a given trajectory on joint n. By iterating the procedure, the manipulator parameters can be identified on the basis of measurements performed joint by joint from the outer link to the base.

The search of an exciting trajectory with this method is simplified for two reasons:

- since the number of parameters to be identified at each step is reduced, the number of columns of matrix $\bar{Y}$ is reduced too;

- since the row vector y_{ij}^T is a function of positions, velocities and accelerations of joints from 1 to j only, the identification of ρ_j can be carried out by fixing joints $j+1$ to n.

1.4. IDENTIFICATION OF DYNAMIC PARAMETERS

A shortcoming of such a method, however, is the risk of accumulating errors due to ill-conditioning of the matrices involved step by step.

1.4.2 Use of energy model

From the principle of conservation of energy, it is

$$\int_{t_a}^{t_b} u^T \dot{q} dt = W(t_b) - W(t_a) \qquad (1.143)$$

where W is the Hamiltonian, and u does not include friction torques; additional terms are obtained on the right-hand side of (1.143), if friction is present. In view of the property of linearity expressed by (1.91), it is

$$W = y'^T(q, \dot{q})\rho \qquad (1.144)$$

where the vector of base dynamic parameters ρ has been considered, and y'^T is a suitable $(1 \times r)$ vector of coefficients which depends on joint positions and velocities and can be computed from the symbolic expressions of kinetic and potential energy. Using (1.144) in (1.143) gives

$$\Delta W = \Delta y'^T(q, \dot{q})\rho \qquad (1.145)$$

where $\Delta W = W(t_b) - W(t_a)$ and $\Delta y' = y'(t_b) - y'(t_a)$.

Eq. (1.145) is the *energy model* which can be used for identification of the base dynamic parameters. If measurements of joint torques, positions and velocities have been obtained at given N couples of time instants $t_{a1}, t_{b1}, \ldots, t_{aN}, t_{bN}$ along a given trajectory, we may write

$$\Delta \bar{W} = \begin{pmatrix} \Delta W(\Delta t_1) \\ \vdots \\ \Delta W(\Delta t_N) \end{pmatrix} = \begin{pmatrix} \Delta y'(\Delta t_1) \\ \vdots \\ \Delta y'(\Delta t_N) \end{pmatrix} \rho = \Delta Y'\rho. \qquad (1.146)$$

where $\Delta t_i = t_{bi} - t_{ai}$ for $i = 1, \ldots, N$, and $N \geq r$. From joint velocity and torque measurements, the vector $\Delta \bar{W}$ can be computed via (1.143).

The model (1.146) has the same form as (1.140), and then the parameters can be identified by using a left pseudoinverse of $\Delta Y'$ as in the solution given by (1.141). The main advantage of this method over the previous one is that it avoids reconstruction of joint accelerations. Nevertheless, for a given trajectory, the condition number of the persistent excitation matrix $\bar{Y}^T \bar{Y}$ is in general smaller than that of the matrix $\Delta Y'^T \Delta Y'$.

1.5 Further reading

Kinematic and dynamic modelling of rigid robot manipulators can be found in any classical robotics textbook, e.g., [72, 19, 21, 86, 83]. Precious reference sources are [41, 13, 1, 68] for kinematics, and [90, 35, 93, 81] for dynamics. Symbolic software packages were developed to derive both kinematic and dynamic models of rigid robot manipulators, e.g., [44].

The Denavit-Hartenberg notation dates back to the original work of [20], which was recently modified in [19, 54]. One advantage of the so-called modified Denavit-Hartenberg notation over the classical one is that it can be used also for tree-structured and closed-chain robot manipulators [54]. The homogeneous transformation representation for direct kinematics of open-chain manipulators was first proposed in [75].

Sufficient conditions for the inverse kinematics problem to have closed-form solutions were given in [75]. These ensure the existence of solutions to 6-degree-of-freedom manipulators provided that there are three revolute joints with intersecting axes or three prismatic joints; in the former case, at most 8 admissible solutions exist, while the number reduces to 2 in the latter case. The kinematic decoupling resulting for spherical-wrist manipulators was developed in [24, 39, 73, 45]. An algebraic approach to the inverse kinematics problem for manipulators having closed-form solutions was presented in [72], which consists of successively post- (or pre-) multiplying both sides of the direct kinematics equation (1.6) by partial transformation matrices so as to isolate the joint variables one after another; the types of equations that can be obtained with this approach were formalized in [21]. Recent methods [62, 78] have found the inverse kinematics solution to general six-revolute-joint manipulators in the form of a polynomial equation of degree 16, i.e, the maximum number of admissible solutions is 16. On the other hand, numerical solution techniques based on iterative algorithms have been proposed, e.g., [91, 34].

The geometric Jacobian of the differential kinematics equation was originally proposed in [95]. The decomposition of the Jacobian into the product of three matrices is due to [80]. The problem of efficient Jacobian computation was addressed in [71]. The analytical Jacobian concept was introduced in [56] in connection with the operational space control problem. A treatment of differential kinematics mapping properties can be found in [83]; the reader is referred to [60, 61, 22] for SVD and QR decomposition.

Pioneering work on dynamics of open-chain manipulators using Lagrange formulation can be found in [92, 8, 87]. An efficient recursive computation of manipulator dynamics was proposed in [38]. The Newton-Euler formulation was used in [70], and its efficient recursive computation was proposed in [64] for inverse dynamics; however, the same formulation can be effectively used also for direct dynamics computation [94]. The two formulations have been compared from a computational viewpoint in [85]. Recursive methods for calculating the direct dynamic model using Newton-Euler equations without calculating firstly the inertia matrix can be found in [6, 25]. On the other hand, the dynamics of tree-structured and closed-loop manipulators was studied in [59, 55, 65]. The set of dynamic parameters was formalized in [50, 53, 57]. Inclusion of actuator inertia and gyroscopic effects was studied in [18, 84]. For friction modelling, the reader is referred to [4].

Efforts to reduce the computational burden of the dynamic model were made, e.g., [42, 55] by suitably customizing the symbolic model to compute. A breakthrough was the symbolic determination of the base dynamic parameters both for open-chain manipulators [28, 30, 48, 67, 63, 46] and for tree-structured and closed-chain manipulators [10, 11, 47]. Alternatively, the base dynamic parameters can be computed numerically, as in [27].

The problem of kinematic calibration is surveyed in [82, 40]. Original work on identification of kinematic parameters was done by [97, 74, 89, 96, 88]. The addition of a fifth parameter to the kinematic model is due to [37], with further refinements in [99]. Symbolic computation of the base kinematic parameters was developed in [52], whereas the reader is referred to [51] for QR decomposition needed for numerical computation of the base parameters. Exciting configurations are given in [12, 52]. Recently, the so-called autonomous methods which do not need an external sensor to measure the end-effector position or orientation have been proposed [9, 23, 98, 49, 69]; these methods construct the identification model by virtually realizing a mechanical contact (point-to-point or frame-to-frame) between the terminal link and the environment.

Identification of inertial parameters using standard least-squares techniques was carried out in [66, 58, 7, 26, 43, 36, 79, 32]. Friction identification was investigated in [3, 16, 17]. As opposed to identification, physical measurement of dynamic parameters was accomplished in [5]. The sequential identification method was developed in [15]. The use of the energy model was proposed in [29] and carried out by sequential method in [76]. The problem of finding exciting trajectories was studied in [2, 31]. Practical implementations of identification methods can be found in [14, 77, 33].

References

[1] J. Angeles, *Spatial Kinematic Chains: Analysis, Synthesis, Optimization*, Springer-Verlag, Berlin, D, 1982.

[2] B. Armstrong, "On finding 'exciting' trajectories for identification experiments involving systems with non-linear dynamics," *Proc. 1987 IEEE Int. Conf. on Robotics and Automation*, Raleigh, NC, pp. 1131–1139, 1987.

[3] B. Armstrong, "Friction: Experimental determination, modeling and compensation," *Proc. 1988 IEEE Int. Conf. on Robotics and Automation*, Philadelphia, PA, pp. 1422–1427, 1988.

[4] B. Armstrong-Hélouvry, *Control of Machines with Friction*, Kluwer Academic Publishers, Boston, MA, 1991.

[5] B. Armstrong, O. Khatib, and J. Burdick, "The explicit dynamic model and inertial parameters of the PUMA 560 Arm," *Proc. 1986 IEEE Int. Conf. on Robotics and Automation*, San Francisco, CA, pp. 510–518, 1986.

[6] W.W. Armstrong, "Recursive solution to the equations of motion of an n-link manipulator," *Proc. 5th World Congr. on Theory of Machines and Mechanisms*, Montreal, CAN, pp. 1343–1346, 1979.

[7] C.G. Atkeson, C.H. An, and J.M. Hollerbach, "Estimation of inertial parameters of manipulator loads and links," *Int. J. of Robotics Research*, vol. 5, no. 3, pp. 101–119, 1986.

[8] A.K. Bejczy, *Robot Arm Dynamics and Control*, memo. TM 33-669, Jet Propulsion Laboratory, California Institute of Technology, 1974.

[9] D.J. Bennet and J.M. Hollerbach, "Self-calibration of single-loop, closed kinematic chains formed by dual or redundant manipulators," *Proc. 27th IEEE Conf. on Decision and Control*, Austin, TX, pp. 627–629, 1988.

[10] F. Bennis and W. Khalil, "Minimum inertial parameters of robots with parallelogram closed-loops," *IEEE Trans. on Systems, Man, and Cybernetics*, vol. 21, pp. 318–326, 1991.

[11] F. Bennis, W. Khalil, and M. Gautier, "The minimum inertial parameters of robots with general closed-loops," *Proc. 1st European Control Conf.*, Grenoble, F, pp. 450–455, 1991.

[12] J.H. Borm and C.H. Menq, "Experimental study of observability of parameter errors in robot calibration," *Proc. 1989 IEEE Int. Conf. on Robotics and Automation*, Scottsdale, AZ, pp. 587–592, 1989.

REFERENCES

[13] O. Bottema and B. Roth, *Theoretical Kinematics*, North Holland, Amsterdam, NL, 1979.

[14] F. Caccavale and P. Chiacchio, "Identification of dynamic parameters and feedforward control for a conventional industrial manipulator," *Control Engineering Practice*, vol. 2, pp. 1039–1050, 1994.

[15] C. Canudas de Wit and A. Aubin, "Parameters identification of robots manipulators via sequential hybrid estimation algorithms," *Prepr. 11th IFAC World Congr.*, Tallinn, Estonia, vol. 9, pp. 178–183, 1990.

[16] C. Canudas de Wit, A. Aubin, B. Brogliato, and P. Drevet, "Adaptive friction compensation in robot manipulators: low-velocities," *Proc. 1989 IEEE Int. Conf. on Robotics and Automation*, Scottsdale, AZ, pp. 1352–1357, 1989.

[17] C. Canudas de Wit and V. Seront, "Robust adaptive friction compensation," *Proc. 1990 IEEE Int. Conf. on Robotics and Automation*, Cincinnati, OH, pp. 1383–1389, 1990.

[18] P. Chedmail, M. Gautier, and W. Khalil, "Automatic modelling of robots including parameters of links and actuators," *Proc. IFAC Symp. on Theory of Robots*, Vienna, A, pp. 295–299, 1986.

[19] J.J. Craig, *Introduction to Robotics: Mechanics and Control*, (2nd ed.), Addison-Wesley, Reading, MA, 1989.

[20] J. Denavit and R.S. Hartenberg, "A kinematic notation for lower-pair mechanisms based on matrices," *ASME J. of Applied Mechanics*, vol. 22, pp. 215–221, 1955.

[21] E. Dombre and W. Khalil, *Modélisation et Commande des Robots*, Hermès, Paris, F, 1988.

[22] J.J. Dongarra, C.B. Moler, J.R. Bunch, and G.W. Stewart, *LINPACK User's Guide*, SIAM, Philadelphia, PA, 1979.

[23] C. Edwards and L. Galloway, "A single-point calibration technique for a six-degree-of-freedom articulated arm," *Int. J. of Robotics Research*, vol. 13, pp. 189–199, 1994.

[24] R. Featherstone, "Position and velocity transformations between robot end-effector coordinates and joint angles," *Int. J. of Robotics Research*, vol. 2, no. 2, pp. 35–45, 1983.

[25] R. Featherstone, *Robot Dynamics Algorithms*, Kluwer Academic Publishers, Boston, MA, 1987.

[26] M. Gautier, "Identification of robots dynamics," *Proc. IFAC Symp. on Theory of Robots*, Vienna, A, pp. 351–356, 1986.

[27] M. Gautier, "Numerical calculation of the base inertial parameters," *J. of Robotic Systems*, vol. 8, pp. 485–506, 1991.

[28] M. Gautier and W. Khalil, "A direct determination of minimum inertial parameters of robots," *Proc. 1988 IEEE Int. Conf. on Robotics and Automation*, Philadelphia, PA, pp. 1682–1687, 1988.

[29] M. Gautier and W. Khalil, "On the identification of the inertial parameters of robots," *Proc. 27th IEEE Conf. on Decision and Control*, Austin, TX, pp. 2264–2269, 1988.

[30] M. Gautier and W. Khalil, "Direct calculation of minimum set of inertial parameters of serial robots," *IEEE Trans. on Robotics and Automation*, vol. 6, pp. 368–373, 1990.

[31] M. Gautier and W. Khalil, "Exciting trajectories for inertial parameters identification," *Int. J. of Robotics Research*, vol. 11, pp. 362–375, 1992.

[32] M. Gautier, W. Khalil, and P.P. Restrepo, "Identification of the dynamic parameters of a closed loop robot," *Proc. 1995 IEEE Int. Conf. on Robotics and Automation*, Nagoya, J, pp. 3045–3050, 1995.

[33] M. Gautier, P.P. Restrepo, and W. Khalil, "Identification of an industrial robot using filtered dynamic model," *Proc. 3rd European Control Conf.*, Rome, I, pp. 2380–2385, 1995.

[34] A.A. Goldenberg, B. Benhabib, and R.G. Fenton, "A complete generalized solution to the inverse kinematics of robots," *IEEE J. of Robotics and Automation*, vol. 1, pp. 14–20, 1985.

[35] H. Goldstein, *Classical Mechanics*, (2nd ed.), Addison-Wesley, Reading, MA, 1980.

[36] I.J. Ha, M.S. Ko, and S.K. Kwon, "An efficient estimation algorithm for the model parameters of robotics manipulators," *IEEE Trans. on Robotics and Automation*, vol. 5, pp. 386–394, 1989.

[37] S.A. Hayati, "Robot arm geometric link parameter estimation," *Proc. 22nd IEEE Conf. on Decision and Control*, San Antonio, TX, pp. 1477–1483, 1983.

[38] J.M. Hollerbach, "A recursive Lagrangian formulation of manipulator dynamics and a comparative study of dynamics formulation complexity," *IEEE Trans. on Systems, Man, and Cybernetics*, vol. 10, pp. 730–736, 1980.

[39] J.M. Hollerbach, "Wrist-partitioned inverse kinematic accelerations and manipulator dynamics," *Int J. of Robotics Research*, vol. 2, no. 4, pp. 61–76, 1983.

[40] J.M. Hollerbach, "A survey of kinematic calibration," *The Robotics Review 1*, O. Khatib, J.J. Craig, and T. Lozano-Pérez (Eds.), MIT Press, Cambridge, MA, pp. 207–242, 1989.

[41] K.H. Hunt, *Kinematic Geometry of Mechanisms*, Clarendon Press, Oxford, UK, 1978.

[42] T. Kanade, P. Khosla, and N. Tanaka, "Real-time control of the CMU direct-drive arm II using customized inverse dynamics," *Proc. 23th IEEE Conf. on Decision and Control*, Las Vegas, NV, pp. 1345–1352, 1984.

[43] H. Kawasaki and K. Nishimura, "Terminal-link parameter estimation and trajectory control of robotic manipulators," *IEEE J. of Robotics and Automation*, vol. 4, pp. 485–490, 1988.

[44] W. Khalil, "A system for generating the symbolic models of robots," *Postpr. 4th IFAC Symp. on Robot Control*, Capri, I, pp. 416–468, 1994.

[45] W. Khalil and F. Bennis, "Automatic generation of the inverse geometric model of robots," *Robotics and Autonomous Systems*, vol. 7, pp. 1–10, 1991.

[46] W. Khalil and F. Bennis, "Comments on "Direct calculation of minimum set of inertial parameters of serial robots"," *IEEE Trans. on Robotics and Automation*, vol. 10, pp. 78–79, 1994.

[47] W. Khalil and F. Bennis, "Symbolic calculation of the base inertial parameters of closed-loop robots," *Int. J. of Robotics Research*, vol. 14, pp. 112–128, 1995.

[48] W. Khalil, F. Bennis, and M. Gautier, "The use of the generalized links to determine the minimal inertial parameters of robots," *J. of Robotic Systems*, vol. 7, pp. 225–242, 1990.

[49] W. Khalil, G. Garcia, and J.F. Delagarde, "Calibration of the geometric parameters of robots without external sensors," *Proc. 1995 IEEE Int. Conf. on Robotics and Automation*, Nagoya, J, pp. 3039–3044, 1995.

[50] W. Khalil and M. Gautier, "On the derivation of the dynamic models of robots," *Proc. 2nd Int. Conf. on Advanced Robotics*, Tokyo, J, pp. 243–250, 1985.

[51] W. Khalil and M. Gautier, "Calculation of the identifiable parameters for robot calibration," *Prepr. 9th IFAC Symp. on Identification and System Parameter Estimation*, Budapest, H, pp. 888–892, 1991.

[52] W. Khalil, M. Gautier, and C. Enguehard, "Identifiable parameters and optimum configurations for robots calibrations," *Robotica*, vol. 9, pp. 63–70, 1991.

[53] W. Khalil, M. Gautier, and J.F. Kleinfinger, "Automatic generation of identification model of robots," *IASTED Int. J. of Robotics and Automation*, vol. 1, pp. 2–6, 1986.

[54] W. Khalil and J.F. Kleinfinger, "A new geometric notation for open and closed-loop robots," *Proc. 1986 IEEE Int. Conf. on Robotics and Automation*, San Francisco, CA, pp. 1174–1180, 1986.

[55] W. Khalil and J.F. Kleinfinger, "Minimum operations and minimum parameters of the dynamic model of tree structure robots," *IEEE J. of Robotics and Automation*, vol. 3, pp. 517–526, 1987.

[56] O. Khatib, "A unified approach for motion and force control of robot manipulators: The operational space formulation," *IEEE J. of Robotics and Automation*, vol. 3, pp. 43–53, 1987.

[57] P.K. Khosla, "Categorization of parameters in the dynamic robot model," *IEEE Trans. on Robotics and Automation*, vol. 5, pp. 261–268, 1989.

[58] P.K. Khosla and T. Kanade, "Parameter identification of robot dynamics," *Proc. 24th IEEE Conf. on Decision and Control*, Fort Lauderdale, FL, pp. 1754–1760, 1985.

[59] J.F. Kleinfinger and W. Khalil, "Dynamic modelling of closed-chain robots," *Proc. 16th Int. Symp. on Industrial Robots*, Bruxelles, B, pp. 401–412, 1986.

REFERENCES

[60] V.C. Klema and A.J. Laub, "The singular value decomposition: Its computation and some applications," *IEEE Trans. on Automatic Control*, vol. 25, pp. 164–176, 1980.

[61] C.L. Lawson and R.J. Hanson, *Solving Least Squares Problems*, Prentice-Hall, Englewood Cliffs, NJ, 1974.

[62] H.Y. Lee and C.G. Liang, "Displacement analysis of the general 7-link 7R mechanism," *Mechanism and Machine Theory*, vol. 23, pp. 219–226, 1988.

[63] S.K. Lin, "An identification method for estimating the inertia parameters of a manipulator," *J. of Robotic Systems*, vol. 9, pp. 505–528, 1992.

[64] J.Y.S. Luh, M.W. Walker, and R.P.C. Paul, "On-line computational scheme for mechanical manipulators," *ASME J. of Dynamic Systems, Measurement, and Control*, vol. 102, pp. 69–76, 1980.

[65] J.Y.S. Luh and Y.F. Zheng, "Computation of input generalized forces for robots with closed kinematic chain mechanisms," *IEEE J. of Robotics and Automation*, vol. 1, pp. 95–103, 1985.

[66] H. Mayeda, K. Osuka, and A. Kangawa, "A new identification method for serial manipulator arms," *Prepr. 9th IFAC World Congr.*, Kyoto, J, vol. 6, pp. 74–79, 1981.

[67] H. Mayeda, K. Yoshida, and K. Osuka, "Base parameters of manipulator dynamic models," *IEEE Trans. on Robotics and Automation*, vol. 6, pp. 312–321, 1990.

[68] J.M. McCarthy, *An Introduction to Theoretical Kinematics*, MIT Press, Cambridge, MA, 1990.

[69] H. Nakamura, T. Itaya, K. Yamamoto, and T. Koyama, "Robot autonomous error calibration method for off line programming system," *Proc. 1995 IEEE Int. Conf. on Robotics and Automation*, Nagoya, J, pp. 1775–1783, 1995.

[70] D.E. Orin, R.B. McGhee, M. Vukobratović, and G. Hartoch, "Kinematic and kinetic analysis of open-chain linkages utilizing Newton-Euler methods," *Mathematical Biosciences*, vol. 43, pp. 107–130, 1979.

[71] D.E. Orin and W.W. Schrader, "Efficient computation of the Jacobian for robot manipulators," *Int. J. of Robotics Research*, vol. 3, no. 4, pp. 66–75, 1984.

[72] R.P. Paul, *Robot Manipulators: Mathematics, Programming, and Control*, MIT Press, Cambridge, MA, 1981.

[73] R.P. Paul and H. Zhang, "Computationally efficient kinematics for manipulators with spherical wrists based on the homogeneous transformation representation," *Int. J. of Robotics Research*, vol. 5, no. 2, pp. 32–44, 1986.

[74] D. Payannet, M.J. Aldon, and A. Liégeois, "Identification and compensation of mechanical errors for industrial robots," *Proc. 15th Int. Symp. on Industrial Robots*, Tokyo, J, pp. 857–864, 1985.

[75] D.L. Pieper, *The Kinematics of Manipulators Under Computer Control*, memo. AIM 72, Stanford Artificial Intelligence Laboratory, 1968.

[76] C. Pressé and M. Gautier, "Identification of robot base parameters via sequential energy method," *Prepr. 3rd IFAC Symp. on Robot Control*, Vienna, A, pp. 117–122, 1991.

[77] M. Prüfer, C. Schmidt, and F. Wahl, "Identification of robot dynamics with differential and integral models: A comparison," *Proc. 1994 IEEE Int. Conf. on Robotics and Automation*, San Diego, CA, pp. 340–345, 1994.

[78] M. Raghavan and B. Roth, "Inverse kinematics of the general 6R manipulator and related linkages," *ASME J. of Mechanical Design*, vol. 115, pp. 502–508, 1990.

[79] B. Raucent, G. Campion, G. Bastin, J.C. Samin, and P.Y. Willems, "Identification of the barycentric parameters of robot manipulators from external measurements," *Automatica*, vol. 28, pp. 1011–1016, 1992.

[80] M. Renaud, "Calcul de la matrice jacobienne necessaire à la commande coordonnee d'un manipulateur," *Mechanism and Machine Theory*, vol. 15, pp. 81–91, 1980.

[81] R.E. Roberson and R. Schwertassek, *Dynamics of Multibody Systems*, Springer-Verlag, Berlin, D, 1988.

[82] Z. Roth, B.W. Mooring, and B. Ravani, "An overview of robot calibration," *IEEE J. of Robotics and Automation*, vol. 3, pp. 377–386, 1987.

[83] L. Sciavicco and B. Siciliano, *Modeling and Control of Robot Manipulators*, McGraw-Hill, New York, NY, 1996.

REFERENCES

[84] L. Sciavicco, B. Siciliano, and L. Villani, "Lagrange and Newton-Euler dynamic modeling of a gear-driven rigid robot manipulator with inclusion of motor inertia effects," *Advanced Robotics*, vol. 10, pp. 317–334, 1996.

[85] D.B. Silver, "On the equivalence of Lagrangian and Newton-Euler dynamics for manipulators," *Int. J. of Robotics Research*, vol. 1, no. 2, pp. 60–70, 1982.

[86] M.W. Spong and M. Vidyasagar, *Robot Dynamics and Control*, Wiley, New York, NY, 1989.

[87] Y. Stepanenko and M. Vukobratović, "Dynamics of articulated open-chain active mechanisms," *Mathematical Biosciences*, vol. 28, pp. 137–170, 1976.

[88] H.W. Stone, *Kinematic Modeling, Identification, and Control of Robotic Manipulators*, Kluwer Academic Publishers, Boston, MA, 1987.

[89] K. Sugimoto and T. Okada, "Compensation of positioning errors caused by geometric deviations in robot systems," *Robotics Research: 2nd Int. Symp.*, H. Hanafusa and H. Inoue (Eds.), MIT Press, Cambridge, MA, pp. 287–298, 1985.

[90] K.R. Symon, *Mechanics* (3rd ed.), Addison-Wesley, Reading, MA, 1971.

[91] L.W. Tsai and A.P. Morgan, "Solving the kinematics of the most general six- and five-degree-of-freedom manipulators by continuation methods," *ASME J. of Mechanisms, Transmission, and Automation in Design* vol. 107, pp. 189–200, 1985.

[92] J.J. Uicker, "Dynamic force analysis of spatial linkages," *ASME J. of Applied Mechanics*, vol. 34, pp. 418–424, 1967.

[93] M. Vukobratović and V. Potkonjak, *Dynamics of Manipulation Robots*, Scientific Fundamentals of Robotics, vol. 1, Springer-Verlag, Berlin, D, 1982.

[94] M.W. Walker and D.E. Orin, "Efficient dynamic computer simulation of robotics mechanism," *ASME J. of Dynamic Systems, Measurement, and Control*, vol. 104, pp. 205–211, 1982.

[95] D.E. Whitney, "Resolved motion rate control of manipulators and human prostheses," *IEEE Trans. on Man-Machine Systems*, vol. 10, pp. 47–53, 1969.

[96] D.E. Whitney, C.A. Lozinski, and J.M. Rourke, "Industrial robot forward calibration method and results," *ASME J. of Dynamic Systems, Measurement, and Control*, vol. 108, pp. 1–8, 1986.

[97] C.H. Wu, "A kinematic CAD tool for the design and control of a robot manipulator," *Int. J. of Robotics Research*, vol. 3, no. 1, pp. 58–67, 1984.

[98] X.-L. Zhong and J.M. Lewis, "A new method for autonomous robot calibration," *Proc. 1995 IEEE Int. Conf. on Robotics and Automation*, Nagoya, J, pp. 1790–1795, 1995.

[99] J. Ziegert and P. Datseris, "Basic considerations for robot calibration," *Proc. 1988 IEEE Int. Conf. on Robotics and Automation*, Philadelphia, PA, pp. 932–938, 1988.

Chapter 2

Joint space control

Traditionally, control design in robot manipulators is understood as the simple fact of tuning a PD (Proportional and Derivative) compensator at the level of each motor driving the manipulator joints. Fundamentally, a PD controller is a position and a velocity feedback that has good closed-loop properties when applied to a double integrator. This controller provides a natural way to stabilize double integrators since it can be understood as an additional mechanical (active) spring and damper which reduces oscillations. To this extent, the control of an n-joint manipulator can be interpreted as the control of n independent chains of double integrators for which a PD controller can be designed. In reality, the manipulator dynamics is much more complex than a simple decoupled second-order linear system. It includes coupling terms and nonlinear components such as gravity, Coriolis and centrifugal forces and friction.

The first experiments conducted with real robots were performed with simple PD compensators and have been, in general, satisfactory as far as stability and middle range performance are concerned. There is some explanation for this. First, as it was understood in the beginning of the 80's, PD controllers can stabilize not only a double integrator structure but also the complete Lagrange robot manipulator dynamics (in the sense of Lyapunov stability) due to the passivity properties enjoyed by the manipulator model. But before this was known, the main reason why the experiments were successful is because the complete robot manipulator dynamics is *locally* equivalent to a linear model described by a set of double integrators and, as such, is locally stabilizable. This local domain enlarges as the non-linearities and the coupling terms become less important, or equivalently as the reduction ratio of the gear boxes (or the harmonic drives) becomes larger. In addition, this region of attraction can be enlarged by increasing the controller gains.

It is instructive for comparative purposes to classify the control objectives into the following two classes:

- *Regulation* which sometimes is also called point-to-point control. A fixed configuration in the joint space is specified; the objective is to regulate the joint variables about the desired position in spite of torque disturbances and independently of the initial conditions. The behaviour of transients and overshooting are, in general, not guaranteed.

- *Tracking control* consists of following a time-varying joint reference trajectory specified within the manipulator workspace. In general, this desired trajectory is assumed to comply with the actuators' capacity. In other words, the joint velocity and acceleration associated with the desired trajectory should respect the maximum velocity and acceleration limits of the manipulator. The control objective is thus to asymptotically track the desired trajectory in spite of disturbances and other unmodelled dynamics.

Although the regulation problem is a particular case of the tracking problem (for which the desired velocity and acceleration are zero), in practice the latter sometimes is artificially solved by regulating the joint variables about a time-varying trajectory. In such situations, tracking errors are not guaranteed to asymptotically vanish but they may remain small and become zero only when the desired velocity is also zero. This will be discussed further in the following sections.

The selection of the controller may depend on the type of task to be performed. For example, tasks only requiring the manipulator to move from one position to another without requiring significant precision during the motion between these two points can be solved by regulators, while tasks like welding, painting, etc. will require tracking controllers.

The material of this chapter is organized as follows. First, the fundamental *properties* of the *dynamic model* of a rigid manipulator are presented; these properties are massively exploited in the subsequent derivation of joint space control algorithms. Then *PD control*, *PID control* and *PD control with gravity compensation* are discussed which are aimed at solving the regulation problem. The tracking control problem is tackled by referring to three classes of control schemes; namely *inverse dynamics control*, *Lyapunov-based control*, and *passivity-based control*. The basic differences among the three solutions are highlighted. In most practical cases model-based compensation is imperfect and robust control is needed. If a *constant bounded disturbance* occurs, it is shown how an *integral action* can be added to the previous controllers so as to reject the disturbance. If *uncertainty* on the *model parameters* occurs, a truly *robust control* additional component is needed so as to counteract the uncertainty. Finally, *adaptive control* schemes are introduced as an alternative solution in the case of model parameter mismatch.

2.1 Dynamic model properties

Typically, control design for rigid robot manipulators is based on the Lagrange dynamics equations of motion

$$H(q)\ddot{q} + C(q,\dot{q})\dot{q} + g(q) = u \qquad (2.1)$$

where q is the $(n \times 1)$ vector of generalized joint coordinates, $H(q)$ is the $(n \times n)$ inertia matrix, $g(q)$ is the $(n \times 1)$ vector of gravity forces, $C(q,\dot{q})\dot{q}$ is the $(n \times 1)$ vector of Coriolis and centrifugal forces and u is the $(n \times 1)$ vector of joint control input torques to be designed; friction has been neglected.

The control algorithms which we will describe in this chapter are based on some very important *properties* of the *dynamic model* of the manipulator in (2.1). Before starting with the detailed analysis of these different control laws, let us therefore give a list of these properties.

Property 2.1 The inertia matrix $H(q)$ is a symmetric positive definite matrix which verifies

$$\lambda_m I \leq H(q) \leq \lambda_M I \qquad (2.2)$$

where λ_m ($\lambda_M < \infty$) denotes the strictly positive minimum (maximum) eigenvalue of H for all configurations q.

□

Property 2.2 The matrix $N(q,\dot{q}) = \dot{H}(q) - 2C(q,\dot{q})$ is skew-symmetric for a particular choice of $C(q,\dot{q})$ (which is always possible), i.e.,

$$z^T N(q,\dot{q}) z = 0 \qquad (2.3)$$

for any $(n \times 1)$ vector z.

□

Property 2.3 As anticipated in (1.115), the input torque vector u can be written as

$$u = Y(q,\dot{q},\ddot{q})\rho \qquad (2.4)$$

where $Y(q,\dot{q},\ddot{q})$ is an $(n \times r)$ matrix and ρ is an $(r \times 1)$ vector of base dynamic parameters.

□

Property 2.4 Given any two $(n \times 1)$ vectors x, y, then

$$C(q, x)y = C(q, y)x. \tag{2.5}$$

□

Property 2.5 The Coriolis and centrifugal terms $C(q, \dot{q})\dot{q}$ verify

$$\|C(q, \dot{q})\dot{q}\| \leq c_0 \|\dot{q}\|^2 \tag{2.6}$$

for some bounded constant $c_0 > 0$.

□

Property 2.6 The matrix $C(q, \dot{q})$ verifies

$$\|C(q, \dot{q})\| \leq k_C \|\dot{q}\| \tag{2.7}$$

for some bounded constant $k_C > 0$.

□

Property 2.7 The gravity vector $g(q)$ verifies

$$\|g(q)\| \leq g_0 \tag{2.8}$$

for some bounded constant $g_0 > 0$.

□

The first two properties were already anticipated in the previous chapter. Furthermore, in the remainder we assume that the state vector $(q, \dot{q})$ is available for measurements.

Remarks

- Positive definiteness of the inertia matrix is a well-known result of classical mechanics for scleronomic systems, i.e., finite dimensional systems having only holonomic constraints and whose Lagrangian does not depend explicitly on time.

- Boundedness of the inertia matrix entries, i.e, $\lambda_M < \infty$, as well as boundedness of the gravity vector norm hold only for manipulators with revolute joints.

- Property 2.2 will be proved to be essential in the development of certain classes of control algorithms and is closely related to passivity properties of the manipulator.

- The vector ρ contains the base dynamic parameters, and the dynamic equations are said to be *linear in the parameters*. Property 2.3 is essential in the development of adaptive controllers.

- Properties 2.4, 2.5, 2.6, 2.7 are very useful because they allow us to establish upper bounds on the dynamic nonlinearities. As we will see further, several control schemes require knowledge of such upper bounds.

- The measurement of $\dot{q}$ can be relaxed via the introduction of nonlinear observers in the control loop, but this is beyond the scopes of the present chapter.

2.2 Regulation

Controllers of PID (Proportional, Integral and Derivative) type are designed in order to solve the regulation problem. They have the advantage of requiring the knowledge of neither the model structure (2.1) nor the model parameters. Let us start the discussion with the simplest stabilizing control structure, i.e., *PD control*.

2.2.1 PD control

A PD controller has the following basic form:

$$u = K_P(q_d - q) - K_D \dot{q} \tag{2.9}$$

where q_d and q respectively denote the desired and actual joint variables, and K_P, K_D are constant, positive definite matrices.

It is well understood why this controller works for a system of double integrators but much less why it can stabilize the Lagrange nonlinear dynamics (2.1). This can be proved on the basis of Property 2.2: $\dot{q}^T N(q, \dot{q}) \dot{q} = 0$ which results from the law of conservation of energy and can be understood as the fact that some internal forces of the system are workless. On the other hand, gravity forces are known to cause positioning errors if no integral action is included. These errors can be analyzed and predicted by studying the equilibrium point(s) of the closed-loop system and then stability of such equilibria can be inferred. Lyapunov analysis and its extension (the La Salle's invariant set theorem) are the suitable tools to do this. Let us first discuss about the equilibria under PD control.

Consider the regulation problem, with q_d *constant*, and let us introduce the error vector, $\tilde{q} = q - q_d$, with $\dot{\tilde{q}} = \dot{q}$ and $\ddot{\tilde{q}} = \ddot{q}$. With this definition, we can now substitute the control law (2.9) into the manipulator dynamics (2.1) so as to obtain the following error equation

$$H(q)\ddot{\tilde{q}} + C(q,\dot{\tilde{q}})\dot{\tilde{q}} + K_D\dot{\tilde{q}} + K_P\tilde{q} + g(q) = 0. \tag{2.10}$$

Notice that the gravity vector is given as a function of the actual state and not as a function of the regulation error. In order to put (2.10) in an autonomous system form equation, we can express $g(q)$ as $g(\tilde{q}+q_d)$ and use trigonometric relations to eliminate the dependence on the constant vector q_d in g. We are thus in a position to determine the possible equilibria, or sets of equilibria, of the above autonomous system. Assume that the initial conditions are given as: $\ddot{\tilde{q}}(0) = 0$, $\dot{\tilde{q}}(0) = 0$ and $\tilde{q}(0) = \tilde{q}_0$. Also, $\tilde{q}_0$ and q_d satisfy the following relation: $K_P\tilde{q}_0 + g(\tilde{q}_0 + q_d) = 0$, and then no motion is possible since $\ddot{\tilde{q}}(t) = \dot{\tilde{q}}(t) = 0$ for all time. According to the definition of invariant sets, the set

$$S = \{(\tilde{q},\dot{\tilde{q}}) : K_P\tilde{q} + g(\tilde{q} + q_d) = 0, \dot{\tilde{q}} = 0\} \tag{2.11}$$

describes a set of all possible equilibria for the system (2.10). To be more specific, let us consider the following example.

Pendulum under PD control

For the sake of simplicity, assume that our objective is to regulate a pendulum about the angle position where gravity forces are maximal, i.e., $q_d = 0$. This will lead us to the following equation error, with $\tilde{q} = q$,

$$m\ddot{q} + k_D\dot{q} + k_P q + gl\cos q = 0 \tag{2.12}$$

where g is the gravity constant, l is the distance from the center of rotation to the center of gravity, m is the motor inertia, and k_P, k_D are the controller gains. Let the system state be described by the vector ξ composed of position $\xi_1 = q$ and velocity $\xi_2 = \dot{q}$, i.e., $\xi = (\xi_1\ \xi_2)^T$. The state space representation of system (2.12) is

$$\begin{aligned}\dot{\xi}_1 &= \xi_2 \\ \dot{\xi}_2 &= \frac{1}{m}(-k_D\xi_2 - k_P\xi_1 - gl\cos\xi_1)\end{aligned} \tag{2.13}$$

In a more compact form we can also refer to system (2.13) as $\dot{\xi} = f(\xi)$. We have that the equilibria are characterized by all points ξ^* such that $f(\xi^*) = 0$. These points can thus be not unique. Indeed, they are given by the invariant set S defined above, which for the considered example simplifies to:

$$S = \{(q,\dot{q}) : k_P q + gl\cos q = 0, \dot{q} = 0\}. \tag{2.14}$$

2.2. REGULATION

Clearly, the desired position $q_d = 0$ does not belong to the set S and hence will never be reached. On the other hand, the equilibria will depend on the physical constants of the systems, i.e., g and l, as well as on the controller gain k_P. Note that if k_P is large enough a unique equilibrium point does exist, whereas if k_P is small enough S will contain more than one equilibrium point —two if $q \in [-\pi, \pi]$, more than two if $q \in R$. It is worth remarking that an estimate of system precision can be obtained from the invariant set S by noticing that the set

$$B = \{(q, \dot{q}) : |q| \leq gl/k_P, \; \dot{q} = 0\} \qquad (2.15)$$

includes the invariant set S, i.e., $S \subseteq B$. Clearly, by increasing the proportional gain k_P, the distance between any point in S to the origin, and hence the positioning error, can be arbitrarily reduced, at least in principle. This preliminary analysis illustrates the concept of invariant set and agrees with the engineering experience of improving system precision by increasing the proportional gain. Stability of the equilibria will now be studied.

There are several ways to perform stability analysis. One possibility is to resort to the Lyapunov direct method and its extension, i.e., the La Salle's invariant set theorem for autonomous nonlinear systems. Another possibility is to use Lyapunov-like functions and to apply Barbalat's lemma. Although this latter approach is more suitable for nonautonomous systems, it can also be applied in connection with the considered problem. Yet another possibility is to consider the gravity forces as a bounded disturbance (this is true in general for revolute joint robots, as remarked for the dynamic model properties) and perform the stability analysis on this basis. This approach is interesting when the explicit determination of the equilibria is difficult to obtain, which renders the stability analysis via formal Lyapunov functions quite complicated and the determination of invariant sets difficult. In such a case, it is more appropriate to look for attractors (a set containing the equilibria) and consider *ultimate boundedness* in case of autonomous systems and *uniform ultimate boundedness* in case of nonautonomous systems, respectively. This notion is sometimes known as *practical stability*. We will come back to these concepts later on in this chapter. Let us carry out the stability analysis of system (2.12) about the equilibria $\xi^\star \in S$ by use of Lyapunov analysis.

The theory given in Appendix A is concerned mainly with systems having the equilibrium point at the origin, i.e., $\xi^\star = 0$; however, these results can be extended to other equilibria different from zero by shifting the origin via a suitable change of coordinates. For instance, let e be the difference between the actual position q and an equilibrium position $q^\star$, i.e.,

$$e = q - q^\star,$$

and thus its time derivative is

$$\dot{e} = \dot{q}.$$

System (2.12) can be rewritten as a function of these new shifted coordinates as

$$m\ddot{e} + k_D\dot{e} + k_P(e + q^\star) + gl\cos(e + q^\star) = 0. \tag{2.16}$$

In the e-coordinates the equilibrium point of system (2.16) is $e^\star = 0$, whereas in the q-coordinates it is given by the set S. In the sequel we assume that k_P is large enough so that a unique equilibrium point does exist (stability arguments are thus global, otherwise the analysis should be performed locally by choosing a particular point in S).

Consider now the following function in the q-coordinates:

$$V = \frac{m}{2}\dot{q}^2 + \frac{k_P}{2}q^2 + gl(1 + \sin q) \tag{2.17}$$

which is a natural choice since it includes the kinetic and potential energy of the pendulum. However, note that this function does not satisfy $V(\xi = \xi^\star) = 0$ as it is required by a Lyapunov function candidate. Notice that its minimum V_0 is obtained at $\xi = \xi^\star$, i.e.,

$$\left(\frac{\partial V}{\partial q}(\xi^\star)\right)^T = k_P q^\star + gl\cos q^\star = 0 \tag{2.18}$$

$$\left(\frac{\partial V}{\partial \dot{q}}(\xi^\star)\right)^T = m\dot{q}^\star = 0 \tag{2.19}$$

and then V_0 is given as

$$V_0 = \frac{k_P}{2}(q^\star)^2 + gl(1 + \sin q^\star). \tag{2.20}$$

Since V is positive with a positive minimum at V_0, we can now define a new shifted function $\bar{V} = V - V_0 \geq 0$, either in the q-coordinates or in the e-coordinates, i.e.,

$$\bar{V} = V(e + q^\star) - V_0$$
$$= \frac{m}{2}\dot{e}^2 + \frac{k_P}{2}(e + q^\star)^2 + gl(\sin q - \sin q^\star) - \frac{k_P}{2}(q^\star)^2 \tag{2.21}$$

so that $\bar{V}$ satisfies the Lyapunov function candidate requirements. Taking the time derivative of $\bar{V}$ along the solutions to (2.16) gives

$$\dot{\bar{V}} = m\dot{e}\ddot{e} + k_P(e + q^\star)\dot{e} + gl\cos(e + q^\star)\dot{e}$$
$$= \dot{e}(-k_D\dot{e} - k_P(e + q^\star) - gl\cos(e + q^\star))$$
$$\quad + k_P(e + q^\star)\dot{e} + gl\cos(e + q^\star)\dot{e}$$
$$= -k_D\dot{e}^2. \tag{2.22}$$

2.2. REGULATION

Since $\dot{V}$ is only negative semi-definite, the equilibrium point ($e = \dot{e} = 0$) is stable in the sense of Lyapunov, but not asymptotically stable. Up to this point, the analysis is incomplete to determine whether or not the equilibrium is reached. The La Salle's invariant set theorem allows us to conclude on this point. This theorem indicates that we have to search for the set of points R in the neighbourhood of the equilibrium (in this case the local condition is not required since, on the assumption of large k_P, the equilibrium point is unique and $\dot{V}$ is globally negative semi-definite) where $\dot{V} = 0$. Then R is given by

$$R = \{(e, \dot{e}) : e = e_0, \; \dot{e} = 0\} \tag{2.23}$$

where e_0 is some constant. Within R we have to look for the largest invariant set, which is given by $S_e = \{(e, \dot{e}) : e = 0, \; \dot{e} = 0\}$ in the e-coordinates, or by S as defined in (2.14) in the q-coordinates. Therefore, by the La Salle's theorem, any solution $\xi(t)$ originating in the neighbourhood of ξ^* will tend to S as $t \to \infty$.

Robot manipulator under PD control

We can now generalize the stability analysis to the complete robot manipulator dynamics (2.1) following the arguments given in the previous example and considering the invariant set S defined in (2.11) with the equilibria $\xi^* = (q^{*T} \;\; \dot{q}^{*T})^T \in S$.

Consider the function

$$V = \frac{1}{2} \dot{\tilde{q}}^T H(q) \dot{\tilde{q}} + \tilde{q}^T K_P \tilde{q} + U(q) + U_0 \tag{2.24}$$

where $U(q)$ represents the potential energy and U_0 is a suitable constant so that V satisfies the Lyapunov candidate requirements (in particular $V(\xi^*) = 0$). Taking the time derivative of V along the solutions to (2.10) gives

$$\dot{V} = -\dot{\tilde{q}}^T K_D \dot{\tilde{q}} + \dot{\tilde{q}}^T \left(\frac{1}{2} \dot{H}(q) - C(q, \dot{q}) \right) \dot{\tilde{q}} - \dot{\tilde{q}}^T g(q) + \dot{\tilde{q}}^T \left(\frac{\partial U(q)}{\partial q} \right)^T$$
$$\leq -\lambda_{\min}(K_D) \|\dot{\tilde{q}}\|^2 \tag{2.25}$$

where $\lambda_{\min}(K_D)$ denotes the smallest eigenvalue of matrix K_D, and we have used Property 2.2 and the fact that $(\partial U(q)/\partial q)^T = g(q)$. The system states $(\dot{\tilde{q}}, \tilde{q})$ are stable in the sense of Lyapunov but, as before, up to here the analysis is inconclusive with respect to the equilibria. The La Salle's theorem can be applied to conclude that the system states will, at least locally, converge to S with S defined as in (2.11); also the system precision will depend on the matrix gain K_P since B will be given by

$$B = \{(\tilde{q}, \dot{\tilde{q}}) : \|\tilde{q}\| \leq \|K_P^{-1}\| g_0, \dot{\tilde{q}} = 0\} \qquad (2.26)$$

where g_0 is defined in Property 2.7. The following lemma summarizes this result.

Lemma 2.1 *Consider the PD controller (2.9) applied to the manipulator dynamics (2.1); then the closed-loop equation has one equilibrium point if K_P is large enough and several equilibria if K_P is small enough. The equilibria are described by the invariant set*

$$S = \{(\tilde{q}, \dot{\tilde{q}}) : K_P \tilde{q} + g(\tilde{q} + q_d) = 0, \dot{\tilde{q}} = 0.\} \qquad (2.27)$$

Furthermore, if K_P and K_D are positive definite matrices, then all the solutions $\xi(t)$ asymptotically converge —globally (if K_P is large) or locally (if K_P is small)— to S as $t \to \infty$.

⋄ ⋄ ⋄

The steady-state error can thus, in principle, be arbitrarily reduced by increasing K_P; nevertheless, measurement noise and other unmodelled dynamics will limit the use of high gains in practice.

2.2.2 PID control

An alternative to high gains over the whole frequency spectrum is to introduce high gains only at those frequencies for which the expected disturbance will have its dominant frequency components. An *integral action* may thus be added to the previous PD controller in order to deal with gravity forces which to some extent can be considered as a constant disturbance (from the local point of view). This leads to a PID control structure. The purpose of this section is to present the (local) stability analysis of the closed-loop robot manipulator dynamics (2.1) under a *PID control*.

Consider the following PID control:

$$u = -K_D \dot{q} + K_P(q_d - q) + K_I \int_0^t (q_d - q) d\tau. \qquad (2.28)$$

Let us describe the robot manipulator dynamics (2.1) in state space form, with $\xi_1 = q$, $\xi_2 = \dot{q}$ and $\xi = (\xi_1^T \ \xi_2^T)^T$, as

$$\begin{aligned} \dot{\xi}_1 &= \xi_2 \\ \dot{\xi}_2 &= H^{-1}(\xi_1)\bigl(-C(\xi_1, \xi_2)\xi_2 - g(\xi_1) + u\bigr) \end{aligned} \qquad (2.29)$$

2.2. REGULATION

and define the state error vectors as

$$\begin{aligned}\tilde{\xi}_1 &= \xi_1 - \xi_{1d} \\ \tilde{\xi}_2 &= \xi_2\end{aligned} \tag{2.30}$$

where ξ_{1d} denotes the position reference vector q_d. Introduce a third state as the integral of the position error plus a bias component

$$\tilde{\xi}_3 = \int_0^t \tilde{\xi}_1 d\tau - K_I^{-1} g(\xi_{1d}). \tag{2.31}$$

Plugging (2.28) into (2.29) and accounting for (2.31) gives

$$\dot{\tilde{\xi}}_1 = \tilde{\xi}_2 \tag{2.32}$$
$$\dot{\tilde{\xi}}_2 = H^{-1}(\xi_1)\left(-C(\xi_1,\tilde{\xi}_2)\tilde{\xi}_2 - g(\xi_1) - K_P\tilde{\xi}_1 - K_D\tilde{\xi}_2 - K_I(\tilde{\xi}_3 + g(\xi_{1d}))\right)$$
$$\dot{\tilde{\xi}}_3 = \tilde{\xi}_1.$$

To study the local stability of the above error equation system, consider its linear tangent approximation about the origin ($\tilde{\xi}^* = 0$)

$$\dot{\tilde{\xi}} = A\tilde{\xi} + o(\tilde{\xi}) \tag{2.33}$$

with

$$A = \begin{pmatrix} 0 & I & 0 \\ (-B_0 K_P + A_0) & -B_0 K_D & -K_I \\ I & 0 & 0 \end{pmatrix} \quad \tilde{\xi} = \begin{pmatrix} \tilde{\xi}_1 \\ \tilde{\xi}_2 \\ \tilde{\xi}_3 \end{pmatrix} \tag{2.34}$$

where $o(\tilde{\xi})$ describes the higher-order terms of Taylor's expansion. Notice that, in view of Properties 2.4, 2.5, 2.6, 2.7, $o(\tilde{\xi})$ satisfies: $\|o(\tilde{\xi})\| \leq c_1 \|\tilde{\xi}\|^2$, where $c_1 > 0$ is a bounded constant. The matrices A_0 and B_0 are given by

$$A_0 = -\left.\frac{\partial(H^{-1}g)}{\partial \xi_1}\right|_{\xi_1 = \xi_{1d}} \qquad B_0 = H^{-1}(\xi_{1d}).$$

If the matrix gains K_P, K_D, K_I are chosen to make the linearized autonomous system (2.33) asymptotically stable, the states of the augmented closed-loop system (2.33) will tend to zero. Since $\tilde{\xi}_1 \to 0$, the steady-state error can be cancelled without any knowledge of the disturbance.

If A is a stable matrix, we can then define V as

$$V = \tilde{\xi}^T P \tilde{\xi} \tag{2.35}$$

where P is a symmetric positive definite matrix satisfying $A^T P + PA = -Q$, for some symmetric $Q > 0$. Then we get

$$\begin{aligned}
\dot{V} &= -\tilde{\xi}^T Q \tilde{\xi} + 2\tilde{\xi}^T P o(\tilde{\xi}) \\
&\leq -\lambda_{\min}(Q)\|\tilde{\xi}\|^2 + 2c_1 p_0 \|\tilde{\xi}\|^3 \\
&= -\|\tilde{\xi}\|^2 (\lambda_{\min}(Q) - 2c_1 p_0 \|\tilde{\xi}\|)
\end{aligned} \quad (2.36)$$

where p_0 is an upper bound on the norm of P. In the neighbourhood of the origin, the negative square term dominates the positive cubic term. The equilibrium point $\tilde{\xi} = 0$ is thus locally stable in the sense of Lyapunov. It is now straightforward to prove, by invoking the La Salle's invariant set theorem, that the error vector $\tilde{\xi}$ converges to zero.

The following lemma gives conditions for ensuring A to be definite negative.

Lemma 2.2 *Consider the PID controller (2.28) with: $K_P = k_P I$, $K_D = k_D I$, $K_I = k_I I$. Assume that the inertia matrix is diagonally dominant, so that $B_0 \approx \text{diag}\{b_j\}$, and that k_P is large enough so that $B_0 k_P + A_0 \approx \text{diag}\{b_j k_P\}$. Then a sufficient condition for A to be a stable matrix —all the eigenvalues of A have strictly negative real parts— is*

$$b_j^2 k_P k_D > k_I \quad \forall j = 1, 2, ..., n \quad (2.37)$$

and then the error system (2.33), under the action of the PID controller, is locally asymptotically stable.

⋄ ⋄ ⋄

Proof. By matrix determinant properties we have that

$$\det(\lambda I - A) = \det\left(\lambda^3 I + \lambda^2 (B_0 K_P - A_0) + \lambda B_0 K_D + K_I\right). \quad (2.38)$$

On the assumption given in the lemma, this expression simplifies to

$$\begin{aligned}
\det(\lambda I - A) &= \det(\lambda^3 I + \lambda^2 \text{diag}\{b_j\} k_P + \lambda \text{diag}\{b_j\} k_D + k_I) \\
&= \prod_{j=1}^{n} (\lambda^3 I + \lambda^2 b_j k_P + \lambda b_j k_D + k_I).
\end{aligned} \quad (2.39)$$

The inequality (2.37) is obtained from Routh's criterion for the above third-order polynomial. If the inertia matrix is not diagonally dominant and if we wish not to have large values for K_P, then we should find values for the matrices K_P, K_D, K_I so that all the roots of the following polynomial

$$\det\left(\lambda^3 I + \lambda^2 (B_0 K_P - A_0) + \lambda B_0 K_D + K_I\right) \quad (2.40)$$

have strictly negative real part.

◇

Notice that the eigenvalues of matrix A will be depending upon the desired position in a nonlinear fashion. More specifically, they depend on the magnitude of the gravity components and on the values of the inertia matrix entries. In some industrial manipulators, gravity can vary from zero up to values that can represent 20% to 30% or more of the maximum permissible torque. This implies that transients in point-to-point control can have substantial differences according to either the type of motion required (going up or going down) or the operational configurations. In other words, if the same transient behaviour is required in the whole manipulator workspace, the controller gains shall be changed according to the value of gravity and the direction of the motion. Another difficulty in using integral action is the presence of dry friction. It can cause oscillations due to the interaction between integral gains and friction nonlinearities. In practice this problem is partially solved by introducing additional piecewise nonlinearities in the integral components, like deadzones. It is customary to activate the integral action only when q is close to q_d to avoid overshooting and deactivate it when the error is too large or when it is below a very small threshold (zero for practical purposes) in order to avoid numerical drift in the integration. These ideas are based on heuristics and in turn on the engineer's common sense. Additional analysis will be required to determine how these modifications affect the transient behaviour and ensure, if possible, a uniform performance in the complete manipulator workspace.

2.2.3 PD control with gravity compensation

The idea of preprogramming the controller gains can be replaced with an exact *gravity* (and eventually friction) *compensation*. To this extent, gravity compensation acts as a bias correction compensating only for the amount of forces that create overshooting and an asymmetric transient behaviour. Formally, this requires that the PD controller be replaced with

$$u = K_P(q_d - q) - K_D \dot{q} + g(q). \qquad (2.41)$$

By following the same analysis as before, the derivative of the function

$$V(t) = \frac{1}{2}\dot{\tilde{q}}^T H(q)\dot{\tilde{q}} + \frac{1}{2}\tilde{q}^T K_P \tilde{q} \qquad (2.42)$$

becomes negative semi-definite for any value of $\dot{q}$, i.e.,

$$\dot{V} = -\dot{\tilde{q}}^T K_D \dot{\tilde{q}} \leq -\lambda_{\min}(K_D)\|\dot{\tilde{q}}\|^2. \qquad (2.43)$$

By invoking the La Salle's invariant set theorem, it can be shown that the largest invariant set S is given by

$$S = \{(q, \dot{q}) : q - q_d = 0, \dot{q} = 0\} \tag{2.44}$$

and hence the regulation error will converge asymptotically to zero, while their high-order derivatives remain bounded. This controller requires knowledge of the gravity components (structure and parameters), though.

In conclusion, the following considerations are in order:

- PD controllers cannot solve the regulation problem but do ensure global stability.

- PID controllers are locally stable and can solve the regulation problem but cannot, in general, guarantee uniformity of the transient behaviour.

- PD controllers with gravity compensation improve over the performance of PID controllers but require exact knowledge of the gravity terms.

None of these control structures is suitable, at least theoretically, to solve the tracking problem. Exact design for tracking control will be discussed next.

2.3 Tracking control

The *tracking control* problem in the joint space consists of following a given time-varying trajectory $q_d(t)$ and its successive derivatives $\dot{q}_d(t)$ and $\ddot{q}_d(t)$ which respectively describe the desired velocity and acceleration. Several schemes for performing these objectives do exist. In this section we will present some of them. They can be classified in the following main groups.

- *Inverse dynamics control* —known in the literature also under the name of computed torque control— is aimed at linearizing and decoupling robot manipulator dynamics; nonlinearities such as Coriolis and centrifugal terms as well as gravity terms can be simply compensated by adding these forces to the control input. However, this type of feedback linearizing controller is more complicated and its conception is no longer straightforward for the robot manipulator dynamics if mechanical (joint and/or link) flexibility is considered; these issues will be further discussed in Part II.

2.3. TRACKING CONTROL

- *Lyapunov-based control* does seek neither to linearize nor to decouple the system nonlinear dynamics. The idea is only to search for asymptotic stability —exponentially, if possible.

- *Passivity-based control* exploits the passivity properties of the robot manipulator dynamics. Like the previous class of control schemes, it does not achieve exact cancellation of the nonlinearities and then it is expected to be more robust than inverse dynamics control.

2.3.1 Inverse dynamics control

Mechanical engineer intuition dictates the idea of cancelling nonlinear terms and decoupling the dynamics of each link by using an *inverse dynamics control* of the form

$$u = H(q)u_0 + C(q,\dot{q})\dot{q} + g(q) \quad (2.45)$$

which applied to (2.1), and in view of the regularity of matrix $H(q)$ as in Property 2.1, yields a set of n decoupled linear systems

$$\ddot{q} = u_0 \quad (2.46)$$

where u_0 is an auxiliary control input to be designed. Typical choices for u_0 are:

$$u_0 = \ddot{q}_d + K_D(\dot{q}_d - \dot{q}) + K_P(q_d - q) \quad (2.47)$$

or with an integral component

$$u_0 = \ddot{q}_d + K_D(\dot{q}_d - \dot{q}) + K_P(q_d - q) + K_I \int_0^t (q_d - q)d\tau, \quad (2.48)$$

respectively leading to the error equation

$$\ddot{\tilde{q}} + K_D\dot{\tilde{q}} + K_P\tilde{q} = 0 \quad (2.49)$$

for control law (2.47), and

$$\tilde{q}^{(3)} + K_D\ddot{\tilde{q}} + K_P\dot{\tilde{q}} + K_I\tilde{q} = 0 \quad (2.50)$$

if control law (2.48) is used. Both error equations (2.49) and (2.50) are exponentially stable by a suitable choice of the matrices K_D, K_P (and K_I). To see this, we can rewrite (2.49) or (2.50) in its state space form, i.e., $\dot{\xi} = A\xi$. It is thus easy to show that if K_D, K_P (and K_I) are suitably chosen then the matrix A is stable. Then we can find, for some symmetric $Q > 0$, a symmetric positive definite matrix P satisfying $A^T P + PA = -Q$.

Global asymptotic Lyapunov stability follows. For any initial position the tracking error will asymptotically converge to zero. The transient behaviour will thus be symmetric and independent of the operation conditions. The matrix gains K_D, K_P (and K_I) should be adjusted once for all.

This controller needs exact knowledge of the model parameters and requires an additional number of numerical computations. The computational burden of u_0 is comparable to that of PID controllers discussed in the previous section. However, to calculate u it is necessary to perform an efficient computation of the inverse dynamics.

Computation time was the main restrictive factor that had prevented this method from having a larger impact. Most commercially available industrial robot manipulators are still equipped with conventional PID controllers. Apart from the difficulties to identify the dynamic parameters and compute the complete model of a robot manipulator, the computer technology until a few years ago did not allow for on-line computation of inverse dynamics.

Nowadays, thanks to the development of efficient inverse dynamics recursive algorithms based on Newton-Euler formulation (see Chapter 1) on one hand, and to the advent of new high-speed microprocessor technology on the other hand, the practical experiments of inverse dynamics control carried out in research laboratories so far can be brought in the industrial scenario. Therefore, it is envisaged that the next generation of control units for industrial robots will embed model-based control algorithms in order to achieve high-performance tracking control. A figure of merit of efficient dynamic model computation for a six-degree-of-freedom robot manipulator using a Pentium processor can be estimated on the order of 0.1 ms.

2.3.2 Lyapunov-based control

Inverse dynamics control is not the only alternative to globally perform exact tracking. *Lyapunov-based control* has been successfully applied to robot manipulators. The method does not seek linearization or decoupling, but only asymptotic convergence of the tracking error.

Consider the following change of coordinates,

$$\dot{\zeta} = \dot{q}_d - \Lambda \tilde{q} \qquad (2.51)$$

$$\sigma = \dot{q} - \dot{\zeta} = \dot{\tilde{q}} + \Lambda \tilde{q} \qquad (2.52)$$

where, as before, $\tilde{q} = q - q_d$ denotes the tracking error in the manipulator joint coordinates and Λ is a constant, positive definite $(n \times n)$ matrix. With these definitions, consider the control law

$$u = H(q)\ddot{\zeta} + C(q,\dot{q})\dot{\zeta} + g(q) - K_D \sigma, \qquad (2.53)$$

2.3. TRACKING CONTROL

where $K_D > 0$ and the matrix $C(q, \dot{q})$ satisfies Property 2.2. This control law applied to the manipulator dynamics gives the closed-loop equation

$$H(q)(\ddot{q} - \dot{\zeta}) + C(q, \dot{q})(\dot{q} - \zeta) + K_D \sigma = 0 \qquad (2.54)$$

that, in view of the above definitions of ζ and σ, can be rewritten as

$$H(q)\dot{\sigma} + C(q, \dot{q})\sigma + K_D \sigma = 0 \qquad (2.55)$$

which describes a first-order differential nonlinear equation in the new variable σ; σ is related to the tracking error by (2.52). Conversely, the tracking error can be seen as a filtered version of the variable σ, or else as the output of a stable linear system having σ as the input, i.e.,

$$\dot{\tilde{q}} = -\Lambda \tilde{q} + \sigma. \qquad (2.56)$$

The solution to (2.56) will be bounded if σ is proved also to be bounded. Furthermore, this solution will tend asymptotically to zero if the input σ also tends to zero. In fact, if $\sigma \in \mathcal{L}_2 \cap \mathcal{L}_\infty$ then the output of a stable and strictly proper linear time-invariant system will have the property $\tilde{q} \in \mathcal{L}_2 \cap \mathcal{L}_\infty$ with $\dot{\tilde{q}} \in \mathcal{L}_\infty$. This is formally stated and demonstrated by the following lemma —see also Appendix A.

Lemma 2.3 *The control law (2.53) applied to the robot manipulator dynamics (2.1) has the following properties:*

(i) $\sigma \in \mathcal{L}_2 \cap \mathcal{L}_\infty$
(ii) $\|\sigma(t)\|^2 \leq \kappa_1 \exp(-\eta_1 t)\|\sigma(0)\|^2$
(iii) $\tilde{q} \in \mathcal{L}_2 \cap \mathcal{L}_\infty,\ \dot{\tilde{q}} \in \mathcal{L}_\infty$
(iv) $\|\tilde{q}(t)\|^2 \leq \kappa_2 \exp(-\eta_2 t)\|\tilde{q}(0)\|^2$
(v) $\|\dot{\tilde{q}}(t)\|^2 \leq \kappa_3 \exp(-\eta_3(\eta_1, \eta_2)t)\| (\tilde{q}(0)^T \quad \dot{\tilde{q}}(0)^T)^T \|$

where κ_1, κ_2 are positive constants and η_1, η_2 are positive constants depending on the controller gains K_D and Λ; also κ_3 is a constant function of $\tilde{q}(0)$, $\dot{\tilde{q}}(0)$, and η_3 is a constant function of η_1, η_2.

◇ ◇ ◇

Proof. *(i)* Consider the following positive definite function

$$V = \frac{1}{2}\sigma^T H(q)\sigma \qquad (2.57)$$

whose time derivative along the solutions to (2.55) is

$$\dot{V} = \sigma^T H(q)\dot{\sigma} + \frac{1}{2}\sigma^T \dot{H}(q)\sigma$$

$$= -\sigma^T K_D \sigma + \sigma^T \left(\frac{1}{2}\dot{H}(q) - C(q,\dot{q})\right)\sigma$$

$$= -\sigma^T K_D \sigma \leq 0 \qquad (2.58)$$

where the last simplification is obtained from Property 2.2. Since $V(t)$ is a decreasing function, we have

$$-V(0) \leq V(t) - V(0) = \int_0^t \dot{V}(\tau)d\tau = -\int_0^t \sigma^T K_D \sigma d\tau \qquad (2.59)$$

from which we obtain

$$\int_0^t \|\sigma(\tau)\|^2 d\tau \leq \frac{V(0)}{\lambda_{\max}(K_D)} < \infty \qquad (2.60)$$

and therefore σ is a bounded square-mean energy signal, i.e., $\sigma \in \mathcal{L}_2$. The boundedness of σ, i.e., $\sigma \in \mathcal{L}_\infty$, follows directly from the boundedness of V.

(ii) It is easy to see that

$$\frac{\dot{V}}{V} = -\frac{\sigma^T K_D \sigma}{\sigma^T H(q)\sigma} \leq -\frac{\lambda_{\min}(K_D)\|\sigma\|^2}{\lambda_M \|\sigma\|^2} = -\frac{\lambda_{\min}(K_D)}{\lambda_M} = \eta_1 \qquad (2.61)$$

where λ_M is defined as in Property 2.1. Integrating both sides gives

$$\ln \frac{V(t)}{V(0)} = \int_0^t \frac{\dot{V}}{V} d\tau \leq -\int_0^t \eta_1 d\tau = -\eta_1 t \qquad (2.62)$$

which leads to

$$V(t) \leq V(0)\exp(-\eta_1 t) \qquad (2.63)$$

and then, using bounds on $V(t)$, we get

$$\|\sigma(t)\|^2 \leq \frac{\lambda_M}{\lambda_{\min}(K_D)}\exp(-\eta_1 t)\|\sigma(0)\|^2 = \kappa_1 \exp(-\eta_1 t)\|\sigma(0)\|^2. \qquad (2.64)$$

(iii) This property is an implication of the following proof, since exponentially decaying signals are in $\mathcal{L}_2 \cap \mathcal{L}_\infty$.

(iv) Since $\Lambda > 0$, then there exist positive constants a_1, a_2, a_3 such that the solution to (2.56), i.e.,

$$\tilde{q}(t) = \exp(-\Lambda t)\tilde{q}(0) + \int_0^t \exp(-\Lambda(t-\tau))\sigma(\tau)d\tau \qquad (2.65)$$

2.3. TRACKING CONTROL

is bounded as follows:

$$\begin{aligned}
\|\tilde{q}(t)\| &\leq a_1 \exp(-a_2 t) + a_3 \exp(-a_2 t) \int_0^t \exp(a_2 \tau) \|\sigma(\tau)\| d\tau \\
&\leq \exp(-a_2 t) \left(a_1 + a_3 \sqrt{\kappa_1} \int_0^t \exp((a_2 - \eta_1/2)\tau) d\tau \right) \\
&= a_1 \exp(-a_2 t) + \frac{a_3 \sqrt{\kappa_1}}{a_2 - \eta_1/2} \left(\exp(-\eta_1/2) - \exp(-a_2 t) \right) \\
&\leq \kappa_2 \exp(-\eta_2 t)
\end{aligned} \quad (2.66)$$

where (2.66) is obtained by using Property *(ii)*, and κ_2, η_2 are given as:

$$\kappa_2 = \max \left\{ \left(a_1 - \frac{a_3 \sqrt{\kappa_1}}{a_2 - \eta_1/2} \right), \frac{a_3 \sqrt{\kappa_1}}{a_2 - \eta_1/2} \right\} \quad (2.67)$$

$$\eta_2 = \min \left\{ a_2, \frac{\eta_1}{2} \right\} \quad (2.68)$$

(v) This property is obtained straightforwardly by combining Properties *(ii)*, *(iv)* and the linear relationship given by the filter equation (2.56).

◇

Remark

- It is possible to obtain a similar proof by strictly following Lyapunov arguments. This can be done by defining the Lyapunov function

$$V = \frac{1}{2} \sigma^T H(q) \sigma + \tilde{q}^T P \tilde{q} \quad (2.69)$$

where $P = P^T = \Lambda^T K_D > 0$. The difference with respect to the previous analysis is that the expression of V here above depends on both σ and $\tilde{q}$ and hence formally defines a Lyapunov function candidate. Simple choices for Λ and K_P are $\Lambda = \lambda I$ and $K_P = k_P I$, with λ and k_P positive. Proceeding as before, it is easy to show that the time derivative of V along the solutions to (2.55) is given by

$$\begin{aligned}
\dot{V} &= -\sigma^T K_D \sigma + \tilde{q}^T P \dot{\tilde{q}} \\
&= -\dot{\tilde{q}}^T K_D \dot{\tilde{q}} - \tilde{q}^T \Lambda^T K_D \Lambda \tilde{q} \leq 0
\end{aligned} \quad (2.70)$$

where the last expression is obtained by using (2.56). Note that now the derivative of V depends on both the position and velocity errors. The proof is completed by invoking Lyapunov direct method.

2.3.3 Passivity-based control

As an alternative to Lyapunov-based control schemes, *passivity-based control* schemes can be designed which explicitly exploit the passivity properties of the Lagrangian model. In comparison with the inverse dynamics method, passivity-based controllers are expected to have better robust properties because they do not rely on the exact cancellation of the manipulator nonlinearities. A general theorem for the analysis of the closed-loop error equation resulting from these schemes is presented next.

Theorem 2.1 *Consider the following differential equation*

$$H(q)\dot{\sigma} + C(q,\dot{q})\sigma + K_D\sigma = \psi \qquad (2.71)$$

where, as before, $H(q)$ is the inertia matrix and $C(q,\dot{q})$ is a matrix chosen so that Property 2.2 holds, K_D is a symmetric positive definite matrix, and the vector σ is given by

$$\sigma = F^{-1}(\cdot)\tilde{q} \qquad (2.72)$$

where $F(\cdot)$ is the transfer function of a strictly proper, stable linear operator (filter), and the mapping $-\sigma \mapsto \psi$ is passive relative to V_1, i.e., $\int_0^t -\sigma(\tau)^T \psi(\tau) d\tau = V_1(t) - V_1(0)$ for all $t \geq 0$. Then $\tilde{q} \in \mathcal{L}_2 \cap \mathcal{L}_\infty$, $\dot{\tilde{q}} \in \mathcal{L}_2$, $\tilde{q}$ is continuous and tends to zero asymptotically. In addition, if ψ is bounded, then σ and consequently $\dot{\tilde{q}}$ also tend to zero asymptotically.

⋄ ⋄ ⋄

Proof. Consider the function

$$V = \frac{1}{2}\sigma^T H(q)\sigma + V_1 \qquad (2.73)$$

which in view of the passivity properties of the mapping $-\sigma \mapsto \psi$ is positive definite. Evaluating the time derivative of V along the solutions to (2.71) gives

$$\begin{aligned}
\dot{V} &= \sigma^T H(q)\dot{\sigma} + \frac{1}{2}\sigma^T \dot{H}(q)\sigma - \sigma^T \psi \\
&= -\sigma^T K_D \sigma + \sigma^T \left(\frac{1}{2}\dot{H}(q) - C(q,\dot{q})\right)\sigma \\
&= -\sigma^T K_D \sigma.
\end{aligned} \qquad (2.74)$$

Therefore, by following the same reasoning as before, we have that $\sigma \in \mathcal{L}_2$, $\tilde{q}$ goes to zero and $\dot{\tilde{q}}$ is bounded. In view of the passivity of the mapping $-\sigma \mapsto \psi$, then $\dot{\tilde{q}}$ also goes to zero if ψ is bounded.

⋄

2.3. TRACKING CONTROL

This theorem will be useful when proving global convergence in the case of input bounded disturbances and/or partial knowledge on the manipulator model parameters, which will be presented in the subsequent sections. In the case of known parameters, the vector ψ is equal to zero and the stability analysis and properties are the same as the Lyapunov-based design in the previous section. The formulation presented in this section then allows unifying most of the proposed control algorithms for rigid robot manipulators enjoying global stability properties.

The control law is then given as in (2.53), i.e.,

$$u = H(q)\ddot{\zeta} + C(q,\dot{q})\dot{\zeta} + g(q) - K_D\sigma. \tag{2.75}$$

Now σ and ζ can be defined as

$$\dot{\zeta} = \dot{q}_d - K(\cdot)\tilde{q} \tag{2.76}$$

$$\sigma = F^{-1}(\cdot)\tilde{q}, \tag{2.77}$$

where $K(\cdot)$ is a linear operator that should be chosen so that $F(\cdot)$ is strictly proper and stable, as required by Theorem 2.1; then $\tilde{q}$ and $\dot{\tilde{q}}$ tend to zero. Note that F and K are related by

$$F^{-1}(s) = sI + K(s) \tag{2.78}$$

where s stands for the Laplace operator.

From this formulation it is easy to recover other choices of passivity-based control laws; e.g., the previous choice of $K(s)$ as

$$K(s) = \Lambda \tag{2.79}$$

where Λ is a symmetric positive definite diagonal matrix. In this case, the operator $F(s)$ becomes

$$F(s) = (sI + \Lambda). \tag{2.80}$$

Another possibility is to select $K(s)$ as

$$K(s) = K_P + \frac{K_I}{s} + \frac{K_R}{s^2}. \tag{2.81}$$

Indeed, it is possible to derive other algorithms by letting $K(s)$ take the following general polynomial form

$$K(s) = \sum_{j=0}^{n} \frac{K_j}{s^j} \tag{2.82}$$

where K_j are the controller gains and n is the order of the desired polynomial. The determination of n relies on other properties apart from stability, e.g., disturbance rejection characteristics.

2.4 Robust control

It can be expected that passivity-based controllers are more robust than the inverse dynamics control law. The argument has been motivated in the context of the fixed-parameter case by the fact that Lyapunov-based controllers or passivity-based controllers do not rely on the exact cancellation of the system nonlinearities, whereas the linearizing schemes do. In the fixed-parameter case, it has been demonstrated that inverse dynamics control may become unstable in the presence of uncertainties. If the model parameters are not exactly known but bounds on these parameters are available, *robust control* design can be applied. Robust design can be classified into *sliding mode (high-gain) control* and *saturating control*; the former being a special case of the latter.

Sliding mode control design consists of defining a *sliding* or *switching surface* which is rendered invariant by the action of switching terms. The invariance of the switching surface implies asymptotic stability of the tracking error in spite of unmodelled dynamics and lack of knowledge of model parameters. Although this method is theoretically appealing, it has the drawback that the produced control input will have high-frequency components and hence will produce "chattering". In practice this provokes the premature fatigue of the motor actuators, thus considerably reducing their lifetime. There are other possibilities for introducing high gains which do not produce high-frequency components but they may provoke unacceptable transient peaks.

A solution for this excess of control authority is given by the robust *saturating control* approach, which substitutes the relay-like switching functions by saturation-like functions. The price paid by the smoothness of the control law is that asymptotic stability of the tracking error is lost. Instead, saturation control gives uniform *ultimate boundedness*, often also called practical stability; that is, the tracking error converges to an error threshold that decreases as the saturation-like function approaches the relay-like function. A trade-off is therefore established between gain magnitude and tracking accuracy.

When the disturbance model is known (e.g., constant disturbance) then there is no need to resort to high-gain design, nor to use saturation controllers. In these cases the model of the disturbance can be explicitly included in the controller so that the disturbance can be asymptotically cancelled.

On the other hand, when the disturbance model is unknown, that is uncertainty occurs on the dynamic model parameters, an additional control input can be introduced which is aimed at providing robustness to parameter uncertainty.

2.4.1 Constant bounded disturbance: integral action

In this section the problem of rejecting a *constant bounded disturbance* is considered. Input torque disturbances, in the context considered here, are characterized by torque loads acting on the torque inputs. By letting d be the constant bounded vector describing the torque input loads, the dynamical model (2.1) becomes

$$H(q)\ddot{q} + C(q,\dot{q})\dot{q} + g(q) = u + d. \tag{2.83}$$

Clearly, if the disturbance d is known, then it is straightforward to cancel d in the same way as Coriolis and gravity vector forces are cancelled. However, in the case consider here, d is unknown. A typical solution for cancelling d in linear systems disturbed by an unknown constant bias is to include in some way the internal model of the disturbance. In the case of a constant disturbance, this internal model corresponds to placing an integral action. This is, to some extent, equivalent to first estimate d and then compensate for it. The integral component can thus be seen as an observer for the unknown constant disturbance.

In this section we will show how to include the *integral action* in all the three foregoing control schemes: inverse dynamics control, Lyapunov-based control, and passivity-based control.

Inverse dynamics control with integral action

Let us first consider the simplest solution, already anticipated above, which consists of adding an integral action to the inverse dynamics control law. Consider the control law (2.45)–(2.48), i.e.,

$$u = H(q)u_0 + C(q,\dot{q})\dot{q} + g(q) \tag{2.84}$$

$$u_0 = \ddot{q}_d + K_D(\dot{q}_d - \dot{q}) + K_P(q_d - q) + K_I \int_0^t (q_d - q)d\tau \tag{2.85}$$

together with the disturbed model (2.83). This gives the following closed-loop equation

$$\ddot{\tilde{q}} + K_D \dot{\tilde{q}} + K_P \tilde{q} + K_I \int_0^t \tilde{q} d\tau = \delta, \tag{2.86}$$

where $\delta = H^{-1}(q)d$. Without loss of generality, we can take the feedback gains as scalar matrices: $K_D = k_D I$, $K_P = k_P I$, $K_I = k_I I$. Then for each joint we have

$$\tilde{q}_i = \frac{s}{s^3 + k_D s^2 + k_P s + k_I} \delta_i. \tag{2.87}$$

Since the static transfer loop gain of this function is equal to zero, the steady-state value of the error will also be zero. In other words, the tracking error $\tilde{q}_i$ will tend to zero in spite of the unknown disturbance δ_i, i.e.,

$$\tilde{q}_i(\infty) = \lim_{s\to 0} \frac{s}{s^3 + k_D s^2 + k_P s + k_I} \delta_i = 0. \tag{2.88}$$

Lyapunov-based control with integral action

Consider now the Lyapunov-based control law (2.53) with the following modification

$$u = H(q)\ddot{\zeta} + C(q,\dot{q})\dot{\zeta} + g(q) - K_D\sigma - \hat{d} \tag{2.89}$$

where $\hat{d}$ is an estimate of d. As before, if $\hat{d} = d$, then the disturbance is cancelled and thereby we recover the same closed-loop properties as in the disturbance-free case —see Lemma 2.3. The question is now how to derive an *estimation* law for $\hat{d}$, so that some of the properties in Lemma 2.3 are still valid (exponential convergence invoked by Properties (ii),(iv),(v) will in this case be replaced with asymptotic convergence). Proceeding as before and substituting (2.89) in (2.83) yields the following error equation

$$H(q)\dot{\sigma} + C(q,\dot{q})\sigma + K_D\sigma = \tilde{d} \tag{2.90}$$

where $\tilde{d} = d - \hat{d}$. Note that this equation is equivalent to the error equation (2.71) with $\psi = \tilde{d}$. Hence from Theorem 2.1 we have that if the mapping $-\sigma \mapsto \tilde{d}$ is passive relative to some function V_1 and in addition $\tilde{d}$ is proved to be bounded, then $\tilde{q}$ is continuous and both $\tilde{q}$ and $\dot{\tilde{q}}$ will asymptotically tend to zero.

Let us first prove the boundedness of $\tilde{d}$ and thereby derive the *estimation* law for $\hat{d}$.

Consider the function

$$V = \frac{1}{2}\sigma^T H\sigma + \frac{1}{2}\tilde{d}^T K_I^{-1}\tilde{d} \tag{2.91}$$

where K_I is some positive definite matrix. Following the proof in Lemma 2.3, we have that the time derivative of V along the solutions to (2.90) is given by

$$\dot{V} = \sigma^T H(q)\dot{\sigma} + \frac{1}{2}\sigma^T \dot{H}(q)\sigma + \sigma^T \tilde{d} + \tilde{d}^T K_I^{-1}\dot{\tilde{d}} \tag{2.92}$$

$$= -\sigma^T K_D\sigma + \sigma^T \left(\frac{1}{2}\dot{H}(q) - C(q,\dot{q})\right)\sigma + \tilde{d}^T \left(\sigma - K_I^{-1}\dot{\tilde{d}}\right)$$

2.4. ROBUST CONTROL

where the last term is obtained by observing that d is constant; hence $\dot{\tilde{d}} = -\dot{\hat{d}}$. To cancel this term, the following estimation law for $\hat{d}$ can be chosen:

$$\dot{\hat{d}} = K_I \sigma \qquad (2.93)$$

giving

$$\dot{V} = -\sigma^T K_D \sigma. \qquad (2.94)$$

The above estimation law relates the estimate of the disturbance to the control vector σ. Therefore, the control law (2.89) can also be rewritten as

$$u = H(q)\ddot{\zeta} + C(q,\dot{q})\dot{\zeta} + g(q) - K_D \sigma - K_I \int_0^t \sigma d\tau \qquad (2.95)$$

which shows how the integral action is introduced. Now, since V is positive definite and $\dot{V}$ negative semi-definite, we have that the states σ and $\tilde{d}$ are bounded. It remains to be proved that the mapping $-\sigma \mapsto \tilde{d}$ is passive relative to some V_1 to complete the requirement on passive systems.

Note that from (2.93) we have $-\sigma^T \tilde{d} = -\tilde{d}^T K_I^{-1} \dot{\tilde{d}}$, and thus

$$-\int_0^t \left(\sigma^T \tilde{d} \right) d\tau = -\left(\int_0^t \tilde{d}^T d\tau \right) K_I^{-1} \dot{\tilde{d}}$$

$$= -\frac{1}{2} \int_0^t \frac{d}{d\tau} \left(\tilde{d}^T K_I^{-1} \tilde{d} \right) d\tau$$

$$= \frac{1}{2} \tilde{d}^T(t) K_I^{-1} \tilde{d}(t) - \frac{1}{2} \tilde{d}^T(0) K_I^{-1} \tilde{d}(0)$$

$$= V_1(t) - V_1(0); \qquad (2.96)$$

therefore $-\sigma \mapsto \tilde{d}$ is a passive mapping relative to $V_1 = \frac{1}{2} \tilde{d}^T K_I^{-1} \tilde{d}$ and then Theorem 2.1 applies.

Notice that the last two terms in the control law (2.89) are a proportional and an integral components in the σ variable. These terms can be interpreted as PID components with respect to the tracking error $\tilde{q}$, by noticing that

$$-K_D \sigma - K_I \int_0^t \sigma d\tau = -K_D \left(\dot{\tilde{q}} + \Lambda \tilde{q} \right) - K_I \int_0^t \left(\dot{\tilde{q}} + \Lambda \tilde{q} \right) d\tau$$

$$= -\left(K_D \Lambda + K_I \right) \tilde{q} - K_D \dot{\tilde{q}} - K_I \Lambda \int_0^t \tilde{q} d\tau$$

where we have assumed that $\tilde{q}(0) = \dot{\tilde{q}}(0) = 0$ for convenience.

Passivity-based control with integral action

The same idea of using an integral action described above can be generalized to the passivity-based control structure described in Section 2.3.3. Consider the control law (2.89) with $\hat{d}$ as in (2.93) and the general definitions for ζ in (2.76) and σ in (2.77), i.e.,

$$u = H(q)\dot{\zeta} + C(q,\dot{q})\zeta + g(q) - K_D\sigma - K_I \int_0^t \sigma d\tau \qquad (2.97)$$

$$\zeta = \dot{q}_d - K(\cdot)\tilde{q} \qquad (2.98)$$

$$\sigma = F^{-1}(\cdot)\tilde{q} \qquad (2.99)$$

where the filters $F(\cdot)$ and $K(\cdot)$ have the same meaning as before. With this control, following exactly the same lines as in the previous section and using Theorem 2.1, we have that

$(i) \quad \sigma \in \mathcal{L}_2 \cap \mathcal{L}_\infty, \; \dot{\sigma} \in \mathcal{L}_2 \cap \mathcal{L}_\infty$

$(ii) \quad \tilde{q} \in \mathcal{L}_2 \cap \mathcal{L}_\infty, \; \dot{\tilde{q}} \in \mathcal{L}_\infty$

$(iii) \quad \lim_{t \to \infty} \begin{pmatrix} \tilde{q}^T & \dot{\tilde{q}}^T & \sigma^T \end{pmatrix}^T = 0.$

2.4.2 Model parameter uncertainty: robust control

In this section the problem of counteracting *model parameter uncertainty* is considered for the same three control schemes as above and their *robust control* versions will be derived.

Robust inverse dynamics control

Consider the rigid manipulator model in (2.1). Let us choose the torque control input vector u as

$$u = H_0(q)(\ddot{q}_d - K_D\dot{\tilde{q}} - K_P\tilde{q}) + C_0(q,\dot{q})\dot{q} + g_0(q) + u_0 \qquad (2.100)$$

where H_0, C_0, g_0 represent nominal values vis-a-vis to parameter uncertainty of H, C, g. Also q_d is the bounded, twice-differentiable reference trajectory, the tracking error is $\tilde{q} = q - q_d$, u_0 is an additional control input that we will define later, and K_D, K_P are constant and positive definite diagonal matrices.

Substituting the control input (2.100) into the dynamic model (2.1) gives the following error equation

$$H(q)(\ddot{\tilde{q}} + K_D\dot{\tilde{q}} + K_P\tilde{q}) = Y(q,\dot{q},q_d,\dot{q}_d,\ddot{q}_d)\tilde{p} + u_0 \qquad (2.101)$$

2.4. ROBUST CONTROL

where we have used the property of linearity of $H(q)$, $C(q,\dot{q})$, $g(q)$ — and therefore of $H_0(q)$, $C_0(q,\dot{q})$, $g_0(q)$ — with respect to a set of constant physical parameters ρ. Note that $Y(\cdot)$ is an $(n \times r)$ matrix and $\tilde{\rho} = \rho_0 - \rho$ is an $(r \times 1)$ vector expressing the error in the parameters.

Therefore we have

$$Y(\cdot)\tilde{\rho} = (H_0(q) - H(q))(\ddot{q}_d - K_D\dot{\tilde{q}} - K_P\tilde{q}) \\ + (C_0(q,\dot{q}) - C(q,\dot{q}))\dot{q} + (g_0(q) - g(q)). \quad (2.102)$$

In the so-called ideal case, i.e., when $\tilde{\rho} \equiv 0$ and $u \equiv 0$, the error equation in (2.101) reduces to the error equation in (2.49). It follows from the positive definiteness of the inertia matrix $H(q)$ that the tracking error asymptotically converges to zero. Due to the mismatch between the nominal and the true system, however, the stability analysis of the perturbed error equation in (2.101) becomes more involved. Indeed the perturbation term on the right-hand side of (2.101) is state-dependent, and therefore it cannot be assumed to be a priori bounded by some positive constant as in the previous section. Moreover, the matrix function $Y(\cdot)$ contains highly nonlinear terms —recall that the centrifugal and Coriolis inertial torques are quadratic functions of $\dot{q}$— that render the equivalent perturbation $Y(\cdot)\tilde{\rho}$ more difficult to deal with in a stability analysis. In the sequel we prove that this problem can be overcome by a suitable choice of the additional input u_0.

First, let us rewrite the error equation in (2.101) in state space form. Choose $\xi_1 = \tilde{q}$ and $\xi_2 = \dot{\tilde{q}}$ as state variables; then $\xi = (\xi_1^T \ \xi_2^T)^T$ is the state vector. We obtain

$$\dot{\xi} = \begin{pmatrix} 0 & I \\ -K_P & -K_D \end{pmatrix} \xi + \begin{pmatrix} 0 \\ H^{-1}(q) \end{pmatrix} (Y(\cdot)\tilde{\rho} + u_0) \quad (2.103)$$

which can be compactly written as

$$\dot{\xi} = A\xi + BH^{-1}(q)u_0(\xi) + Be(\xi) \quad (2.104)$$

where

$$A = \begin{pmatrix} 0 & I \\ -K_P & -K_D \end{pmatrix} \quad B = \begin{pmatrix} 0 \\ I \end{pmatrix}$$

and $e(\xi) = H^{-1}(q)Y(\cdot)\tilde{\rho}$.

We can recognize from (2.104) that the so-called matching conditions are verified in this case, i.e., the uncertainty enters into the system in the same place as the input does. One slight difficulty comes from the fact that H^{-1} premultiplies u and is unknown. We will see that the a priori knowledge of an upper bound of the maximum eigenvalue of $H(q)$ is sufficient to overcome this problem.

Consider the following positive definite function

$$V = \xi^T P \xi \qquad (2.105)$$

where P is a symmetric positive definite matrix satisfying $A^T P + PA = -Q$, with Q symmetric and positive definite too. Taking the time derivative of V along the trajectories of the error system (2.101) yields

$$\dot{V} = -\xi^T Q \xi + 2\xi^T PBH^{-1}(q)(Y(\cdot)\tilde{\rho} + u_0). \qquad (2.106)$$

From (2.106) it follows that

$$\dot{V} \leq -\xi^T Q \xi + 2\|\xi^T PB\| \, \|H^{-1}(q)Y(\cdot)\tilde{\rho}\| + 2\xi^T PBH^{-1}(q)u_0. \qquad (2.107)$$

We now make the assumption that there exists a known function $\beta(\cdot) : R^{2n} \times R \to R^{\ell}$ and a constant vector $\alpha^* \in R^{\ell}$ such that

$$\begin{aligned} \beta_i(\xi,t) &\geq 0 \\ \alpha_i^* &\geq 0 \qquad 1 \leq i \leq \ell \\ \|H^{-1}(q,t)Y(\cdot)\tilde{\rho}\| &\leq \beta^T(\xi,t)\alpha^* \end{aligned} \qquad (2.108)$$

for all $(\xi,t) \in R^{2n} \times R$.

From Properties 2.1, 2.5, and 2.6 of the dynamic model, a possible choice for $\beta^T(\xi,t)\alpha^*$ when revolute joints only are considered is $\beta^T(\xi,t)\alpha^* = \alpha_1^* + \alpha_2^*\|q\| + \alpha_3^*\|\dot{q}\| + \alpha_4^*\|\dot{q}\|^2$, where the α_i^*'s depend on λ_M, g_0, c_0, K_D, K_P, q_d, $\dot{q}_d$, $\ddot{q}_d$.

The above assumption clearly implies some a priori knowledge on the system parameters ρ.

Introducing (2.108) into (2.107) gives

$$\dot{V} \leq -\xi^T Q \xi + 2\|\xi^T PB\|\beta^T(\xi,t)\alpha^* + 2\xi^T PBH^{-1}(q)u_0. \qquad (2.109)$$

Choose the additional control torque input u_0 as

$$u_0 = -\frac{(\beta^T(\xi,t)\alpha^*)^2 \bar{\lambda}}{\varepsilon} B^T P \xi \qquad (2.110)$$

with $\varepsilon > 0$ and $\bar{\lambda} \geq \lambda_M$. Hence, plugging (2.110) into (2.109) yields

$$\dot{V} \leq -\xi^T Q \xi + \frac{2}{\varepsilon}\|\xi^T PB\|\beta^T(\xi,t)\alpha^* \left(\varepsilon - \|\xi^T PB\|\beta^T(\xi,t)\alpha^*\right) \qquad (2.111)$$

where we have used the fact that $\lambda_{\min}(H^{-1}) = 1/\lambda_M$.

2.4. ROBUST CONTROL

Therefore, as long as the following inequality is verified

$$\|\xi^T PB\|\beta^T(\xi,t)\alpha^* \geq \varepsilon, \tag{2.112}$$

we get

$$\dot{V} < -\xi^T Q \xi. \tag{2.113}$$

This inequality is strict; indeed we can easily verify that if $\xi = 0$ then (2.112) is not satisfied.

On the other hand, if $\|\xi^T PB\|\beta^T(\xi,t)\alpha^* \leq \varepsilon$, we obtain

$$\dot{V} < -\xi^T Q \xi + 2\|\xi^T PB\|\beta^T(\xi,t)\alpha^* \tag{2.114}$$

so that

$$\dot{V} \leq -\xi^T Q \xi + \varepsilon_1 \tag{2.115}$$

with $\varepsilon_1 = 2\varepsilon$.

From (2.113) and (2.115) we conclude that the additional torque input in (2.110) allows us to establish that in all cases the following inequality is true:

$$\dot{V} \leq -\lambda_{\min}(Q)\|\xi\|^2 + \varepsilon_1. \tag{2.116}$$

At this point, let S_ε denote a compact set around the origin $\xi = 0$; the subscript indicates that the size of S_ε is directly related to ε, and $S_\varepsilon \to \{\xi = 0\}$ as $\varepsilon \to 0$. The following result can be established.

Theorem 2.2 *The state ξ is globally ultimately uniformly bounded with respect to the compact set S_ε, i.e., given any $\varepsilon > 0$ there exists a finite time $T > 0$ —which does not depend on the initial time but may depend on $\|\xi(0)\|$— such that $\xi(t)$ enters S_ε and remains inside for all $t > T$.*

⋄ ⋄ ⋄

Proof. Inequality (2.116) implies that as long as the state vector ξ lies outside a compact set of the state space, then V is strictly decreasing. We will use this fact to demonstrate the ultimate boundedness of ξ. Assume that $\xi(0)$ lies outside the ball B_r defined as

$$B_{r(\varepsilon)} = \left\{\xi : \|\xi\| \leq \left(\frac{\varepsilon_1}{\lambda_{\min}(Q)} + h\right)^{\frac{1}{2}} = r(\varepsilon),\ h > 0\right\}. \tag{2.117}$$

Then from (2.116) we see that, outside $B_{r(\varepsilon)}$, $\dot{V}$ is strictly negative and therefore V strictly decreases. From the positive definiteness of V we infer that there exists a finite time T_1 such that $\|\xi(T_1)\| = r(\varepsilon)$. Thus we have

$$V(T_1) - V(0) \leq \int_0^{T_1} -\lambda_{\min}(Q)h\,d\tau = -T_1 \lambda_{\min}(Q)h \tag{2.118}$$

where we assume, without loss of generality, that $t_0 = 0$. Note also that ξ is uniformly bounded in $[0, T_1]$. This is easily proved by using the same inequality as in (2.118) and replacing T_1 with t. Then

$$V(T_1) \leq V(0) - T_1 \lambda_{\min}(Q) h. \tag{2.119}$$

Defining $\gamma_1(\cdot)$ and $\gamma_2(\cdot)$ as

$$\begin{aligned} \gamma_1(\cdot) &= \lambda_{\min}(P) \|\cdot\|^2 \\ \gamma_2(\cdot) &= \lambda_{\max}(P) \|\cdot\|^2, \end{aligned} \tag{2.120}$$

it follows that

$$\gamma_1(r(\varepsilon)) \leq V(T_1) \leq \gamma_2(\|\xi(0)\|) - T_1 \lambda_{\min}(Q) h. \tag{2.121}$$

Then from (2.121) we have that

$$T_1 \leq \frac{\gamma_2(\|\xi(0)\|) - \gamma_1(r(\varepsilon))}{\lambda_{\min}(Q) h}. \tag{2.122}$$

It is now clear that as long as h is strictly positive then T_1 is finite. Assume that $\xi(t)$ lies inside $B_{r(\varepsilon)}$ for some $t > T_1$. Nothing guarantees that ξ will remain inside $B_{r(\varepsilon)}$, because $\dot{V}$ is no longer ensured to be negative inside $B_{r(\varepsilon)}$ —see (2.116). Assume then that ξ leaves $B_{r(\varepsilon)}$ at $t = T_2 > T_1$. Thus for some $t > T_2$, $\dot{V}$ is strictly negative and we know that ξ must reenter $B_{r(\varepsilon)}$ at $t = T_3 > T_2$. Now for all $T_2 < t \leq T_3$ we have

$$\gamma_1(\|\xi\|) \leq V(t) < V(T_2) \leq \gamma_2(r(\varepsilon)) \tag{2.123}$$

so that

$$\|\xi\| \leq \gamma_1^{-1}(\gamma_2(r(\varepsilon))). \tag{2.124}$$

Hence, we have that ξ remains inside the ball centered at the origin with radius $\gamma_1^{-1}(\gamma_2(r(\varepsilon))) = (\lambda_{\max}(P)/\lambda_{\min}(P))^{\frac{1}{2}} r(\varepsilon) = \bar{r}$.

Now it is clear that, given any prespecified $\bar{r} > 0$, we can find some $\varepsilon > 0$ and $h > 0$ such that ξ eventually lies in the ball $B_{\bar{r}}$. Conversely if ε in u_0 tends to zero, ξ is ultimately bounded with respect to a ball of radius that also tends to zero, as we can always find some h in (2.117) which tends to zero; however, in this latter case note that T_1 in (2.122) is not guaranteed to be bounded, i.e., the origin will be reached asymptotically.

The above analysis can be further explained as follows. First observe that the level sets of the positive definite function V in (2.105) are compact sets of the state space around the origin $\xi = 0$; namely, they are $2n$-dimensional ellipsoidal sets. Thus strict negativeness of $\dot{V}$ outside $B_{r(\varepsilon)}$

2.4. ROBUST CONTROL

implies that ξ eventually remains in the smallest level set of V which contains $B_{r(\varepsilon)}$. This is easily visualized in the plane —or in the 3-dimensional Euclidean space— where the level sets are defined as

$$V^{-1}(c) = \{\xi : \xi^T P \xi \leq c^2\} \tag{2.125}$$

which can be written as

$$V^{-1}(c) = \{\xi_p : \xi_p^T D \xi_p \leq c^2\} \tag{2.126}$$

where $P = LDL^T$, $D = \text{diag}(\lambda_{\min}(P), \lambda_{\max}(P))$, and $\xi_p = L^T \xi$. Thus

$$V^{-1}(c) = \left\{\xi_p : \xi_{p1}^2 \frac{\lambda_{\min}(P)}{c^2} + \xi_{p2}^2 \frac{\lambda_{\max}(P)}{c^2} \leq 1\right\}. \tag{2.127}$$

The smallest level set containing $B_{r(\varepsilon)}$ is therefore $V^{-1}\left((\lambda_2)^{\frac{1}{2}}r(\varepsilon)\right)$. It can be easily verified that $V^{-1}\left((\lambda_2)^{\frac{1}{2}}r(\varepsilon)\right)$ is contained in $B_{\bar{r}}$ with $\bar{r} = (\lambda_{\max}(P)/\lambda_{\min}(P))^{\frac{1}{2}}r(\varepsilon)$.

◇

Remarks

- The input in (2.110) belongs to the class of functions which allow us to draw conclusions on ultimate boundedness of the state. This is not the only one. Indeed let us consider the following function

$$u_0 = \begin{cases} -\dfrac{\beta^T(\xi,t)\alpha^*\bar{\lambda}}{\|\xi^T PB\|} B^T P \xi & \text{if } \|\xi^T PB\|\beta^T(\xi,t)\alpha^* \geq \varepsilon \\[2mm] -\dfrac{(\beta^T(\xi,t)\alpha^*)^2\bar{\lambda}}{\varepsilon} B^T P \xi & \text{if } \|\xi^T PB\|\beta^T(\xi,t)\alpha^* < \varepsilon. \end{cases} \tag{2.128}$$

Notice that this input is a continuous function of time as long as ε is strictly positive. Substituting (2.128) into (2.109) gives the following results.

- If $\left\|\dot{\tilde{q}} + \frac{k_D}{2}\tilde{q}\right\| \beta^T(\xi,t)\alpha^* \geq \varepsilon$, then

$$\dot{V} = -\xi^T Q \xi + 2\left\|\dot{\tilde{q}} + \frac{k_D}{2}\tilde{q}\right\| \beta^T(\xi,t)\alpha^*$$

$$-\frac{2}{\left\|\dot{\tilde{q}} + \frac{k_D}{2}\tilde{q}\right\|} \xi^T P B H^{-1}(q)\left(\dot{\tilde{q}} + \frac{k_D}{2}\tilde{q}\right)\beta^T(\xi,t)\alpha^*\bar{\lambda}$$

(2.129)

so that we obtain

$$\dot{V} = -\xi^T Q \xi + 2\left\|\dot{\tilde{q}} + \frac{k_D}{2}\tilde{q}\right\| \beta^T(\xi,t)\alpha^*$$

$$-2\left\|\dot{\tilde{q}} + \frac{k_D}{2}\tilde{q}\right\| \lambda_{\min}(H^{-1}(q))\beta^T(\xi,t)\alpha^*\bar{\lambda} \leq -\xi^T Q \xi.$$

(2.130)

- If $\left\|\dot{\tilde{q}} + \frac{k_D}{2}\tilde{q}\right\| \beta^T(\xi,t)\alpha^* < \varepsilon$, then

$$\dot{V} \leq -\xi^T Q \xi + 2\|\xi^T P B\|\beta^T(\xi,t)\alpha^*$$

$$-2\|\xi^T P B\|^2 \lambda_{\min} H^{-1}(q)(\rho^T(\xi,t)\alpha^*)^2 \frac{\bar{\lambda}}{\varepsilon}$$

(2.131)

and we finally get

$$\dot{V} \leq -\xi^T Q \xi + 2\varepsilon.$$

(2.132)

Therefore, we can see that the same conclusions can be drawn with u_0 in (2.110) or in (2.128). However it is important to note that, in spite of the fact that both controllers in (2.110) and (2.128) theoretically lead to the same stability result, they are of very different nature. Indeed u_0 in (2.110) is *continuous*, whereas u in (2.128) is *discontinuous* since the saturation function aims at smoothing the sign function around the discontinuity at zero. This difference has very important practical consequences; if ε is decreased to improve tracking performance, the control in (2.128) will lead to the phenomenon of *chattering*, whereas the one in (2.110) will not. On the other hand, the smaller ε the larger u_0 in (2.110), especially during the transient period. At this stage, we may think of decreasing ε proportionally to $\|\xi\|$ until a prespecified threshold value is reached, so as to limit the magnitude of the additional input u_0.

2.4. ROBUST CONTROL

- It is expected that this class of inputs, introduced to counteract the effects of a wrong identification of the system parameters, will somewhat "shake" the system in order to force the state to eventually reach a small neighbourhood of the origin. In fact, numerical simulations confirm that the controller in (2.110) lead to smooth transients more than the saturation function in (2.128) does. However, the rate of convergence seems slower. Notice that both inputs tend to zero when the tracking errors tend to zero. This means that the asymptotic value of such a controller will generally be small in magnitude.

- The above analysis applies to robot manipulators with revolute joints only. Indeed in this case the matrix $H^{-1}(q)$ is positive definite, because λ_M is bounded. When prismatic joints are considered, it is no longer guaranteed that $\lambda_{\min}(H^{-1}(q)) = 1/\lambda_M$ is strictly positive; however, for practically limited joint ranges, the property can be assumed to hold as well.

- If $H_0 = 0$, $C_0 = 0$, $g_0 = 0$, i.e., $\rho_0 = 0$, then the above analysis still applies. Nevertheless, we may expect the controller to behave better if the nominal parameter vector ρ_0 is close to the true parameter vector ρ.

- This type of stability (*ultimate boundedness* with respect to a small neighbourhood of the origin) shall not be confused with Lyapunov stability. Indeed the feedback gain depends directly on the size of the ball to which the state is driven. Moreover the origin is guaranteed to be neither an equilibrium point of the closed-loop system nor Lyapunov stable for $\varepsilon > 0$. Consider $r' > 0$ with $r' < \bar{r}$; then nothing guarantees that there exists some $\eta(r') > 0$ such that $\|\xi(t_0)\| < \eta$ implies $\|\xi(t)\| \leq r'$ for all $t \geq t_0$. We only know that $\|\xi(t)\| < \bar{r}$ for all $t \geq T_1$.

Robust passivity-based control

We study now the extension of the passivity-based control algorithms presented in Section 2.3.3, computed with fixed (but wrong or uncertain) parameters, following the same philosophy as for the above two kinds of controllers. As stated in Section 2.3.3, there exists a whole class of control laws (see (2.75)) which all lead to the same error equation (see (2.71)), the only difference being the way the variable σ is defined (see (2.72)). For the sake of clarity, we will use σ as defined in (2.52).

Consider the following control input

$$u = H_0(q)\ddot{\zeta} + C_0(q,\dot{q})\dot{\zeta} + g_0(q) - K_D \sigma + u_0 \qquad (2.133)$$

where H_0, C_0, g_0 have the same meaning as in (2.100), K_D is a positive definite, diagonal matrix, and σ is defined in (2.72); the additional control input u_0 will be defined later and will have the same structure as u_0 in (2.110) or (2.128).

Substituting (2.133) into the manipulator dynamics leads to the following error equation

$$H(q)\dot{\sigma} + C(q,\dot{q})\sigma + K_D\sigma = Y(q,\dot{q},q_d,\dot{q}_d,\ddot{q}_d)\tilde{\rho} + u_0 \qquad (2.134)$$

where

$$Y(\cdot)\tilde{\rho} = (H_0(q) - H(q))\ddot{\zeta} + (C_0(q,\dot{q}) - C(q,\dot{q}))\dot{\zeta} + (g_0(q) - g(q)).$$

The stability analysis is based on the choice of the following positive definite function of σ:

$$V = \frac{1}{2}\sigma^T H(q)\sigma. \qquad (2.135)$$

Taking the time derivative of V along the solutions to the error equation in (2.134) gives

$$\dot{V} = -\sigma^T K_D \sigma + \sigma^T Y(\cdot)\tilde{\rho} + \sigma^T u_0. \qquad (2.136)$$

Proceeding as we did for inverse dynamics control, we can bound the uncertain term $Y(\cdot)\tilde{\rho}$ from above and obtain

$$\dot{V} \leq -\lambda_{\min}(K_D)\|\sigma\|^2 + \|\sigma\|\beta^T(\sigma,t)\alpha^* + \sigma^T u_0. \qquad (2.137)$$

By choosing u_0 as

$$u_0 = -\frac{(\beta^T(\sigma,t)\alpha^*)^2}{\varepsilon}\sigma, \qquad (2.138)$$

we obtain

$$\dot{V} \leq -\lambda_{\min}(K_D)\|\sigma\|^2 + \frac{\|\sigma\|}{\varepsilon}\beta^T(\sigma,t)\alpha^* \left(\varepsilon - \|\sigma\|\beta^T(\sigma,t)\alpha^*\right). \qquad (2.139)$$

Thus, if $\|\sigma\|\beta^T(\sigma,t)\alpha^* \geq \varepsilon$, we get

$$\dot{V} \leq -\lambda_{\min}(K_D)\|\sigma\|^2, \qquad (2.140)$$

while if $\|\sigma\|\beta^T(\sigma,t)\alpha^* < \varepsilon$ we get

$$\dot{V} \leq -\lambda_{\min}(K_D)\|\sigma\|^2 + 2\varepsilon. \qquad (2.141)$$

We conclude that for all values of $\|\sigma\|$ the following inequality holds:

$$\dot{V} \leq -\lambda_{\min}(K_D)\|\sigma\|^2 + 2\varepsilon. \qquad (2.142)$$

2.4. ROBUST CONTROL

We have not used here the fact that the stability analysis is conducted with the Lyapunov-like function V in (2.135) which is a positive definite function of σ, i.e., the variable from which u_0 is computed. This was not the case for the inverse dynamics control algorithm —compare V in (2.135) with V in (2.105). Indeed we can slightly simplify the computation of u_0 in (2.138) by choosing

$$u = -\frac{\beta^T(\sigma, t)\alpha^*}{\varepsilon}\sigma \quad (2.143)$$

as an additional input. The reason for doing this is mainly to reduce the input magnitude by choosing uncertainty upper bounds as less conservative as possible.

According to the same reasoning as above, the following result can be established.

Theorem 2.3 *Given some $r > 0$, there exist some $\varepsilon > 0$ and $h > 0$ such that σ is ultimately uniformly bounded with respect to the ball B_r centered at the origin with radius $r = \left(\frac{2\varepsilon}{\lambda_{\min}(K_D)} + h\right)^{\frac{1}{2}} \frac{\lambda_M}{\lambda_m}$.*

◇ ◇ ◇

Proof. It is easy to see that each time σ is outside the ball B_ε then

$$\dot{V} \leq -\lambda_{\min}(K_D)\|\sigma\|^2 \leq -\lambda_{\min}(K_D)\varepsilon^2. \quad (2.144)$$

Thus σ eventually converges to the ball $B_{\bar{r}}$ of radius $\bar{r} = (\lambda_M/\lambda_m)\varepsilon$. This time, we do not need to consider any constant $h > 0$ that defines a neighbourhood of the boundary of the ball outside of which $\dot{V}$ is strictly negative, since inequality (2.144) holds even on the boundary of B_ε.

We conclude the proof by noticing that $\tilde{q}$ can be considered as the output of a first-order linear system with transfer function $1/(s + \lambda)$ and with input $\sigma(t)$. Therefore if for some $T_1 > 0$ we have $\|\sigma(t)\| \leq r$ for all $t \geq T_1$, then asymptotically $\|\tilde{q}(t)\| \leq r/\lambda$ and $\|\dot{\tilde{q}}(t)\| \leq 2r$. In fact, we have

$$\tilde{q}(t) = \exp(-\lambda t)\tilde{q}(0) + \int_0^t \exp(-\lambda(t-\tau))\sigma(\tau)d\tau \quad (2.145)$$

so that

$$\tilde{q}(t) \leq \exp(-\lambda t)\left(\tilde{q}(0) + \int_0^{T_1} \exp(\lambda\tau)\sigma(\tau)d\tau\right) + \frac{r}{\lambda}(1 - \exp(-\lambda(t-T_1))). \quad (2.146)$$

Since we know that T_1 is finite for a strictly positive ε, the conclusion follows.

◇

Remarks

- It is important to emphasize that if the nominal parameters ρ_0 are equal to the true ones, then we still get asymptotic convergence of the tracking error to zero, because the additional input always acts in the right direction. However it is expected that the closer ρ to ρ_0, the smaller the upper bound on the uncertainty.

- An alternative way to define the additional input in (2.138) or (2.143) is to bound the uncertain term in (2.134) from above as

$$\|\sigma^T Y(\cdot)\tilde{\rho}\| \leq \|\sigma^T Y(\cdot)\|\|\tilde{\rho}\| \leq \|\sigma^T Y(\cdot)\|\alpha^* \qquad (2.147)$$

where $\|\tilde{\rho}\| \leq \alpha^*$, α^* being known. Thus the variable leading the input is no longer σ alone, but instead $Y^T \sigma$. Indeed we can replace u_0 in (2.133) with $\bar{u}_0 = Y(\cdot)u_0$; this is possible as $Y(\cdot)$ is a known matrix and is assumed to be available on line. Choose u_0 as

$$u_0 = -\frac{\alpha^{*2}}{\varepsilon} Y^T(\cdot)\sigma. \qquad (2.148)$$

Then inequality (2.139) becomes

$$\dot{V} \leq -\lambda_{\min}(K_D)\|\sigma\|^2 + \frac{\|\sigma^T Y(\cdot)\|\alpha^*}{\varepsilon}\left(\varepsilon - \|\sigma^T Y(\cdot)\|\alpha^*\right) \qquad (2.149)$$

and the above reasoning applies. It is not clear whether significant closed-loop behaviour differences exist between the control laws in (2.138) and (2.148). However the input in (2.148) is simpler to compute as it does not involve any matrix norm computation, being α^* independent of the desired trajectory and the feedback gains. Moreover, as this kind of robust method involves upper bounds on the control input, it is challenging to try finding the "best" upper bound in order to reduce the input magnitude as much as possible.

- In the above analysis we have always considered a strictly positive constant ε in the additional input u_0. As remarked in a previous comment, there are strong practical reasons for doing this; namely, chattering phenomena when saturation type functions are used or very large values of the input when high-gain-type functions are used. However, the motivation for choosing $\varepsilon > 0$ also arises from theoretical considerations. Take the saturation function with $\varepsilon = 0$; the input looks like a sign function with a discontinuity at 0. It follows that the set of differential equations describing the closed-loop system has a right-hand side containing a discontinuous function of the state. Therefore, the usual mathematical tools allowing us to conclude on existence and uniqueness of solutions to ordinary differential equations with state-continuous right-hand side no longer apply. We have to introduce other concepts developed for differential equations with discontinuous right-hand side such as Filippov's definition of solutions. This problem is well known in the control literature in the field of the so-called sliding mode controllers (i.e., control laws which are intentionally designed to be discontinuous on a certain hyperplane of the state space). Concerning the high-gain-type of controllers, we can straightforwardly see that taking $\varepsilon = 0$ makes no sense since this introduces a division by zero in the input which, consequently, is no longer defined.

2.5 Adaptive control

Controllers that can handle regulation and tracking problems without the need of knowledge of the process parameters are by themselves an appealing procedure. Such control schemes belong to the class of *adaptive control*.

The global convergence properties of the existing adaptive schemes are basically due to Property 2.3 that model (2.1) enjoys. This property, together with Property 2.2, has been successfully exploited to derive Lyapunov-based adaptive controllers.

Complexity of these controllers and the computational time involved have been the main difficulties for their application in practice. Nevertheless, they present an interesting alternative for tracking control when a high degree of performance is required. The first experiments using adaptive techniques have just started but, at the moment, it seems too early to decide whether or not this method will be introduced into the industrial set-ups. On the other hand, the constant progress in the new generation of microprocessors, such as the DSP generation, may suggest their use in the near future, as well as other control alternatives.

2.5.1 Adaptive gravity compensation

Stable adaptive algorithms have been proposed in connection with the PD control + gravity compensation control structure in order to avoid the problem of identifying the parameters associated with the gravity vector.

Below, we show how a PD controller with *adaptive gravity compensation* can be designed. This controller yields the global asymptotic stability of the whole system even if the inertia and gravity parameters are unknown, provided that upper and lower bounds of the inertia matrix are available.

The convergence is ensured for any value of the proportional and derivative matrix gains, assumed to be symmetric and positive definite. The only constraint is in the adaptation gain which has to be greater than a lower bound. In the common case in which only the robot manipulator payload is unknown, one integrator is sufficient to implement this controller, while a PID algorithm requires as many integrators as the number of the joints.

Since the gravity vector $g(q)$ is linear in terms of robot manipulator parameters, it can be expressed as $g(q) = Y_g(q)\rho_g$, where ρ_g is the $(r_g \times 1)$ vector of constant unknown parameters, and $Y_g(q)$ is an $(n \times r_g)$ known matrix. Even if the inertia matrix is supposed to be unknown, we assume known upper and lower bounds on the magnitude of its eigenvalues as in Property 2.1; also we assume that Property 2.6 holds. Consider the control law

$$u = -K_P \tilde{q} - K_D \dot{q} + Y_g(q)\hat{\rho}_g \qquad (2.150)$$

with the parameter adaptation dynamics

$$\dot{\hat{\rho}}_g = -\nu Y^T(q)\left(\gamma \dot{q} + \frac{2\tilde{q}}{1 + 2\tilde{q}^T \tilde{q}}\right) \qquad (2.151)$$

in which $\tilde{q} = q - q_d$ is the position error, K_P and K_D are symmetric positive definite matrices, ν is a positive constant and γ is such that

$$\gamma > \max\{\gamma_1, \gamma_2\} \qquad (2.152)$$

where

$$\gamma_1 = \frac{2\lambda_M}{\sqrt{\lambda_m \lambda_{\min}(K_P)}}$$

$$\gamma_2 = \frac{1}{\lambda_{\min}(K_D)}\left(\frac{\lambda_{\max}^2(K_D)}{2\lambda_{\min}(K_P)} + 4\lambda_M + \frac{k_C}{\sqrt{2}}\right).$$

Theorem 2.4 *Consider the system in (2.150) and (2.151). If γ is taken as in (2.152), then $\tilde{q}(t)$, $\dot{q}(t)$ and $\hat{\rho}$ are bounded for any $t \geq 0$. Moreover*

$$\lim_{t \to \infty}\left\|(\tilde{q}^T \quad \dot{q}^T)^T\right\| = 0. \qquad (2.153)$$

◇ ◇ ◇

Proof. We select as Lyapunov function candidate ($\tilde{\rho}_g = \rho_g - \hat{\rho}_g$)

$$V = \gamma\left(\frac{1}{2}\dot{q}^T H(q)\dot{q} + \frac{1}{2}\tilde{q}^T K_P \tilde{q}\right) + \frac{2\tilde{q}^T H(q)\tilde{q}}{1 + 2\tilde{q}^T \tilde{q}} + \frac{1}{2\nu}\tilde{\rho}_g^T \tilde{\rho}_g \qquad (2.154)$$

2.5. ADAPTIVE CONTROL

which is positive definite since by assumption $\gamma > \gamma_1$. The time derivative of (2.154) is given by

$$\dot{V} = -\gamma \dot{q}^T K_D \dot{q} - \frac{2\tilde{q}^T K_P \tilde{q}}{1 + 2\tilde{q}^T \tilde{q}} + \frac{2\dot{q}^T H(q)\dot{q}}{1 + 2\tilde{q}^T \tilde{q}} + \frac{2\dot{q}^T C(q,\dot{q})\tilde{q}}{1 + 2\tilde{q}^T \tilde{q}}$$

$$- \frac{2\dot{q}^T K_D \tilde{q}}{1 + 2\tilde{q}^T \tilde{q}} - \frac{8\dot{q}^T H(q)\tilde{q}\dot{q}^T \tilde{q}}{(1 + 2\tilde{q}^T \tilde{q})^2}.$$
(2.155)

At this point, note that the following inequalities hold:

$$2\frac{\dot{q}^T H(q)\dot{q}}{1 + 2\tilde{q}^T \tilde{q}} \leq 2\lambda_M \|\dot{q}\|^2 \tag{2.156}$$

$$\frac{2\dot{q}^T C(q,\dot{q})\tilde{q}}{1 + 2\tilde{q}^T \tilde{q}} \leq k_C \|\dot{q}\|^2 \frac{2\|\tilde{q}\|}{1 + 2\|\tilde{q}\|^2} \leq \frac{k_C}{\sqrt{2}} \|\dot{q}\|^2 \tag{2.157}$$

$$\frac{8\dot{q}^T H(q)\tilde{q}\dot{q}^T \tilde{q}}{(1 + 2\tilde{q}^T \tilde{q})^2} \leq \lambda_M \|\dot{q}\|^2 \frac{8\|\tilde{q}\|^2}{1 + 4\|\tilde{q}\|^4} \leq 2\lambda_M \|\dot{q}\|^2 \tag{2.158}$$

which imply that

$$\dot{V} \leq -\gamma \lambda_{\min}(K_D) \|\dot{q}\|^2 - 2\lambda_{\min}(K_P)\frac{\|\tilde{q}\|^2}{1 + 2\|\tilde{q}\|^2} + \left(4\lambda_M + \frac{k_C}{\sqrt{2}}\right)\|\dot{q}\|^2$$

$$+ \frac{2\lambda_{\max}(K_D)\|\dot{q}\|\|\tilde{q}\|}{1 + 2\|\tilde{q}\|^2}.$$
(2.159)

Since

$$\gamma > \frac{1}{\lambda_{\min}(K_D)}\left(\frac{\lambda_{\max}^2(K_D)}{2\lambda_{\min}(K_P)} + 4\lambda_M + \frac{k_C}{\sqrt{2}}\right), \tag{2.160}$$

the function $\dot{V}$ is negative semi-definite and vanishes if and only if $\tilde{q} = 0$, $\dot{q} = 0$. The conclusion follows by applying the La Salle's theorem.

◇

Remarks

- Since the constants on the right-hand side of (2.152) are bounded, we can always choose γ so that (2.152) is satisfied.

- The above stability analysis is based on the Lyapunov function in (2.154). Note that the choice of V in (2.154) is much less "natural" than the one when gravity is ignored.

2.5.2 Adaptive inverse dynamics control

The inverse dynamics control method described in Section 2.3.1 is quite appealing, since it allows the designer to transform a multi-input multi-output highly coupled nonlinear system into a very simple decoupled linear system of order two, whose control design is a well-established problem. However this feedback linearizing method relies on the perfect knowledge of system parameters. Indeed we saw in the foregoing section that parameter errors result in a mismatch term in the error equation, which can be interpreted as an equivalent state-dependent nonlinear perturbation acting at the input of the closed-loop system. A way to deal with this kind of uncertainty has already been described: roughly speaking, it uses a fixed-parameter nominal control law with an additional high-gain input which aims at cancelling the mismatch perturbation term in the error equation. We have also seen that the tracking accuracy is directly related to the magnitude of the additional input. Therefore it is expected that if high accuracy is desired, the methods in Section 2.4.1 will be hardly implementable, due either to very high gains in the control law or to chattering phenomena. A solution which allows us to retrieve asymptotic stability of the tracking errors when the parameters are unknown without requiring high gain or discontinuous inputs is to replace the true control parameters in the nominal input with some time-varying parameters, which are usually called the estimates of the true parameters —even if they do not provide in general any accurate estimate of these parameters, unless some conditions on the desired trajectory are verified. One major problem that we encounter in this *adaptive inverse dynamics control* scheme is the design of a suitable update law for the estimates that guarantees boundedness of all the signals in the closed-loop system, and convergence of the tracking error to zero. Let us then replace H, C, g in (2.45) with their estimates, i.e.,

$$u = \hat{H}(q)(\ddot{q}_d - K_D \dot{\tilde{q}} - K_P \tilde{q}) + \hat{C}(q, \dot{q})\dot{q} + \hat{g}(q). \tag{2.161}$$

We assume here that $\hat{H}$, $\hat{C}$, $\hat{g}$ have the same functional form as H, C, g with estimated parameters $\hat{p}$. From Property 2.3 of the dynamic model we can rewrite (2.161) as

$$u = Y(q, \dot{q}, \ddot{q})\hat{p} \tag{2.162}$$

where $Y(q, \dot{q}, \ddot{q})$ is a known $(n \times r)$ matrix. Substituting (2.161) into the manipulator dynamics gives the following closed-loop error equation

$$\hat{H}(\ddot{\tilde{q}} + K_D \dot{\tilde{q}} + K_P \tilde{q}) = Y(q, \dot{q}, \ddot{q})\tilde{p} \tag{2.163}$$

where

$$Y(q, \dot{q}, \ddot{q})\tilde{p} = (\hat{H}(q) - H(q))\ddot{q} + (\hat{C}(q, \dot{q}) - C(q, \dot{q}))\dot{q} + (\hat{g}(q) - g(q)). \tag{2.164}$$

2.5. ADAPTIVE CONTROL

Before going further on with the stability analysis, let us introduce two assumptions that we will need in the sequel:

Assumption 2.1 The acceleration $\ddot{q}$ is measurable,

□

Assumption 2.2 The estimate of the generalized inertia matrix $\hat{H}(q)$ is full-rank for all q.

□

The error equation in (2.163) can be rewritten as

$$\ddot{\tilde{q}} + K_D \dot{\tilde{q}} + K_P \tilde{q} = \hat{H}^{-1}(q) Y(q, \dot{q}, \ddot{q}) \tilde{\rho} = \Phi(q, \dot{q}, \ddot{q}, \hat{\rho}) \tilde{\rho}. \quad (2.165)$$

This equation can be cast in state space form by choosing $\xi_1 = \tilde{q}$, $\xi_2 = \dot{\tilde{q}}$, $\xi = (\xi_1^T \ \xi_2^T)^T$, i.e.,

$$\dot{\xi} = A\xi + B\Phi\tilde{\rho} \quad (2.166)$$

with

$$A = \begin{pmatrix} 0 & I \\ -K_P & -K_D \end{pmatrix} \quad B = \begin{pmatrix} 0 \\ I \end{pmatrix}.$$

The following result can be established.

Lemma 2.4 *If Assumptions 2.1 and 2.2 hold and the parameter estimate is updated as*

$$\dot{\hat{\rho}} = -\Gamma^{-1} \Phi^T B^T P \xi \quad (2.167)$$

where Γ is a symmetric positive definite matrix, then the state ξ of system (2.166) asymptotically tends to zero.

◊ ◊ ◊

Proof. Choose the following Lyapunov function candidate

$$V = \xi^T P \xi + \tilde{\rho}^T \Gamma \tilde{\rho} \quad (2.168)$$

where P is the unique symmetric positive definite solution to the equation $A^T P + PA = -Q$, for a given symmetric positive definite matrix Q. Taking the time derivative of V along the trajectories of (2.166) gives

$$\dot{V} = -\xi^T Q \xi + 2\tilde{\rho}^T (\Phi^T B^T P \xi + \Gamma \dot{\hat{\rho}}). \quad (2.169)$$

Choosing the update law (2.167) yields

$$\dot{V} = -\xi^T Q \xi. \quad (2.170)$$

We can therefore conclude that $\xi \in L_2 \cap L_\infty$, $\hat{\rho} \in L_\infty$, and then the control input u in (2.161) is bounded. It follows that $\ddot{q} \in L_\infty$ so that $\dot{\xi} \in L_\infty$. Then ξ is uniformly continuous and, since $\xi \in L_2$, it can be concluded that ξ asymptotically converges to zero.

◊

Remarks

- We may wonder why we have chosen to derive an error equation in (2.163) that renders Assumptions 2.1 and 2.2 necessary. Indeed, following what we did in Section 2.3.2, we could have written

$$\ddot{\tilde{q}} + K_D \dot{\tilde{q}} + K_P \tilde{q} = H^{-1}(q) Y(\cdot) \tilde{\rho}. \tag{2.171}$$

But since $H^{-1}(q)$ is neither known nor linear with respect to some set of physical parameters, it seems impossible to use such an equivalent error equation.

- The algorithm we have presented here hinges on both Assumptions 2.1 and 2.2. For both practical and theoretical reasons, these assumptions are hardly acceptable. In most cases, indeed, it is not easy to obtain an accurate measure of acceleration; robustness of the above adaptive scheme with respect to such a disturbance has to be established. Moreover, from a pure theoretical viewpoint, measuring q, $\dot{q}$ and $\ddot{q}$ means that not only do we need the whole system state vector —an assumption which is generally considered in control theory as being somewhat stringent in itself— but we also need its derivative! Concerning Assumption 2.2, it is claimed sometimes that the estimates can be modified in order to ensure positive definiteness of $\hat{H}(q)$. The modification uses the fact that if a compact region Ω in the parameter space within which $\hat{H}(q)$ remains full rank and which contains the true parameter vector ρ is known, then the gradient update law in (2.167) can be modified in such a way that the estimates do not leave Ω. Besides the fact that the region Ω has to be a priori known —which may require in certain cases a very good a priori knowledge of the system parameters— and although this kind of projection algorithm is familiar to researchers in the field of adaptive control, its practical feasibility is not yet settled.

- It is possible to improve the above method by relaxing either Assumption 2.1 or Assumption 2.2, but not 2.1 and 2.2 together! For the sake of clarity of the presentation, and since our aim was to highlight the difficulties associated with the adaptive implementation of the inverse dynamics control method rather than to provide an exhaustive overview of the literature, we have preferred to restrict ourselves to this basic algorithm.

2.5.3 Adaptive passivity-based control

As we have already pointed out, adaptive implementation of a fixed-parameter controller may not always be obvious, even if the structure of the fixed-parameter scheme seems at first sight very simple. This is mainly due to the fact that, in general, it may not be easy to find a suitable error equation from the process dynamics and the designed output; by "suitable" we mean an error equation that contains a term linear in the parameter error vector, that is the mismatch measure between the true system and its estimate. If we would be able to find a Lyapunov function for the true fixed-parameter controller, it is expected that the same function with an additional term quadratic in the parameter error vector will enable us to find an update law for the estimates that renders the whole closed-loop system asymptotically stable. Such "two-terms" Lyapunov functions have been proved to work well in the case of adaptive control of linear systems. This idea is applied below directly to the previous passivity-based control schemes, which have been shown to embed the Lyapunov-based control schemes as a particular case.

Consider the error equation in (2.134) with $u_0 \equiv 0$; also in (2.133) replace $H_0(q)$, $C_0(q,\dot{q})$, $g_0(q)$ with $\hat{H}(q)$, $\hat{C}(q,\dot{q})$, $\hat{g}(q)$ as we did for the adaptive inverse dynamics controller in the preceding section. We therefore obtain a set of closed-loop differential equations similar to (2.71), with $\psi = Y(\cdot)\tilde{\rho}$. At this point, the following result can be established for the *adaptive passivity-based control*.

Lemma 2.5 *If the parameter estimate is updated as*

$$\dot{\hat{\rho}} = -\Gamma Y^T(\cdot)\sigma, \qquad (2.172)$$

where Γ is a symmetric positive definite matrix, then $\tilde{q}$, $\dot{\tilde{q}}$, σ asymptotically tend to zero and $\tilde{\rho}$ is bounded.

⋄ ⋄ ⋄

Proof. It is easily verified that the system $-\sigma \mapsto \psi$ with state $\hat{\rho}$ is passive relative to the function $V_2 = \frac{1}{2}\hat{\rho}^T\Gamma^{-1}\hat{\rho}$. Indeed for all $T \geq 0$ we have

$$\int_0^T -\sigma^T Y(\cdot)\hat{\rho}\,d\tau = \int_0^T \dot{\hat{\rho}}^T \Gamma^{-1}\hat{\rho}\,d\tau$$

$$= \frac{1}{2}\hat{\rho}^T(T)\Gamma^{-1}\hat{\rho}(T) - \frac{1}{2}\hat{\rho}^T(0)\Gamma^{-1}\hat{\rho}(0). \quad (2.173)$$

Since Γ is a bounded positive definite matrix, the result follows. The positive definite function V in (2.73) is thus given by

$$V = \frac{1}{2}\sigma^T H(q)\sigma + \frac{1}{2}\tilde{\rho}^T\Gamma^{-1}\tilde{\rho}. \qquad (2.174)$$

Therefore the conclusions of Theorem 2.1 in Section 2.3.3 hold, i.e., $\tilde{q}$, $\dot{\tilde{q}}$ and σ asymptotically tend to zero. But we cannot conclude that $\tilde{\rho}$ converges to zero, i.e., the estimated parameters are not guaranteed to converge to the true parameters; it is easy to conclude, however, that $\tilde{\rho}$ remains bounded.

$\diamond$

Remarks

- Consider the error equation in (2.71) together with the update law in (2.172). Then it is possible to decompose this closed-loop system in two subsystems as follows:

    (H1) Input $u_1 = Y(\cdot)\tilde{\rho}$, output $y_1 = \sigma$, state σ.

    (H2) Input $u_2 = \sigma$, output $y_2 = -Y(\cdot)\tilde{\rho}$, state $\hat{\rho}$.

    It is then easy to see that equation (2.71) corresponds to the interconnection of Subsystem (H1) with Subsystem (H2). As shown just above, Subsystem (H2) is passive relative to V_2. Also we have

    $$< u_1 | y_1 >_t = \int_0^t \sigma^T \left(H(q)\dot{\sigma} + C(q,\dot{q})\sigma + K_D\sigma \right) d\tau$$
    $$= \frac{1}{2} \left(s^T H(q)s \right)(t) - \frac{1}{2} \left(\sigma^T H(q)\sigma \right)(0) +$$
    $$+ \int_0^t \sigma^T K_D \sigma d\tau. \qquad (2.175)$$

    From (2.175) we conclude that Subsystem (H1) is strictly passive relative to $V_1 = \frac{1}{2}\sigma^T H(q)\sigma$, and $W_1 = \sigma^T K_D \sigma$. Therefore we conclude that the passivity theorem applies to this case.

- Notice that although such a passivity interpretation does not bring us any new conclusion on stability, it is however an interesting way to understand the underlying properties of the closed-loop system. For example, we may see that, by replacing the update law in (2.172) with any other estimation algorithm, stability properties are preserved provided the new update law is passive. This is exactly the same with the term $K_D\sigma$ in Subsystem (H1), which can be changed into any other function of σ provided this does not destroy the strict passivity of the subsystem.

- The positive definite function in (2.57) can be slightly modified to $V = \frac{1}{2}\sigma^T H(q)\sigma + \tilde{q}^T \Lambda^T K_D \tilde{q}$ which represents a Lyapunov function for the system in (2.71) with $\psi = 0$, as pointed out in (2.69). It is noteworthy that we can easily modify the feedback interconnection above so that it fits with this new function V. Indeed consider the following three subsystems

(J1) Input $u_1 = Y(\cdot)\tilde{\rho} - K_D\sigma$, output $y_1 = \sigma$, state σ.

(J2) Input $u_2 = \sigma$, output $y_2 = K_D\sigma$, state $\tilde{q}$.

(J3) Input $u_3 = \sigma$, output $y_3 = -Y(\cdot)\tilde{\rho}$, state $\hat{\rho}$.

Eq. (2.71) corresponds to the negative feedback interconnection of Subsystem (J1) with Subsystems (J2) and (J3). Then we have

$$\begin{aligned} <u_1|y_1>_t &= \int_0^t \sigma^T \left(H(q)\dot{\sigma} + C(q,\dot{q})\sigma\right) d\tau \\ &= \frac{1}{2}\left(\sigma^T H(q)\sigma\right)(t) - \frac{1}{2}\left(\sigma^T H(q)\sigma\right)(0) \quad (2.176) \end{aligned}$$

and

$$\begin{aligned} <u_2|y_2>_t &= \int_0^t \sigma^T K_D \sigma d\tau \\ &= \int_0^t \dot{\tilde{q}}^T K_D \dot{\tilde{q}} d\tau + \int_0^t \tilde{q}^T \Lambda^T K_D \Lambda \tilde{q} d\tau \\ &\quad + \left(\tilde{q}^T \Lambda^T K_D \tilde{q}\right)(t) - \left(\tilde{q}^T \Lambda^T K_D \tilde{q}\right)(0). (2.177) \end{aligned}$$

From (2.176) and (2.177) we conclude that Subsystem (J1) is passive relative to $V_1 = \frac{1}{2}\sigma^T H(q)\sigma$, whereas Subsystem (J2) is strictly passive relative to $V_2 = \tilde{q}^T \Lambda^T K_D \tilde{q}$ and $W_2 = \dot{\tilde{q}}^T K_D \dot{\tilde{q}} + \tilde{q}^T \Lambda^T K_D \tilde{q}$. Therefore we conclude that the passivity theorem applies again to this case. In fact, we are able here to associate a passive subsystem with each positive definite term of the positive definite function V used to prove stability.

2.6 Further reading

A survey of dynamic model properties used for control purposes can be found in [37]. Passivity properties enjoyed by the manipulator model were first studied in [74]. They were used in [5] to show stability of PD control. This property results from the law of conservation of energy, see [7]. It can be understood as the fact that some internal forces of the system are workless. Further studies on the stability of PID controllers in connection with the dynamics (2.1) can also be found in [5].

Eq. (2.1) was studied in this chapter without regard to friction forces. It can be expected that simplified friction model, such as viscous friction, will enhance dissipative properties. However, friction models are not yet completely understood. Some friction phenomena such as hysteresis, the Dalh's effect (nonlinear dynamical friction properties) and the Stribeck's effect (positive damping at low velocities) require further investigation in connection with the dissipative system properties. Some references concerning friction modelling, identification and adaptive compensation can be found in [20], [24], [28]. The use of notch filters to cope with friction over-compensation is studied in [21]. Mechanisms for friction identification are discussed in [22]. A more complete friction model including friction dynamics (hysteresis, stick-slip effect and pre-displacement motion) has been recently proposed in [29]. Such a model suitably describes most of the observed physical phenomena and can be used for simulation and control. An adaptive version that uses partial knowledge of this dynamic model is reported in [23]. For a complete discussion on modelling, identification and control of systems with friction, see [6].

The inverse dynamics control scheme was developed at the beginning of the 70's under the name of computed torque control. One of the first experiments carried out with the inverse dynamics control was done by [52]. Methods seeking to simplify the dynamic model and hence the computational burden involved with the inverse dynamics control method were studied in [8], among others. An efficient alternative to inverse dynamics computation is represented by Newton-Euler recursive algorithms presented in Chapter 1.

The control algorithm presented in Section 2.3.2 was first introduced in [67]. Although the analysis presented by the authors did not formally describe a Lyapunov-based control approach (rather a Lyapunov-like control), it is easy to prove that a simple modification in the Lyapunov candidate can lead to a formal Lyapunov function. This was pointed out in [71].

Stability analysis using passive system theory was carried out in [59] and [16], where a unified presentation of fixed and adaptive parameter control schemes from the passivity point of view was given. In [18] it is shown that under some restrictions on the Lyapunov function, the results in [59] and [16] generalize. Theorem 2.1 was first presented in [59]. In this chapter this theorem was modified in order to assess the dissipative system ideas rather than the passive system definitions. More reading and discussion about these different system definitions can be found in Appendix A. The lemma concerning the input–output relations when the input–output mapping is linear, and used in the proof of Lemma 2.3, is discussed at the end of

Appendix A as well as in [35]. In the fixed-parameter case, it has been demonstrated [2] that inverse dynamics control may become unstable in the presence of uncertainty, whereas by making the matrix gains $K_P(q)$ and $K_D(q)$ state-dependent the PD controller becomes passive and stability is ensured.

The main method of robust control we have presented above is mainly based on the work in [31]. Many robust controllers have been proposed in the literature which use this idea. Among them let us cite the works in [39, 64, 49, 72]. Other controllers guaranteeing practical stability, but using a high-gain type input rather than the saturation function can be found in [60, 34, 73, 63]. One of the first references where both approaches are compared is [40].

The method proposed in Section 2.4.2 differs from the one exposed in [72]. Although both philosophies are quite similar, the method used here to design the additional input u_0 in (2.110) or (2.128) is quite different from the one in [72]. The main difference is that in [72] u_0 is defined through an implicit equation which has a solution if the matrix $H_0(q)$ is such that $\|H^{-1}(q)H_0(q) - I\| \leq \alpha < 1$, for some α and for all $q \in R^n$. No such an assumption has been done here. In [63] a general family of nonlinear feedback gains is presented, which guarantees global uniform ultimate boundedness of the tracking errors on the basis of weak assumptions on the nominal inertia matrix $H_0(q)$; in particular, $H_0(q)$ is required to remain full-rank, and $H^{-1}(q)H_0(q)$ is positive-definite, with $\lambda_{\min}(H^{-1}(q)H_0(q)+H_0(q)H^{-1}(q)/2) > \delta > 0$. A detailed stability analysis is done using two time-scale system techniques. The idea of premultiplying the additional input u in (2.148) by $Y(\cdot)$ is taken from [70].

The above method relies on the a priori knowledge of an upper bound of the uncertain term in the error equation. An important question is whether or not it is possible to relax this assumption, i.e., are we able to design additional control inputs u if α^* is unknown? A solution consists of replacing α^* in u_0 with a time-varying term $\hat{\alpha}(t)$, which has to be updated through a suitable update law. Some preliminary results have been presented in the literature. An extension of the work in [31] when α^* is unknown was proposed in [32]; similar ideas are to be found in [78]; the application of the approach to the case of manipulators has been presented in [65]. However, boundedness of the signals in the closed-loop system is guaranteed under the condition that the boundary layer size ε of the saturation function exponentially converges to zero, so that the input becomes discontinuous and chattering phenomena may occur. Another solution has been proposed in [19], that can be considered as the extension of the scheme in [31]. The input is a continuous function of time, so that chattering phenomena are avoided. In [48] the authors, inspired by the work in [10], presented an algorithm that applies to static state feedback linearizable nonlinear systems with uncertainties. However the method in [48] applies to very particular classes of manipulators when the physical parameters are not known (see the example in [48]).

Finally, note that robust control of rigid manipulators was also considered in the pioneering work [66] with a sliding-mode philosophy. A survey of the robust approaches can be found in [1].

One of the very first algorithms dealing with adaptive control of rigid manipulators was presented in [36]. The stability analysis was done assuming that the joint dynamics be decoupled, each joint being considered as an independent second-order linear system. Other pioneering works in the field can for example be found in [9, 55]; although none of the fundamental dynamic model properties are used, the complete dynamic equations are taken into account, but the control input is discontinuous and may lead to chattering. Positive definiteness of the inertia matrix is explicitly used in [38]; it is however assumed that some time-varying quantities are constant during the adaptation. It is worth noticing that all these algorithms were based on the concept of Model Reference Adaptive Systems (MRAS) developed in [47] for linear systems. Therefore they are conceptually very different from the truly nonlinear algorithms presented here (and much less elegant).

Adaptive compensation of gravity was presented in [76]. The choice of the Lyapunov function in (2.154) was motivated by the work in [46]. The adaptive version of the inverse dynamics control scheme presented here was studied in [33]. Several different adaptive implementations of inverse dynamics control have been proposed in the literature [54, 44] (see, e.g., [59] for a review).

As we already noticed, there is an infinite number of ways to define the variable σ in what we called the passivity-based controllers. The choice we made here ($\sigma = \dot{\tilde{q}} + \Lambda \tilde{q}$) was in fact the one proposed in [67, 68]. As this algorithm has represented a major breakthrough in the field of manipulator control, we have preferred to keep this choice here. A number of other controllers using the same underlying philosophy have been proposed in the literature [62, 45, 69, 53] (see [16] for a review and a unification of those schemes from a passivity point of view). Theorem 2.1 is taken from [59].

A passive modified version of the least-squares estimation algorithm has been proposed in [51] and [17] which, according to the remarks at the end of the last section, guarantees closed-loop stability of the scheme. Other schemes are to be found in [41] (no use is made of the skew-symmetry property), [42] (it is shown that an optimal control law can be explicitly derived for rigid manipulators by minimizing some quadratic criterion), [11] (a variety of controllers are proposed and compared from a theoretical point of view), and [77] (the recursive Newton-Euler formulation is used instead of the Lagrange one to derive the manipulator dynamics, thus simplifying computation in view of practical implementations).

Even though adaptive control provides a solution which is robust to parameter uncertainty, robustness of adaptive controllers is in turn a topic that has interested researchers in the field. Indeed, disturbances at the output (e.g., measurement noise) or unmodelled dynamics (e.g., flexibility) may result in unbounded closed-loop signals. In particular the estimated parameters may diverge; this well-known phenomenon in adaptive control is called *parameter drift*. Solutions inspired from adaptive control of linear systems have been studied [61, 75], where a modified estimation ensures boundedness of the estimates. In [14], the controller in [67, 68] is modified so as to enhance its robustness. A general review of robust control (adaptive or not) of rigid manipulators is given in [1].

Practical experiments with comparisons between PD control, inverse dynamics control and adaptive control are presented in [68] and [57]; the algorithm presented in [67] is implemented by using a recursive Newton-Euler formulation.

Although not treated in this chapter, the problem of designing nonlinear controllers using only information of joint positions has been extensively studied, since in practice many industrial robots are not endowed with joint velocity transducers. High-resolution sensors allow, within a reasonable accuracy, estimating velocity by a crude interpolation. However, stability issues related to the impact of such an approximation on feedback design are often neglected. These reasons have motivated substantial research in the area of manipulator motion control via observed velocities. State feedback controllers for both regulation and tracking problems are used in [56] in connection with a nonlinear model-based observer in the feedback loop, ensuring local asymptotic stability. An observer based on sliding mode concept has been proposed in [30], whereas in [27] both a sliding and a smooth observer are considered which guarantee local exponential stability for the case where the system parameters are exactly known. Robustness issues and an adaptive solution for the case of model parameter inaccuracies have been proposed in [25] and in [26], respectively. Passivity-based observers are designed in [13], both for the regulation and the tracking problems. The regulation case allows for further simplification in the control design and often yields global or semi-global stability results. Simple PD controllers with gravity compensation where the velocity in the derivative term is replaced with a low-pass filtered joint position have been presented in [12], as well as in [43] leading to global asymptotic stability. Gravity compensation can be removed by means of a PID controller with an extra integral term in the filtered position [58].

To conclude the further reading section on the classical topic of manipulator motion control in the joint space, we would like to mention an alternative approach which has received the attention of some researchers; namely, learning control. The underlying idea is to use the error information caused by unmodelled dynamic effects in order to generate input torque compensation terms aimed at reducing such an error over repeated trials. The method does not enjoy the same formal elegance of the methods whose theory has been widely discussed in this chapter, but it proves itself quite useful for practical implementation of control algorithms on robot manipulators executing repetitive tasks, as typical in certain industrial applications. We refer the interested reader to the early works [4, 15], including both PD and PID types of controller, and to subsequent papers which are referenced in a brief recent survey on learning control [3].

References

[1] C. Abdallah, D. Dawson, P. Dorato, and M. Jamshidi, "Survey of robust control for rigid robots," *IEEE Control Systems Mag.*, vol. 11, no. 2, pp. 24–30, 1991.

[2] R.J. Anderson, "Passive computed torque algorithms for robots," *Proc. 28th IEEE Conf. on Decision and Control*, Tampa, FL, pp. 1638–1644, 1989.

[3] S. Arimoto, "Learning control," in *Robot Control*, M.W. Spong, F.L. Lewis, and C.T. Abdallah (Eds.), IEEE Press, New York, NY, pp. 185–188, 1993.

[4] S. Arimoto, S. Kawamura, and F. Miyazaki, "Bettering operation of robots by learning," *Advanced Robotics*, vol. 1, pp. 123–140, 1984.

[5] S. Arimoto and F. Miyazaki, "Stability and robustness of PID feedback control of robot manipulators," in *Robotics Research: 1st Int. Symp.*, M. Brady and R.P. Paul (Eds.), MIT Press, Cambridge, MA, pp. 783–789, 1983.

[6] B. Armstrong-Hélouvry, P. Dupont, and C. Canudas de Wit, "A survey of analysis tools and compensation methods for the control of machines with friction," *Automatica*, vol. 10, pp. 1083–1138, 1994.

[7] V.I. Arnold, *Mathematical Methods of Classical Mechanics*, Springer-Verlag, New York, NY, 1974.

[8] A. Aubin, C. Canudas de Wit, and H. Sidaoui, "Dynamic model simplification of industrial manipulators," *Prepr. 3rd IFAC Symp. on Robot Control*, Vienna, A, pp. 9–14, 1991.

[9] A. Balestrino, G. De Maria, and L. Sciavicco, "An adaptive model following control for robotic manipulators," *ASME J. of Dynamic Systems, Measurement, and Control*, vol. 105, pp. 143–151, 1983.

[10] B.R. Barmish, M. Corless, and G. Leitmann, "A new class of stabilizing controllers for uncertain dynamical systems," *SIAM J. on Control and Optimization*, vol. 21, pp. 246–255, 1983.

[11] D.S. Bayard and J.T. Wen, "New class of control laws for robotic manipulators — Part 2. Adaptive case," *Int. J. of Control*, vol. 47, pp. 1387–1406, 1988.

[12] H. Berghuis and H. Nijmeijer, "Global regulation of robots using only position measurements," *Systems & Control Lett.*, vol. 21, pp. 289–293, 1993.

[13] H. Berghuis and H. Nijmeijer, "A passivity approach to controller-observer design for robots," *IEEE Trans. on Robotics and Automation*, vol. 9, pp. 740–754, 1993.

[14] H. Berghuis, R. Ortega, and H. Nijmeijer, "A robust adaptive controller for robot manipulators," *Proc. 1992 IEEE Int. Conf. on Robotics and Automation*, Nice, F, pp. 1876–1881, 1992.

[15] P. Bondi, G. Casalino, and L. Gambardella, "On the iterative learning control theory for robotic manipulators," *IEEE J. of Robotics and Automation*, vol. 4, pp. 14–22, 1988.

[16] B. Brogliato, I.D. Landau, and R. Lozano, "Adaptive motion control of robot manipulators: A unified approach based on passivity," *Int. J. of Robust and Nonlinear Control*, vol. 1, pp. 187–202, 1991.

[17] B. Brogliato and R. Lozano, "Passive least squares type estimation algorithm for direct adaptive control," *Int. J. of Adaptive Control and Signal Processing*, vol. 6, pp. 35–44, 1992.

[18] B. Brogliato, R. Lozano, I.D. Landau, "New relationships between Lyapunov functions and the passivity theorem," *Int. J. of Adaptive Control and Signal Processing*, vol. 7, pp. 353–365, 1993.

[19] B. Brogliato and A. Trofino-Neto, "Practical stabilization of a class of nonlinear systems with partially known uncertainties," *Automatica*, vol. 31, pp. 145–150, 1995.

[20] C. Canudas de Wit, *Adaptive Control of Partially Known Systems: Theory and Applications*, Elsevier Science Publishers, Amsterdam, NL, 1988.

[21] C. Canudas de Wit, "Robust control for servo-mechanisms under inexact friction compensation," *Automatica*, vol. 29, pp. 757–761, 1993.

[22] C. Canudas de Wit, "Application of a bounded error on-line estimation algorithm to robotics systems," *Int. J. of Adaptive Control and Signal Processing*, vol. 8, pp. 73–84, 1994.

[23] C. Canudas de Wit, "Adaptive friction compensation with partially known dynamic friction model," *Int. J. of Adaptive Control and Signal Processing*, vol. 11, pp. 65–80, 1997.

[24] C. Canudas de Wit, K.J. Åström, and K. Braun, "Adaptive friction compensation in DC motors drives," *IEEE Trans. on Robotics and Automation*, vol. 3, pp. 681–685, 1987.

[25] C. Canudas de Wit and N. Fixot, "Robot control via robust estimated state feedback," *IEEE Trans. on Automatic Control*, vol. 36, pp. 1497–1501, 1991.

[26] C. Canudas de Wit and N. Fixot, "Adaptive control of robot manipulators via velocity estimated state feedback," *IEEE Trans. on Automatic Control*, vol. 37, pp. 1234–1237, 1992.

[27] C. Canudas de Wit, N. Fixot, and K.J. Åström, "Trajectory tracking in robot manipulators via nonlinear estimated state feedback," *IEEE Trans. on Robotics and Automation*, vol. 8, pp. 138–144, 1992.

[28] C. Canudas de Wit, P. Noel, A. Aubin, and B. Brogliato, "Adaptive friction compensation in robot manipulators: low velocities," *Int. J. of Robotics Research*, vol. 10, no. 3, pp. 189–199, 1991.

[29] C. Canudas de Wit, H. Olsson, K.J. Åström, and P. Lischinsky, "A new model for control of systems with friction," *IEEE Trans. on Automatic Control*, vol. 40, pp. 419–425, 1995.

[30] C. Canudas de Wit and J.J.-E. Slotine, "Sliding observers for robot manipulators," *Automatica*, vol. 27, pp. 859–864, 1991.

[31] M. Corless and G. Leitmann, "Continuous state feedback guaranteeing uniform ultimate boundedness for uncertain dynamic systems," *IEEE Trans. on Automatic Control*, vol. 26, pp. 1139–1144, 1981.

[32] M. Corless and G. Leitmann, "Adaptive control of systems containing uncertain functions and unknown functions with uncertain bounds," *J. of Optimization Theory and Application*, vol. 41, pp. 155–168, 1983.

[33] J.J. Craig, *Adaptive Control of Mechanical Manipulators*, Addison-Wesley, Reading, MA, 1988.

[34] D.M. Dawson and Z. Qu, "On the global uniform ultimate boundedness of a DCAL-like controller," *IEEE Trans. on Robotics and Automation*, vol. 8, pp. 409–413, 1992.

[35] C.A. Desoer and M. Vidyasagar, *Feedback Systems: Input–Output Properties*, Academic Press, New York, NY, 1975.

[36] S. Dubowsky and D.T. DesForges, "The application of model-referenced adaptive control to robotic manipulators," *ASME J. of Dynamic Systems, Measurement, and Control*, vol. 101, pp. 193–200, 1979.

[37] W.M. Grimm, "Robot non-linearity bounds evaluation techniques for robust control," *Int. J. of Adaptive Control and Signal Processing*, vol. 4, pp. 501–522, 1990.

[38] R. Horowitz and M. Tomizuka, "An adaptive control scheme for mechanical manipulators — Compensation of nonlinearity and decoupling control," *ASME J. of Dynamic Systems, Measurement, and Control*, vol. 108, pp. 127–135, 1986.

[39] L.-C. Fu and T.-L. Liao, "Globally stable robust tracking of nonlinear systems using variable structure control with an application to a robotic manipulator," *IEEE Trans. on Automatic Control*, vol. 35, pp. 1345–1350, 1990.

[40] S. Jayasuriya and C.N. Hwang, "Tracking controllers for robot manipulators: A high gain perspective," *ASME J. of Dynamic Systems, Measurement, and Control*, vol. 110, pp. 39–45, 1988.

[41] R. Johansson, "Adaptive control of robot manipulator motion," *IEEE Trans. on Robotics and Automation*, vol. 6, pp. 483–490, 1990.

[42] R. Johansson, "Quadratic optimization of motion coordination and control," *IEEE Trans. on Automatic Control*, vol. 35, pp. 1197–1208, 1990.

[43] R. Kelly, "A simple set-point robot controller by using only position measurements," *Prepr. 13th IFAC World Congress*, Sydney, AUS, vol. 6, pp. 173–176, 1993.

[44] R. Kelly and R. Carelli, "Unified approach to adaptive control of robotic manipulators," *Proc. 27th IEEE Conf. on Decision and Control*, Austin, TX, pp. 699–703, 1988.

[45] R. Kelly and R. Ortega, "Adaptive motion control design of robot manipulators: An input–output approach," *Proc. 1988 IEEE Int. Conf. on Robotics and Automation*, Philadelphia, PA, pp. 699–703, 1988.

[46] D.E. Koditschek, "Natural motion for robot arms," *Proc. 23th IEEE Conf. on Decision and Control*, Las Vegas, NV, pp. 733–735, 1984.

[47] I.D. Landau, *Adaptive Control: The Model Reference Approach*, Dekker, New York, NY, 1979.

[48] T.-L. Liao, L.-C. Fu, and C.-F. Hsu, "Adaptive robust tracking of nonlinear systems with an application to a robotic manipulator," *Systems & Control Lett.*, vol. 15, pp. 339–348, 1990.

[49] K.Y. Lim and M. Eslami, "Robust adaptive controller designs for robot manipulator systems," *IEEE J. of Robotics and Automation*, vol. 3, pp. 54–66, 1987.

[50] R. Lozano, B. Brogliato, and I.D. Landau, "Passivity and global stabilization of cascaded nonlinear systems," *IEEE Trans. on Automatic Control*, vol. 37, pp. 1386–1388, 1992.

[51] R. Lozano and C. Canudas de Wit, "Passivity based adaptive control for mechanical manipulators using LS type estimation," *IEEE Trans. on Automatic Control*, vol. 35, pp. 1363–1365, 1990.

[52] B.R. Markiewicz, *Analysis of the Computed Torque Drive Method and Comparison with Conventional Position Servo for a Computer-Controlled Manipulator*, memo. TM 33-601, JPL, Pasadena, CA, 1973.

[53] W. Messner, R. Horowitz, W.-W. Kao, and M. Boals, "A new adaptive learning rule," *Proc. 1990 IEEE Int. Conf. on Robotics and Automation*, Cincinnati, OH, pp. 1522–1527, 1990.

REFERENCES

[54] R. Middleton and G.C. Goodwin, "Adaptive computed torque control for rigid link manipulators," *Systems & Control Lett.*, vol. 10, pp. 9–16, 1988.

[55] S. Nicosia and P. Tomei, "Model reference adaptive control algorithms for industrial robots," *Automatica*, vol. 20, pp. 635–644, 1984.

[56] S. Nicosia and P. Tomei, "Robot control by using only joint position measurements," *IEEE Trans. on Automatic Control*, vol. 35, pp. 1058–1061, 1990.

[57] G. Niemeyer and J.-J.E. Slotine, "Performance in adaptive manipulator control," *Int. J. of Robotics Research*, vol. 10, pp. 149–161, 1991.

[58] R. Ortega, A. Loria, and R. Kelly, "A semiglobally state output feedback PI^2D regulator for robot manipulators," *IEEE Trans. on Automatic Control*, vol. 40, pp. 1432–1436, 1995.

[59] R. Ortega and M.W. Spong, "Adaptive motion control of rigid robots: a tutorial," *Automatica*, vol. 25, pp. 877–888, 1989.

[60] Z. Qu, J.F. Dorsey, X. Zhang, and D. Dawson, "Robust control of robots by the computed torque law," *Systems & Control Lett.*, vol. 16, pp. 25–32, 1991.

[61] J.S. Reed and P.A. Ioannou, "Instability analysis and robust adaptive control of robotic manipulators," *IEEE Trans. on Robotics and Automation*, vol. 5, pp. 381–386, 1989.

[62] N. Sadegh and R. Horowitz, "Stability and robustness analysis of a class of adaptive controllers for robotic manipulators," *Int. J. of Robotics Research* vol. 9, no. 3, pp. 74–92, 1990.

[63] C. Samson, "Robust control of a class of nonlinear systems and applications to robotics," *Int. J. of Adaptive Control and Signal Processing*, vol. 1, pp. 49–68, 1987.

[64] R. Shoureshi, M.E. Momot, and M.D. Roesler, "Robust control for manipulators with uncertain dynamics," *Automatica*, vol. 26, pp. 353–359, 1990.

[65] S.N. Singh, "Adaptive model following control of nonlinear robotic systems," *IEEE Trans. on Automatic Control*, vol. 30, pp. 1099–1100, 1985.

[66] J.-J.E. Slotine, "The robust control of robot manipulators," *Int. J. of Robotics Research*, vol. 4, no. 2, pp. 49–64, 1985.

[67] J.-J.E. Slotine and W. Li, "On the adaptive control of robot manipulators," *Int. J. of Robotics Research*, vol. 6, no. 3, pp. 49–59, 1987.

[68] J.-J.E. Slotine and W. Li, "Adaptive manipulator control: A case study," *IEEE Trans. on Automatic Control*, vol. 33, pp. 995–1003, 1988.

[69] J.-J.E. Slotine and W. Li, "Composite adaptive control of robot manipulators," *Automatica*, vol. 25, pp. 509–519, 1989.

[70] M.W. Spong, "On the robust control of robot manipulators," *IEEE Trans. on Automatic Control*, vol. 37, pp. 1782–1786, 1993.

[71] M.W. Spong, R. Ortega, and R. Kelly, "Comments on 'Adaptive manipulator control: A case study'," *IEEE Trans. on Automatic Control*, vol. 35, pp. 761–762, 1990.

[72] M.W. Spong and M. Vidyasagar, *Robot Dynamics and Control*, Wiley, New York, NY, 1989.

[73] Y. Stepanenko and J. Yuan, "Robust adaptive control of a class of nonlinear mechanical systems with unbounded and fast varying uncertainties," *Automatica*, vol. 28, pp. 265–276, 1992.

[74] M. Takegaki and S. Arimoto, "A new feedback method for dynamic control of manipulators" *ASME J. of Dynamic Systems, Measurement, and Control*, vol. 102, pp. 119–125, 1981.

[75] G. Tao, "On robust adaptive control of robot manipulators," *Automatica*, vol. 28, pp. 803–807, 1992.

[76] P. Tomei, "Adaptive PD controller for robot manipulators," *IEEE Trans. on Robotics and Automation*, vol. 7, pp. 565–570, 1991.

[77] M.W. Walker, "Adaptive control of manipulators containing closed kinematic loops," *IEEE Trans. on Robotics and Automation*, vol. 6, pp. 10–19, 1990.

[78] D.S. Yoo and M.J. Chung, "A variable structure control with simple adaptation laws for upper bounds on the norm of the uncertainty," *IEEE Trans. on Automatic Control*, vol. 37, pp. 860–864, 1992.

Chapter 3

Task space control

In the above joint space control schemes, it was assumed that the reference trajectory is available in terms of the time history of joint positions, velocities and accelerations. On the other hand, robot manipulator motions are typically specified in the task space in terms of the time history of end-effector position, velocity and acceleration. This chapter is devoted to control of rigid robot manipulators in the task space.

The natural strategy to achieve task space control goes through two successive stages; namely, kinematic inversion of the task space variables into the corresponding joint space variables, and then design of a joint space control. Hence this approach, termed *kinematic control*, is congenial to analyze the important properties of kinematic mappings: *singularities* and *redundancy*.

A different strategy consists of designing a control scheme directly in the task space that utilizes the kinematic mappings to reconstruct task space variables from measured joint space variables. This approach has the advantage to operate directly on the task space variables. However, it does not allow an easy management of the effects of singularities and redundancy, and may become computationally demanding if, besides positions, also velocities and accelerations are of concern.

The material of this chapter is organized as follows. The *inversion* of *differential kinematics* is discussed in terms of both the pseudoinverse and the damped least-squares inverse of the Jacobian. *Inverse kinematics algorithms* are proposed which are aimed at generating the reference trajectories for joint space control schemes; velocity resolution schemes are presented based on the use of either the pseudoinverse or the transpose of the Jacobian matrix, and the extension to acceleration resolution is also discussed. As opposed to the above kinematic control schemes, two kinds of *direct task space control* schemes are presented which are analogous to those analyzed in the joint space; namely, a PD control with gravity compensation scheme that achieves end-effector *regulation*, and an inverse dynamics control scheme that allows end-effector trajectory *tracking*.

3.1 Kinematic control

Control of robot manipulators is naturally achieved in the joint space, since the control inputs are the joint torques. Nevertheless, the user specifies a motion in the task space, and thus it is important to extend the control problem to the task space. This can be achieved by following two different strategies. Let us start by illustrating the more natural one, *kinematic control*, which consists of inverting the kinematics of the manipulator to compute the joint motion corresponding to the given end-effector motion. In view of the difficulties in finding closed-form solutions to the inverse kinematics problem, it is worth considering the problem of *differential kinematics inversion* which is well posed for any manipulator kinematic structure and allows a natural treatment of singularities and redundancy.

3.1.1 Differential kinematics inversion

The differential kinematics equation, in terms of either the geometric or the analytical Jacobian establishes a linear mapping between joint space velocities and task space velocities, even if the Jacobian is a function of the joint configuration. This feature suggests the use of the differential kinematics equation

$$v = J(q)\dot{q} \tag{3.1}$$

to solve the inverse kinematics problem.

Assume that a task space trajectory is given $(x(t), v(t))$. The goal is to find a feasible joint space trajectory $(q(t), \dot{q}(t))$ that reproduces the given trajectory.

Joint velocities can be obtained by solving the differential kinematics equation for $\dot{q}$ at the current joint configuration; then, joint positions $q(t)$ can be computed by integrating the velocity solution over time with known initial conditions. This approach is based on the knowledge of the manipulator Jacobian and thus is applicable to any manipulator structure, on condition that a suitable inverse for the matrix J can be found.

3.1. KINEMATIC CONTROL

Pseudoinverse

With reference to the geometric Jacobian, the basic inverse solution to (3.1) is obtained by using the *pseudoinverse* $J^\dagger$ of the matrix J; this is a unique matrix satisfying the Moore-Penrose conditions

$$JJ^\dagger J = J \qquad J^\dagger J J^\dagger = J^\dagger$$
$$(JJ^\dagger)^T = JJ^\dagger \qquad (J^\dagger J)^T = J^\dagger J \qquad (3.2)$$

or, alternatively, the equivalent conditions

$$\begin{aligned} J^\dagger a &= a & \forall a \in \mathcal{N}^\perp(J) \\ J^\dagger b &= 0 & \forall b \in \mathcal{R}^\perp(J) \\ J^\dagger(a+b) &= J^\dagger a + J^\dagger b & \forall a \in \mathcal{R}(J), \forall b \in \mathcal{R}^\perp(J). \end{aligned} \qquad (3.3)$$

The inverse solution can then be written as

$$\dot{q} = J^\dagger(q)v \qquad (3.4)$$

that provides a least-squares solution with minimum norm to equation (3.1); in detail, solution (3.4) satisfies the condition

$$\min_{\dot{q}} \|\dot{q}\| \qquad (3.5)$$

of all $\dot{q}$ that fulfill

$$\min_{\dot{q}} \|v - J\dot{q}\|. \qquad (3.6)$$

If the Jacobian matrix is full-rank, the right pseudoinverse of J can be computed as

$$J^\dagger = J^T(JJ^T)^{-1}, \qquad (3.7)$$

and (3.4) provides an exact solution to (3.1); further, if J square, the pseudoinverse (3.7) reduces to the standard inverse Jacobian matrix J^{-1}.

To gain insight into the properties of the inverse mapping described by (3.4), it is useful to consider the singular value decomposition of J, and thus

$$J^\dagger = V\Sigma^\dagger U^T = \sum_{i=1}^{r} \frac{1}{\sigma_i} v_i u_i^T \qquad (3.8)$$

where r denotes the rank of J. The following properties hold:

- $\sigma_1 \geq \sigma_2 \geq \ldots \geq \sigma_r > \sigma_{r+1} = \ldots = \sigma_m = 0$,

- $\mathcal{R}(J^\dagger) = \mathcal{N}^\perp(J) = \text{span}\{v_1, \ldots, v_r\}$,
- $\mathcal{N}(J^\dagger) = \mathcal{R}^\perp(J) = \text{span}\{u_{r+1}, \ldots, u_n\}$.

The null space $\mathcal{N}(J^\dagger)$ is the set of task velocities that yield null joint space velocities at the current configuration; these task velocities belong to the orthogonal complement of the feasible space task velocities. Hence, one effect of the pseudoinverse solution (3.4) is to filter the infeasible components of the given task velocities while allowing exact tracking of the feasible components; this is due to the minimum norm property (3.5).

The range space $\mathcal{R}(J^\dagger)$ is the set of joint velocities that can be obtained as a result of all possible task velocities. Since these joint velocities belong to the orthogonal complement of the null space joint velocities, the pseudoinverse solution (3.4) satisfies the least-squares condition (3.6).

If a task velocity is assigned along u_i, the corresponding joint velocity computed via (3.4) lies along v_i and is magnified by the factor $1/\sigma_i$. When a *singularity* is approached, the r-th singular value tends to zero and a fixed task velocity along u_r requires large joint velocities. At a singular configuration, the u_r direction becomes infeasible and v_r adds to the set of null space velocities of the manipulator.

Redundancy

For a kinematically *redundant* manipulator a nonempty null space $\mathcal{N}(J)$ exists which is available to set up systematic procedures for an effective handling of redundant degrees of freedom. The general inverse solution can be written as

$$\dot{q} = J^\dagger(q)v + (I - J^\dagger(q)J(q))\dot{q}_0 \tag{3.9}$$

which satisfies the least-squares condition (3.6) but loses the minimum norm property (3.5), by virtue of the addition of the homogeneous term $(I - J^\dagger J)\dot{q}_0$; the matrix $(I - J^\dagger J)$ is a projector of the joint vector $\dot{q}_0$ onto $\mathcal{N}(J)$.

In terms of the singular value decomposition, solution (3.9) can be written in the form

$$\dot{q} = \sum_{i=1}^{r} v_i u_i^T v + \sum_{i=r+1}^{m} v_i v_i^T \dot{q}_0 + \sum_{i=m+1}^{n} v_i v_i^T \dot{q}_0. \tag{3.10}$$

Three contributions can be recognized in (3.10); namely, the least-squares joint velocities, the null space joint velocities due to singularities (if $r < m$), and the null space joint velocities due to redundant degrees of freedom (if $m < n$).

3.1. KINEMATIC CONTROL

This result is of fundamental importance for redundancy resolution, since solution (3.9) evidences the possibility of choosing the vector $\dot{q}_0$ so as to exploit the redundant degrees of freedom. In fact, the contribution of $\dot{q}_0$ is to generate null space motions of the structure that do not alter the task space configuration but allow the manipulator to reach postures which are more dexterous for the execution of the given task.

A typical choice of the null space joint velocity vector is

$$\dot{q}_0 = \alpha \left(\frac{\partial w(q)}{\partial q} \right)^T \tag{3.11}$$

with $\alpha > 0$; $w(q)$ is a scalar objective function of the joint variables and $(\partial w(q)/\partial q)^T$ is the vector function representing the *gradient* of w. In this way, we seek to *locally* optimize w in accordance with the kinematic constraint expressed by (3.1). Usual objective functions are:

- the *manipulability measure* defined as

$$w(q) = \sqrt{\det(J(q) J^T(q))}, \tag{3.12}$$

which vanishes at a singular configuration, and thus redundancy may be exploited to escape singularities;

- the *distance from mechanical joint limits* defined as

$$w(q) = -\frac{1}{2n} \sum_{i=1}^{n} \left(\frac{q_i - \bar{q}_i}{q_{iM} - q_{im}} \right)^2, \tag{3.13}$$

where q_{iM} (q_{im}) denotes the maximum (minimum) limit for q_i and $\bar{q}_i$ the middle of the joint range, and thus redundancy may be exploited to keep the manipulator off joint limits;

- the *distance from an obstacle* defined as

$$w(q) = \min_{p,o} \|p(q) - o\|, \tag{3.14}$$

where o is the position vector of an opportune point on the obstacle and p is the position vector of the closest manipulator point to the obstacle, and thus redundancy may be exploited to avoid collisions with obstacles.

Damped least-squares inverse

In the neighbourhood of singular configurations the use of a pseudoinverse is not adequate and a numerically robust solution is achieved by the *damped least-squares inverse* technique. This is based on the solution to the modified differential kinematics equation

$$J^T v = (J^T J + \lambda^2 I)\dot{q} \tag{3.15}$$

in place of equation (3.1); in (3.15) the scalar λ is the so-called *damping factor*. Note that, when $\lambda = 0$, equation (3.15) reduces to (3.1).

The solution to (3.15) can be written in either of the equivalent forms

$$\dot{q} = J^T(JJ^T + \lambda^2 I)^{-1} v \tag{3.16}$$
$$\dot{q} = (J^T J + \lambda^2 I)^{-1} J^T v. \tag{3.17}$$

The computational load of (3.16) is lower than that of (3.17), being usually $n \geq m$. Let then

$$\dot{q} = J^\#(q) v \tag{3.18}$$

indicate the damped least-squares inverse solution computed with either of the above forms. Solution (3.18) satisfies the condition

$$\min_{\dot{q}} \|v - J\dot{q}\|^2 + \lambda^2 \|\dot{q}\|^2 \tag{3.19}$$

that gives a trade-off between the least-squares condition (3.6) and the minimum norm condition (3.5). In fact, condition (3.19) accounts for both accuracy and feasibility in choosing the joint space velocity $\dot{q}$ required to produce the given task space velocity v. In this regard, it is essential to select a suitable value for the damping factor; small values of λ give accurate solutions but low robustness in the neighbourhood of singular configurations, while large values of λ result in low tracking accuracy even if feasible and accurate solutions would be possible.

Resorting to the singular value decomposition, the damped least-squares inverse solution (3.18) can be written as

$$\dot{q} = \sum_{i=1}^{r} \frac{\sigma_i}{\sigma_i^2 + \lambda^2} v_i u_i^T v. \tag{3.20}$$

Remarkably, it is:

- $\mathcal{R}(J^\#) = \mathcal{R}(J^\dagger) = \mathcal{N}^\perp(J) = \mathrm{span}\{v_1, \ldots, v_r\}$,
- $\mathcal{N}(J^\#) = \mathcal{N}(J^\dagger) = \mathcal{R}^\perp(J) = \mathrm{span}\{u_{r+1}, \ldots, u_n\}$,

3.1. KINEMATIC CONTROL

that is, the structural properties of the damped least-squares inverse solution are analogous to those of the pseudoinverse solution.

It is clear that, with respect to the pure least-squares solution (3.4), the components for which $\sigma_i \gg \lambda$ are little influenced by the damping factor, since in this case it is

$$\frac{\sigma_i}{\sigma_i^2 + \lambda^2} \approx \frac{1}{\sigma_i}. \tag{3.21}$$

On the other hand, when a singularity is approached, the smallest singular value tends to zero while the associated component of the solution is driven to zero by the factor σ_i/λ^2; this progressively reduces the joint velocity to achieve near-degenerate components of the commanded task velocity. At the singularity, solutions (3.18) and (3.4) behave identically as long as the remaining singular values are significantly larger than the damping factor. Note that an upper bound of $1/2\lambda$ is set on the magnification factor relating the task velocity component along u_i to the resulting joint velocity along v_i; this bound is reached when $\sigma_i = \lambda$.

The damping factor λ determines the degree of approximation introduced with respect to the pure least-squares solution; then, using a constant value for λ may turn out to be inadequate for obtaining good performance over the entire manipulator workspace. An effective choice is to adjust λ as a function of some measure of closeness to the singularity at the current configuration of the manipulator; to this purpose, manipulability measures or estimates of the smallest singular value can be adopted. Remarkably, currently available microprocessors even allow real-time computation of full singular value decomposition.

A singular region can be defined on the basis of the estimate of the smallest singular value of J; outside the region the exact solution is used, while inside the region a configuration-varying damping factor is introduced to obtain the desired approximate solution. The factor must be chosen so that continuity of joint velocity $\dot{q}$ is ensured in the transition at the border of the singular region.

Without loss of generality, for a 6-degree-of-freedom manipulator, we can select the damping factor according to the following law:

$$\lambda^2 = \begin{cases} 0 & \hat{\sigma}_6 \geq \epsilon \\ \left(1 - \left(\frac{\hat{\sigma}_6}{\epsilon}\right)^2\right) \lambda_{\max}^2 & \hat{\sigma}_6 < \epsilon, \end{cases} \tag{3.22}$$

where $\hat{\sigma}_6$ is the estimate of the smallest singular value, and ϵ defines the size of the singular region; the value of $\lambda_{\max}$ is at user's disposal to suitably shape the solution in the neighbourhood of a singularity.

Eq. (3.22) requires computation of the smallest singular value. In order to avoid a full singular value decomposition, we can resort to a recursive algorithm to find an estimate of the smallest singular value. Suppose that an estimate $\hat{v}_6$ of the last input singular vector is available, so that $\hat{v}_6 \approx v_6$ and $\|\hat{v}_6\| = 1$. This estimate is used to compute the vector $\hat{v}_6'$ from

$$(J^T J + \lambda^2 I)\hat{v}_6' = \hat{v}_6. \tag{3.23}$$

Then the square of the estimate $\hat{\sigma}_6$ of the smallest singular value can be found as

$$\hat{\sigma}_6^2 = \frac{1}{\|\hat{v}_6'\|} - \lambda^2, \tag{3.24}$$

while the estimate of v_6 is updated using

$$\hat{v}_6 = \frac{\hat{v}_6'}{\|\hat{v}_6'\|}. \tag{3.25}$$

The above estimation scheme is based on the assumption that v_6 is slowly rotating, which is normally the case. However, if the manipulator is close to a double singularity (e.g., a shoulder and a wrist singularity for the anthropomorphic manipulator), the vector v_6 will instantaneously rotate if the two smallest singular values cross. The estimate of the smallest singular value will then track σ_5 initially, before $\hat{v}_6$ converges again to v_6. Therefore, it is worth extending the scheme by estimating not only the smallest but also the second smallest singular value. Assume that the estimates $\hat{v}_6$ and $\hat{\sigma}_6$ are available and define the matrix

$$M = J^T J + \lambda^2 I - (\hat{\sigma}_6^2 + \lambda^2)\hat{v}_6 \hat{v}_6^T. \tag{3.26}$$

With this choice, the second smallest singular value of J plays in

$$M\hat{v}_5' = \hat{v}_5 \tag{3.27}$$

the same role as σ_6 in (3.23) and then will provide a convergent estimate of $\hat{v}_5$ to v_5 and $\hat{\sigma}_5$ to σ_5.

At this point, suppose that $\hat{v}_5$ is an estimate of v_5 so that $\hat{v}_5 \approx v_5$ and $\|\hat{v}_5\| = 1$. This estimate is used to compute $\hat{v}_5'$ from (3.27). Then, an estimate of the square of the second smallest singular value of J is found from

$$\hat{\sigma}_5^2 = \frac{1}{\|\hat{v}_5'\|} - \lambda^2, \tag{3.28}$$

and the estimate of v_5 is updated using

$$\hat{v}_5 = \frac{\hat{v}_5'}{\|\hat{v}_5'\|}. \tag{3.29}$$

3.1. KINEMATIC CONTROL

On the basis of this modified estimation algorithm, crossing of singularities can be effectively detected; also, by switching the two singular values and the associated estimates $\hat{v}_5$ and $\hat{v}_6$, the estimation of the smallest singular value will be accurate even when the two smallest singular values cross.

User-defined accuracy

The above damped least-squares inverse method achieves a compromise between accuracy and robustness of the solution. This is performed without specific regard to the components of the particular task assigned to the manipulator end-effector. The *user-defined accuracy* strategy based on the weighted damped least-squares inverse method allows us to discriminate between directions in the task space where higher accuracy is desired and directions where lower accuracy can be tolerated. This is the case, for instance, of spot welding or spray painting in which the tool angle about the approach direction is not essential to the fulfillment of the task.

Let a weighted end-effector velocity vector be defined as

$$\bar{v} = Wv \tag{3.30}$$

where W is the $(m \times m)$ task-dependent weighting matrix taking into account the anisotropy of the task requirements. Substituting (3.30) into (3.1) gives

$$\bar{v} = \bar{J}(q)\dot{q} \tag{3.31}$$

where $\bar{J} = WJ$. It is worth noticing that if W is full-rank, solving (3.1) is equivalent to solving (3.31), but with different conditioning of the system of equations to solve. This suggests selecting only the strictly necessary weighting action in order to avoid undesired ill-conditioning of $\bar{J}$.

Eq. (3.31) can be solved by using the weighted damped least-squares inverse technique, i.e.,

$$\bar{J}^T(q)\bar{v} = \left(\bar{J}^T(q)\bar{J}(q) + \lambda^2 I\right)\dot{q}. \tag{3.32}$$

Again, the singular value decomposition of the matrix $\bar{J}$ is helpful, i.e.,

$$\bar{J} = \sum_{i=1}^{r} \bar{\sigma}_i \bar{u}_i \bar{v}_i^T \tag{3.33}$$

and the solution to (3.32) can be written as

$$\dot{q} = \sum_{i=1}^{r} \frac{\bar{\sigma}_i}{\bar{\sigma}_i^2 + \lambda^2} \bar{v}_i \bar{u}_i^T \bar{v}. \tag{3.34}$$

It is clear that the singular values $\bar{\sigma}_i$ and the singular vectors $\bar{u}_i$ and $\bar{v}_i$ depend on the choice of the weighting matrix W. While this has no effect on the solution $\dot{q}$ as long as $\bar{\sigma}_r \gg \lambda$, close to singularities where $\bar{\sigma}_r \ll \lambda$, for some $r < m$, the solution can be shaped by properly selecting the matrix W.

For a 6-degree-of-freedom manipulator with spherical wrist, it is worthwhile to devise a special handling of the wrist singularity, since such a singularity is difficult to predict at the planning level in the task space. It can be recognized that, at the wrist singularity, there are only two components of the angular velocity vector that can be generated by the wrist itself. The remaining component might be generated by the inner joints, at the expense of loss of accuracy along some other task space directions, though. For this reason, lower weight should be put on the angular velocity component that is infeasible to the wrist. For the anthropomorphic manipulator, this is easily expressed in the frame attached to link 4; let R_4 denote the rotation matrix describing orientation of this frame with respect to the base frame, so that the infeasible component is aligned with the x-axis. We propose then to choose the weighting matrix as

$$W = \begin{pmatrix} I & 0 \\ 0 & R_4 \text{diag}\{w, 1, 1\} R_4^T \end{pmatrix}. \tag{3.35}$$

Similarly to the choice of the damping factor as in (3.22), the weighting factor w is selected according to the following expression:

$$(1-w)^2 = \begin{cases} 0 & \hat{\sigma}_6 \geq \epsilon \\ \left(1 - \left(\frac{\hat{\sigma}_6}{\epsilon}\right)^2\right)(1 - w_{\min})^2 & \hat{\sigma}_6 < \epsilon, \end{cases} \tag{3.36}$$

where $w_{\min} > 0$ is a design parameter.

3.1.2 Inverse kinematics algorithms

In the previous section the differential kinematics equation has been utilized to solve for joint velocities. Open-loop reconstruction of joint variables through numerical integration unavoidably leads to solution drift and then to task space errors. This drawback can be overcome by devising a closed-loop *inverse kinematics algorithm* based on the task space error between the desired and actual end-effector locations x_d and x, i.e., $x_d - x$. It is worth considering also the differential kinematics equation in the form

$$\dot{x} = J_a(q)\dot{q} \tag{3.37}$$

where the definition of the task error has required consideration of the analytical Jacobian J_a in lieu of the geometric Jacobian.

3.1. KINEMATIC CONTROL

Jacobian pseudoinverse

At this point, the joint velocity vector has to be chosen so that the task error tends to zero. The simplest algorithm is obtained by using the *Jacobian pseudoinverse*

$$\dot{q} = J_a^\dagger(q)\left(\dot{x}_d + K(x_d - x)\right), \tag{3.38}$$

which plugged into (3.37) gives

$$(\dot{x}_d - \dot{x}) + K(x_d - x) = 0. \tag{3.39}$$

If K is an $(m \times m)$ positive definite (diagonal) matrix, the linear system (3.39) is *asymptotically stable*; the tracking error along the given trajectory converges to zero with a rate depending on the eigenvalues of K.

If it is desired to exploit redundant degrees of freedom, solution (3.38) can be generalized to

$$\dot{q} = J_a^\dagger(q)\left(\dot{x}_d + K(x_d - x)\right) + \left(I - J_a^\dagger(q)J_a(q)\right)\dot{q}_0 \tag{3.40}$$

that logically corresponds to (3.9). In case of numerical problems in the neighbourhood of singularities, the pseudoinverse can be replaced with a suitable damped least-squares inverse.

Jacobian transpose

A computationally efficient inverse kinematics algorithm can be derived by considering the *Jacobian transpose* in lieu of the pseudoinverse.

Lemma 3.1 *Consider the joint velocity vector*

$$\dot{q} = J_a^T(q)K(x_d - x) \tag{3.41}$$

where K is a symmetric positive definite matrix. If x_d is constant and J_a is full-rank for all joint configurations q, then $x = x_d$ is a globally asymptotically stable equilibrium point for the system (3.37) and (3.41).

◇ ◇ ◇

Proof. A simple Lyapunov argument can be used to analyze the convergence of the algorithm. Consider the positive definite function candidate

$$V = \frac{1}{2}(x_d - x)^T K(x_d - x); \tag{3.42}$$

its time derivative along the trajectories of the system (3.37) and (3.41) is

$$\dot{V} = (x_d - x)^T K \dot{x}_d - (x_d - x)^T K J_a(q) J_a^T(q) K(x_d - x). \tag{3.43}$$

If $\dot{x}_d = 0$, $\dot{V}$ is negative definite as long as J_a is full-rank.

◇

Remarks

- If $\dot{x}_d \neq 0$, only boundedness of tracking errors can be established; an estimate of the bound is given by

$$\|x_d - x\|_{\max} = \frac{\|\dot{x}_d\|_{\max}}{k\sigma_r^2(J_a)} \tag{3.44}$$

where K has been conveniently chosen as a diagonal matrix $K = kI$. It is anticipated that k can be increased to diminish the errors, but in practice upper bounds exist due to discrete-time implementation of the algorithm.

- When a singularity is encountered, $\mathcal{N}(J_a^T)$ is non-empty and $\dot{V}$ is only semi-definite; $\dot{V} = 0$ for $x \neq x_d$ with $K(x_d - x) \in \mathcal{N}(J_a^T)$, and the algorithm may get stuck. It can be shown, however, that such an equilibrium point is unstable as long as $\dot{x}_d$ drives $K(x_d - x)$ outside $\mathcal{N}(J_a^T)$. An enhancement of the algorithm can be achieved by rendering the matrix $J_a^T K$ less sensitive to variations of joint configurations along the task trajectory; this is accomplished by choosing a configuration-dependent K that compensates for variations of J_a.

The most attractive feature of the Jacobian transpose algorithm is certainly the need of computing only direct kinematics functions $k(q)$ and $J_a(q)$. Further insight into the performance of solution (3.41) can be gained by considering the singular value decomposition of the Jacobian transpose, and thus

$$J^T = \sum_{i=1}^{m} \sigma_i v_i u_i^T \tag{3.45}$$

which reveals a continuous, smooth behaviour of the solution close and through singular configurations; note that in (3.45) the geometric Jacobian has been considered and it has been assumed that no representation singularities are introduced.

Use of redundancy

In case of redundant degrees of freedom, it is possible to combine the Jacobian pseudoinverse solution with the Jacobian transpose solution as illustrated below. This is carried out in the framework of the so-called *augmented task space* approach to exploit redundancy in robotic systems. The idea is to introduce an additional constraint task by specifying a $(p \times 1)$ vector x_c as a function of the manipulator joint variables, i.e.,

$$x_c = k_c(q), \tag{3.46}$$

3.1. KINEMATIC CONTROL

with $p \leq n-m$ so as to constrain at most all the available redundant degrees of freedom. The constraint task vector x_c can be chosen by embedding scalar objective functions of the kind introduced in (3.12)–(3.14).

Differentiating (3.46) with respect to time gives

$$\dot{x}_c = J_c(q)\dot{q}. \qquad (3.47)$$

The result is an augmented differential kinematics equation given by (3.37) and (3.47), based on a Jacobian matrix

$$J' = \begin{pmatrix} J_a \\ J_c \end{pmatrix}. \qquad (3.48)$$

When a constraint task is specified independently of the end-effector task, there is no guarantee that the matrix J' remains full-rank along the entire task path. Even if $\text{rank}(J_a) = m$ and $\text{rank}(J_c) = p$, then $\text{rank}(J') = m + p$ if and only if $\mathcal{R}(J_a^T) \cap \mathcal{R}(J_c^T) = \{\emptyset\}$; singularities of J' are termed *artificial singularities* and it can be shown that those are given by singularities of the matrix $J_c(I - J_a^\dagger J_a)$.

The above discussion suggests that, when solving for joint velocities, a *task priority strategy* is advisable so as to avoid conflicting situations between the end-effector task and the constraint task. This can be achieved by computing the null space joint velocity in (3.40) as

$$\dot{q}_0 = \left(J_c(q)(I - J_a^\dagger(q)J_a(q)) \right)^\dagger \left(\dot{x}_c - J_c(q)J_a^\dagger(q)\dot{x} + K_c(x_{cd} - x_c) \right), \qquad (3.49)$$

where x_{cd} denotes the desired value of the constraint task and K_c is a positive definite matrix. The operator $(I - J_a^\dagger J_a)$ projects the secondary velocity contribution $\dot{q}_0$ on the null space $\mathcal{N}(J_a)$, guaranteeing correct execution of the primary end-effector task while the secondary constraint task is correctly executed as long as it does not interfere with the end-effector task. Obviously, if desired, the order of priority can be switched, e.g., in an obstacle avoidance task when an obstacle comes to be along the end-effector path.

On the other hand, the occurrence of artificial singularities suggests that pseudoinversion of the matrix $J_c(I - J_a^\dagger J_a)$ has to be avoided. Hence, by recalling the Jacobian transpose solution for the end-effector task (3.41), we can conveniently choose the null space joint velocity vector as

$$\dot{q}_0 = J_c^T(q)K_c(x_{cd} - x_c). \qquad (3.50)$$

The avoidance of the pseudoinversion of the matrix $J_c(I - J_a^\dagger J_a)$ allows the algorithm to work even at an artificial singularity. In detail, if $\text{rank}(J_c) = p$

but $\mathcal{R}(J_a^T) \cap \mathcal{R}(J_c^T) \neq \{\emptyset\}$, when $J_c^T K_c(x_{cd} - x_c) \in \mathcal{R}(J_a^T)$, $\dot{q}_0$ vanishes with $x_c \neq x_{cd}$; the higher-priority task is still tracked ($x = x_d$) and errors occur for the lower-priority task. However, it may be observed that the desired constraint task is often constant over time ($\dot{x}_{cd} = 0$) and the actual errors will be smaller.

Position and orientation errors

The above inverse kinematics algorithms make use of the analytical Jacobian since they operate on error variables (position and orientation) which are defined in the task space.

For what concerns the position error, it is obvious that its expression is given by

$$e_p = p_{ed} - p_e, \qquad (3.51)$$

where p_{ed} and p_e respectively denote the desired and actual end-effector positions. Further, its derivative is

$$\dot{e}_p = \dot{p}_{ed} - \dot{p}_e. \qquad (3.52)$$

On the other hand, for what concerns the orientation error, some considerations are in order. If $R_{ed} = (\, n_{ed} \quad s_{ed} \quad a_{ed} \,)$ denotes the desired rotation matrix of the end-effector frame, the orientation error with actual end-effector frame R_e is given by

$$e_o = u_r \sin \vartheta; \qquad (3.53)$$

the (3×1) unit vector u_r and the angle ϑ describe the equivalent rotation

$$R_r(\vartheta) = R_{ed} R_e^T \qquad (3.54)$$

needed to align R_e with R_{ed}. It can be easily shown that the expression of this rotation matrix is

$$R_r(\vartheta) = \begin{pmatrix} u_{rx}^2(1-c_\vartheta) + c_\vartheta & u_{rx}u_{ry}(1-c_\vartheta) - u_{rz}s_\vartheta \\ u_{rx}u_{ry}(1-c_\vartheta) + u_{rz}s_\vartheta & u_{ry}^2(1-c_\vartheta) + c_\vartheta \\ u_{rx}u_{rz}(1-c_\vartheta) - u_{ry}s_\vartheta & u_{ry}u_{rz}(1-c_\vartheta) + u_{rx}s_\vartheta \end{pmatrix}$$

$$\begin{matrix} u_{rx}u_{rz}(1-c_\vartheta) + u_{ry}s_\vartheta \\ u_{ry}u_{rz}(1-c_\vartheta) - u_{rx}s_\vartheta \\ u_{rz}^2(1-c_\vartheta) + c_\vartheta \end{matrix} \Big), \qquad (3.55)$$

where $u_r = (\, u_{rx} \quad u_{ry} \quad u_{rz} \,)^T$, and standard abbreviations have been used for sine and cosine. Notice that (3.53) gives a unique solution for $-\pi/2 < \vartheta < \pi/2$, but this interval is not limiting for a convergent inverse

3.1. KINEMATIC CONTROL

kinematics algorithm. It can be shown that a computational expression of the orientation error is given by

$$e_o = \frac{1}{2}(n_e \times n_{ed} + s_e \times s_{ed} + a_e \times a_{ed}); \qquad (3.56)$$

the limitation on ϑ sets the conditions $n_e^T n_{ed} \geq 0$, $s_e^T s_{ed} \geq 0$, $a_e^T a_{ed} \geq 0$.

Differentiation of (3.56) gives

$$\dot{e}_o = L^T \omega_d - L\omega \qquad (3.57)$$

where

$$L = -\frac{1}{2}\left(S(n_{ed})S(n_e) + S(s_{ed})S(s_e) + S(a_{ed})S(a_e)\right) \qquad (3.58)$$

and S is the skew-symmetric matrix performing the vector product between vectors a and b: $S(a)b = a \times b$.

Finally, the end-effector task error dynamics is given by combining (3.52) and (3.57), i.e.,

$$\dot{x}_d - \dot{x} = \begin{pmatrix} \dot{p}_{ed} \\ L^T \omega_d \end{pmatrix} - \begin{pmatrix} I & 0 \\ 0 & L \end{pmatrix} J\dot{q}. \qquad (3.59)$$

This equation evidences the possibility of using the geometric Jacobian in lieu of the analytical Jacobian in all the above inverse kinematics algorithms.

3.1.3 Extension to acceleration resolution

All the above schemes solve inverse kinematics at the velocity level. In order to achieve dynamic control of a manipulator in the task space, it is necessary to compute not only the joint velocity solutions but also the accleration solutions to the given task space motion trajectory. The second-order kinematics can be obtained by further differentiating equation (3.37), i.e.,

$$\ddot{x} = J_a(q)\ddot{q} + \dot{J}_a(q)\dot{q}. \qquad (3.60)$$

At this point, it would be quite natural to solve (3.60) for the joint accelerations by regarding the second term on the right-hand side as associated with $\ddot{x}$. Thus, the Jacobian pseudoinverse solution corresponding to (3.4) is —referring to the analytical Jacobian—

$$\ddot{q} = J_a^\dagger(q)(\ddot{x} - \dot{J}_a(q)\dot{q}) \qquad (3.61)$$

that can be integrated over time, with known initial conditions, to find $\dot{q}(t)$ and $q(t)$.

In terms of the inverse kinematics algorithms presented above, the acceleration solution logically corresponding to (3.38) is

$$\ddot{q} = J_a^\dagger(q)(\ddot{x} - \dot{J}_a(q)\dot{q} + K_D(\dot{x}_d - \dot{x}) + K_P(x_d - x)) + (I - J_a^\dagger(q)J_a(q))\ddot{q}_0, \tag{3.62}$$

where K_P and K_D are positive definite (diagonal) matrices that shape the convergence of the task space error dynamics

$$(\ddot{x}_d - \ddot{x}) + K_D(\dot{x}_d - \dot{x}) + K_P(x_d - x) = 0. \tag{3.63}$$

Note that the $(n \times 1)$ vector $\ddot{q}_0$ in (3.62) indicates arbitrary joint accelerations that can be utilized for redundancy resolution. Thus, $\ddot{q}_0$ can be chosen either as —see (3.11)—

$$\ddot{q}_0 = \alpha \left(\frac{\partial w(q)}{\partial q}\right)^T, \tag{3.64}$$

or as —see (3.50)—

$$\ddot{q}_0 = J_c^T \left(K_{Dc}(\dot{x}_{cd} - \dot{x}_c) + K_{Pc}(x_{cd} - x_c)\right) \tag{3.65}$$

with obvious meaning of K_{Pc} and K_{Dc}.

One shortcoming of the above procedure is that solving redundancy at the acceleration level may generate internal instability of joint velocities. The occurrence of this phenomenon, which in fact sets kinetic limitations on the use of redundancy, is basically related to the instability of the *zero dynamics* of the second-order system described by (3.60) under solutions of the kind (3.62).

A first remedy is to compute acceleration solutions by symbolic differentiation of velocity solutions so as to inherit all the properties of a first-order solution and then avoid the above inconvenience of joint velocity instability. This technique, however, may be too computationally demanding since it requires the calculation of the symbolic expression of the derivative of the Jacobian pseudoinverse.

A simple convenient solution is to modify the null space joint accelerations by the addition of a damping term on the joint velocities, that is,

$$\ddot{q} = J_a^\dagger(q)(\ddot{x} - \dot{J}_a(q)\dot{q} + K_D(\dot{x}_d - \dot{x}) + K_P(x_d - x)) \tag{3.66}$$
$$+ (I - J_a^\dagger(q)J_a(q))(\ddot{q}_0 - K_v\dot{q}),$$

3.2. DIRECT TASK SPACE CONTROL

being K_v an $(n \times n)$ positive definite matrix. When $\ddot{q}_0 = 0$, the added contribution guarantees exponential stability of joint velocities in the null space and then overcomes the above drawback of instability. On the other hand, when $\ddot{q}_0 \neq 0$, suitable values of K_v have to be chosen in order not to interfere with the additional constraints specified by $\ddot{q}_0$.

3.2 Direct task space control

The above inverse kinematics algorithms can be used to solve the task space control problem with a kinematic control strategy, that is, transformation of task space variables into joint space variables which constitute the reference inputs to some joint space control scheme.

A different strategy can be pursued by designing a *direct task space control* scheme. The task space variables are usually reconstructed from the joint space variables, via the kinematic mappings. In fact, it is quite rare to have sensors to directly measure end-effector positions and velocities.

From the discussion of the above sections it should be clear that the manipulator Jacobian plays a central role in the relationship between joint space and task space variables. And in fact, the two task space control schemes illustrated in the following are based respectively on the Jacobian transpose and the Jacobian pseudoinverse; the schemes are analogous to the PD control with gravity compensation and the inverse dynamics controllers developed in the joint space. It is worth remarking that the analytical Jacobian is utilized since the control schemes operate directly on task space quantities, e.g., end-effector position and orientation.

3.2.1 Regulation

Consider the dynamic model of an n-degree-of-freedom robot manipulator in the joint space

$$H(q)\ddot{q} + C(q,\dot{q})\dot{q} + g(q) = u, \qquad (3.67)$$

with obvious meaning of the symbols. Let then x_d indicate an $(m \times 1)$ desired *constant* end-effector location vector. It is wished to find a control vector u such that the error $x_d - x$ tends asymptotically to zero.

Theorem 3.1 *Consider the PD control law with gravity compensation*

$$u = J_a^T(q)K_P(x_d - x) - J_a^T(q)K_D J_a(q)\dot{q} + g(q), \qquad (3.68)$$

where K_P and K_D are $(m \times m)$ symmetric positive definite matrices. If J_a is full-rank for all joint configurations q, then

$$x = x_d \qquad \dot{x}_d = 0$$

is a globally asymptotically stable equilibrium point for the closed-loop system (3.67) and (3.68).

◊ ◊ ◊

Proof. Following the standard Lyapunov approach, choose the positive definite function candidate

$$V = \frac{1}{2}\dot{q}^T H(q)\dot{q} + \frac{1}{2}(x_d - x)^T K_P(x_d - x), \quad (3.69)$$

with K_P a positive definite matrix. Differentiating (3.69) with respect to time gives

$$\dot{V} = \dot{q}^T H(q)\ddot{q} + \frac{1}{2}\dot{q}^T \dot{H}(q,\dot{q})\dot{q} + (\dot{x}_d - \dot{x})^T K_P(x_d - x). \quad (3.70)$$

Observing that $\dot{x}_d = 0$ and recalling the skew-symmetry of the matrix $N = \dot{H} - 2C$, eq. (3.70) along the trajectories of the system (3.67) becomes

$$\dot{V} = \dot{q}^T \left(u - g(q) - J_a^T(q) K_P(x_d - x) \right). \quad (3.71)$$

At this point, the choice of the control law as in (3.68) with K_D positive definite gives

$$\dot{V} = -\dot{q} J_a^T(q) K_D J_a(q)\dot{q} \leq 0. \quad (3.72)$$

By analogy with the joint space stability analysis, the equilibrium points of the system are described by

$$J_a^T(q) K_P(x_d - x) = 0. \quad (3.73)$$

Therefore, on the assumption of full rank of J_a for any joint configuration q, end-effector regulation is achieved, i.e., $x = x_d$.

◊

Remarks

- Regarding the damping term in (3.68), if $\dot{x}$ is reconstructed from $\dot{q}$ measurements, the quantity $-K_D \dot{x}$ can be simplified into $-K_D \dot{q}$.

- Note that, even if task space variables x and $\dot{x}$ are directly measured, the joint variables q are needed to evaluate $J_a(q)$ and $g(q)$.

3.2.2 Tracking control

If tracking of a desired time-varying trajectory $x_d(t)$ is desired, an inverse dynamics control can be designed as in the joint space, i.e.,

$$u = H(q)u_0 + C(q,\dot{q})\dot{q} + g(q) \qquad (3.74)$$

which transforms the system (3.67) into the linear system

$$\ddot{q} = u_0. \qquad (3.75)$$

The second-order differential kinematics equation (3.60) comes into support to design the new control input u_0. In fact, according to the pseudoinverse solution (3.62), the choice

$$u_0 = J_a^\dagger(q)\left(\ddot{x}_d - \dot{J}_a(q)\dot{q} + K_D(\dot{x}_d - \dot{x}) + K_P(x_d - x)\right), \qquad (3.76)$$

with K_P and K_D positive definite matrices, leads to the same error dynamics as in equation (3.63). Note that this time, also $\dot{q}$ is needed by the controller, besides q, x and $\dot{x}$.

The control law (3.74) and (3.76) is the task space counterpart of the joint space inverse dynamics control scheme derived in the previous chapter. Other model-based control schemes can be devised with do not seek linearization or decoupling, but only asymptotic convergence of the tracking error. With reference to the Lyapunov-based control scheme already developed in the joint space, it is possible to formalize a task space version by introducing a reference velocity

$$\dot{\zeta}_x = \dot{x}_d + \Lambda(x_d - x) \qquad (3.77)$$

and modify the control law (3.74) as

$$u = H(q)\ddot{\zeta} + C(q,\dot{q})\dot{\zeta} + g(q) + J_a^T(q)K_D J_a(q)(\dot{\zeta} - \dot{q}), \qquad (3.78)$$

where

$$\dot{\zeta} = J_a^\dagger(q)\dot{\zeta}_x \qquad (3.79)$$

$$\ddot{\zeta} = J_a^\dagger(q)(\ddot{\zeta}_x - \dot{J}_a(q)\dot{\zeta}). \qquad (3.80)$$

In a similar fashion, it is possible to derive the task space counterpart of the joint space passivity-based control scheme.

The above solutions require the matrix J_a not to be singular. Also, if the manipulator is redundant, an additional null space acceleration term can be added to (3.76) that produces a null space torque contribution via (3.74);

likewise, redundancy resolution can be embedded in the control scheme based on (3.78). It is argued, however, that handling singularities and redundancy with a direct task space control is not as straightforward as with the above two-stage control approach. In fact, when the manipulator Jacobian is embedded in the dynamic control loop, its distinctive features (singularities and redundancy) affect the performance of the overall control scheme and may also overload the computational burden of the controller. On the other hand, when the manipulator Jacobian is handled within the inverse kinematics level, singularities and redundancy have already been solved when the joint quantities are presented as reference inputs to the joint space controller.

Nonetheless, a direct task space control solution becomes advisable when studying contact between the manipulator and environment, e.g., in motion and force control problems, as shown in the following chapter. In particular, the above direct task space control schemes will be reformulated on the basis of a task space dynamic model to better highlight the effects of interaction between the end effector and the environment.

3.3 Further reading

Kinematic control as a task space control scheme is presented in [37], and a survey on kinematic control of redundant manipulators can be found in [38].

The inversion of differential kinematics dates back to [43] under the name of resolved motion rate control. The adoption of the pseudoinverse of the Jacobian is due to [20]. More on the properties of the pseudoinverse can be found in [2]. The use of null-space joint velocities for redundancy resolution was proposed in [21], and further refined in [44, 25] as concerns the choice of objective functions. The reader is referred to [27] for a complete treatment of redundant manipulators.

The adoption of the damped least-squares inverse was independently presented in [28] and [42]. More about kinematic control in the neighbourhood of kinematic singularities can be found in [6]. The technique for estimating the smallest singular value of the Jacobian is due to [26], and its modification to include the second smallest singular value was achieved by [7]. The use of the damped least-squares inverse for redundant manipulators was presented in [17]. The user-defined accuracy strategy was proposed in [8] and further refined in [9]. A review of the damped least-squares inverse kinematics with experiments on an industrial robot manipulator was recently presented in [12].

Closed-loop inverse kinematics algorithms are discussed in [35]. The original Jacobian transpose inverse kinematics algorithm was proposed in

[32]; the choice of suitable gains for achieving robustness to singularities was discussed in [4]. Singular value decomposition of the Jacobian transpose is due to [10]. Combination of the Jacobian transpose solution with the pseudoinverse solution was proposed in [5]. References on the augmented task space approach are [16, 33, 36, 31]. The occurrence of artificial singularities was pointed out in [1], and their properties were studied in [3]. The task priority strategy is due to [29]. The use of the Jacobian transpose for the constraint task was presented in [11, 39]. The expression of the end-effector orientation error (3.56) is due to [23], and its properties were studied in [45, 22].

The extension to acceleration resolution was proposed in [37]. The issue of internal velocity instability was raised by [18] and further investigated in [24, 14]; the instability of zero dynamics of robotic systems is analyzed in [13]. Symbolic differentiation of joint velocity solutions to obtain stable acceleration solutions was proposed in [40], while the addition of a velocity damping term in the null space of the Jacobian is due to [30]. More about redundancy resolution at the acceleration level can be found in [15].

The PD control with gravity compensation was originally proposed by [41]. Direct task space control via second-order differential kinematics equation was developed in [34], while the reader is referred to [19] for the so-called operational space control which is based on a task space dynamic model of the manipulator.

References

[1] J. Baillieul, "Kinematic programming alternatives for redundant manipulators," *Proc. 1985 IEEE Int. Conf. on Robotics and Automation*, St. Louis, MO, pp. 722–728, 1985.

[2] T.L. Boullion and P.L. Odell, *Generalized Inverse Matrices*, Wiley, New York, NY, 1971.

[3] P. Chiacchio, S. Chiaverini, L. Sciavicco, and B. Siciliano, "Closed-loop inverse kinematics schemes for constrained redundant manipulators with task space augmentation and task priority strategy," *Int. J. of Robotics Research*, vol. 10, pp. 410–425, 1991.

[4] P. Chiacchio and B. Siciliano, "Achieving singularity robustness: An inverse kinematic solution algorithm for robot control," in *Robot Control: Theory and Applications*, IEE Control Engineering Series 36, K. Warwick and A. Pugh (Eds.), Peter Peregrinus, Herts, UK, pp. 149–156, 1988.

[5] P. Chiacchio and B. Siciliano, "A closed-loop Jacobian transpose scheme for solving the inverse kinematics of nonredundant and redundant robot wrists," *J. of Robotic Systems*, vol. 6, pp. 601–630, 1989.

[6] S. Chiaverini, "Inverse differential kinematics of robotic manipulators at singular and near-singular configurations," *1992 IEEE Int. Conf. on Robotics and Automation — Tutorial on 'Redundancy: Performance Indices, Singularities Avoidance, and Algorithmic Implementations'*, Nice, F, May 1992.

[7] S. Chiaverini, "Estimate of the two smallest singular values of the Jacobian matrix: Application to damped least-squares inverse kinematics," *J. of Robotic Systems*, vol. 10, pp. 991–1008, 1993.

[8] S. Chiaverini, O. Egeland, and R.K. Kanestrøm, "Achieving user-defined accuracy with damped least-squares inverse kinematics," *Proc. 5th Int. Conf. on Advanced Robotics*, Pisa, I, pp. 672–677, 1991.

[9] S. Chiaverini, O. Egeland, J.R. Sagli, and B. Siciliano, "User-defined accuracy in the augmented task space approach for redundant manipulators," *Laboratory Robotics and Automation*, vol. 4, pp. 59–67, 1992.

[10] S. Chiaverini, L. Sciavicco, and B. Siciliano, "Control of robotic systems through singularities," in *Advanced Robot Control*, Lecture Notes in Control and Information Science, vol. 162, C. Canudas de Wit (Ed.), Springer-Verlag, Berlin, D, pp. 285–295, 1991.

[11] S. Chiaverini, B. Siciliano, and O. Egeland, "Redundancy resolution for the human-arm-like manipulator," *Robotics and Autonomous Systems*, vol. 8, pp. 239–250, 1991.

[12] S. Chiaverini, B. Siciliano, and O. Egeland, "Review of the damped least-squares inverse kinematics with experiments on an industrial robot manipulator," *IEEE Trans. on Control Systems Technology*, vol. 2, pp. 123–134, 1994.

[13] A. De Luca, "Zero dynamics in robotic systems," in *Nonlinear Synthesis*, Progress in Systems and Control Series, C.I. Byrnes and A. Kurzhanski (Eds.), Birkhäuser, Boston, MA, pp. 68–87, 1991.

[14] A. De Luca and G. Oriolo, "Issues in acceleration resolution of robot redundancy," *Prepr. 3rd IFAC Symp. on Robot Control*, Wien, A, pp. 665–670, 1991.

[15] A. De Luca, G. Oriolo, and B. Siciliano, "Robot redundancy resolution at the acceleration level," *Laboratory Robotics and Automation*, vol. 4, pp. 97–106, 1992.

[16] O. Egeland, "Task-space tracking with redundant manipulators," *IEEE J. of Robotics and Automation*, vol. 3, pp. 471–475, 1987.

[17] O. Egeland, J.R. Sagli, I. Spangelo, and S. Chiaverini, "A damped least-squares solution to redundancy resolution," *Proc. 1991 IEEE Int. Conf. on Robotics and Automation*, Sacramento, CA, pp. 945–950, 1991.

[18] P. Hsu, J. Hauser, and S. Sastry, "Dynamic control of redundant manipulators," *J. of Robotic Systems*, vol. 6, pp. 133–148, 1989.

[19] O. Khatib, "A unified approach for motion and force control of robot manipulators: The operational space formulation," *IEEE J. of Robotics and Automation*, vol. 3, pp. 43–53, 1987.

[20] C.A. Klein and C.H. Huang, "Review of pseudoinverse control for use with kinematically redundant manipulators," *IEEE Trans. on Systems, Man, and Cybernetics*, vol. 13, pp. 245–250, 1983.

[21] A. Liégeois, "Automatic supervisory control of the configuration and behavior of multibody mechanisms," *IEEE Trans. on Systems, Man, and Cybernetics*, vol. 7, pp. 868–871, 1977.

[22] S.K. Lin, "Singularity of a nonlinear feedback control scheme for robots," *IEEE Trans. on Systems, Man, and Cybernetics*, vol. 19, pp. 134–139, 1989.

[23] J.Y.S. Luh, M.W. Walker, and R.P.C. Paul, "Resolved-acceleration control of mechanical manipulators," *IEEE Trans. on Automatic Control*, vol. 25, pp. 468–474, 1980.

[24] A.A. Maciejewski, "Kinetic limitations on the use of redundancy in robotic manipulators," *IEEE Trans. on Robotics and Automation*, vol. 7, pp. 205–210, 1991.

[25] A.A. Maciejewski and C.A. Klein, "Obstacle avoidance for kinematically redundant manipulators in dynamically varying environments," *Int. J. of Robotics Research*, vol. 4, no. 3, pp. 109–117, 1985.

[26] A.A. Maciejewski and C.A. Klein, "Numerical filtering for the operation of robotic manipulators through kinematically singular configurations," *J. of Robotic Systems*, vol. 5, pp. 527–552, 1988.

[27] Y. Nakamura, *Advanced Robotics: Redundancy and Optimization*, Addison-Wesley, Reading, MA, 1991.

[28] Y. Nakamura and H. Hanafusa, "Inverse kinematic solutions with singularity robustness for robot manipulator control," *ASME J. of Dynamic Systems, Measurement, and Control*, vol. 108, pp. 163–171, 1986.

[29] Y. Nakamura, H. Hanafusa, and T. Yoshikawa, "Task-priority based redundancy control of robot manipulators," *Int. J. of Robotics Research*, vol. 6, no. 2, pp. 3–15, 1987.

[30] Z. Novaković and B. Siciliano, "A new second-order inverse kinematics solution for redundant manipulators," in *Advances in Robot Kinematics*, S. Stifter and J. Lenarčič (Eds.), Springer-Verlag, Wien, A, pp. 408–415, 1991.

[31] C. Samson, M. Le Borgne, and B. Espiau, *Robot Control: The Task Function Approach*, Oxford Engineering Science Series, vol. 22, Clarendon Press, Oxford, UK, 1991.

[32] L. Sciavicco and B. Siciliano, "Coordinate transformation: A solution algorithm for one class of robots," *IEEE Trans. on Systems, Man, and Cybernetics*, vol. 16, pp. 550–559, 1986.

[33] L. Sciavicco and B. Siciliano, "A solution algorithm to the inverse kinematic problem for redundant manipulators," *IEEE J. of Robotics and Automation*, vol. 4, pp. 403–410, 1988.

[34] L. Sciavicco and B. Siciliano, "The augmented task space approach for redundant manipulator control," *Proc. 2nd IFAC Symp. on Robot Control*, Karlsruhe, D, pp. 125–129, 1988.

[35] L. Sciavicco and B. Siciliano, *Modeling and Control of Robot Manipulators*, McGraw-Hill, New York, NY, 1996.

[36] H. Seraji, "Configuration control of redundant manipulators: Theory and implementation," *IEEE Trans. on Robotics and Automation*, vol. 5, pp. 472–490, 1989.

[37] B. Siciliano, "A closed-loop inverse kinematic scheme for on-line joint based robot control," *Robotica*, vol. 8, pp. 231–243, 1990.

[38] B. Siciliano, "Kinematic control of redundant robot manipulators: A tutorial," *J. of Intelligent and Robotic Systems*, vol. 3, pp. 201–212, 1990.

REFERENCES

[39] B. Siciliano, "Solving manipulator redundancy with the augmented task space method using the constraint Jacobian transpose," *1992 IEEE Int. Conf. on Robotics and Automation — Tutorial on 'Redundancy: Performance Indices, Singularities Avoidance, and Algorithmic Implementations,'* Nice, F, May 1992.

[40] B. Siciliano and J.-J.E. Slotine, "A general framework for managing multiple tasks in highly redundant robotic systems," *Proc. 5th Int. Conf. on Advanced Robotics*, Pisa, I, pp. 1211–1216, 1991.

[41] M. Takegaki and S. Arimoto, "A new feedback method for dynamic control of manipulators," *ASME J. of Dynamic Systems, Measurement, and Control*, vol. 102, pp. 119–125, 1981.

[42] C.W. Wampler, "Manipulator inverse kinematic solutions based on vector formulations and damped least-squares methods," *IEEE Trans. on Systems, Man, and Cybernetics*, vol. 16, pp. 93–101, 1986.

[43] D.E. Whitney, "Resolved motion rate control of manipulators and human prostheses," *IEEE Trans. on Man-Machine Systems*, vol. 10, pp. 47–53, 1969.

[44] T. Yoshikawa, "Manipulability of robotic mechanisms," *Int. J. of Robotics Research*, vol. 4, no. 2, pp. 3–9, 1985.

[45] J.S.-C. Yuan, "Closed-loop manipulator control using quaternion feedback," *IEEE J. of Robotics and Automation*, vol. 4, pp. 434–440, 1988.

Chapter 4

Motion and force control

In this chapter we deal with the motion control problem for situations in which the robot manipulator end effector is in *contact with the environment*. Many robotic tasks involve intentional interaction between the manipulator and the environment. Usually, the end effector is required to follow in a stable way the edge or the surface of a workpiece while applying prescribed forces and torques. The specific feature of robotic problems such as polishing, deburring, or assembly, demands control also of the exchanged forces at the contact. These forces may be explicitly set under control or just kept limited in a indirect way, by controlling the end-effector position. In any case, *force* specification is often complemented with a requirement concerning the end-effector *motion*, so that the control problem has in general *hybrid* (mixed) objectives.

In setting up the proper framework for analysis, an essential role is played by the model of the environment. The predicted performance of the overall system will depend not only on the robot manipulator dynamics but also on the assumptions made for the interaction between the manipulator and environment. On one hand, the environment may behave as a simple mechanical system undergoing small but finite deformations in response to applied forces. When contact occurs, the arising forces will be dictated by the *dynamic balancing* of two coupled systems, the robot manipulator and the environment. On the other hand, if the environment is stiff enough and the manipulator is continuously in contact, part of its degrees of freedom will be actually lost, the motion being locally constrained to given directions. Contact forces are then viewed as an attempt to violate the imposed *kinematic constraints*.

Three control strategies are discussed; namely, *impedance control*, *parallel control*, and *hybrid force/motion control*. Impedance control tries to assign desired dynamic characteristics to the interaction with rather unmodelled objects in the workspace. Parallel control provides the additional feature of regulating the contact force to a desired value. Hybrid force/motion control exploits the partition of the task space into directions of admissible motion and of reaction forces, both arising from the existence of a rigid constraint.

All strategies can incorporate the best available model of the environment; however, the achieved performance will suffer anyway from uncertainty in the location and geometric characteristics (orientation, curvature) of the contact surfaces. Moreover, for impedance and hybrid force/motion control, the measure of contact forces may or may not be needed in nominal conditions. Thus, the main differences between these two strategies rely in the control objectives rather than in the implementation requirements of a specific control law. On the other hand, parallel control is aimed at closing a force control loop around a pre-existing position (impedance) control loop and, as such, it naturally makes use of contact force measurements.

We will investigate these three frameworks, each allowing the design of different types of controllers. A stability analysis of the most interesting ones will be provided.

4.1 Impedance control

The underlying idea of impedance control is to assign a prescribed dynamic behaviour for a robot manipulator while its end effector is interacting with the environment. The desired performance is specified by a *generalized dynamic impedance*, i.e., by a complete set of linear or nonlinear second-order differential equations representing a mass-spring-damper system. However, the way in which the environment generates forces in reaction to deformation is hardly modelled.

In general, impedance control is suitable for those tasks where contact forces must be kept small, typically to avoid jamming among parts in assembly or insertion operations, while accurate regulation of forces is not required. In fact, an explicit force error loop is absent in this approach, so that it is often stated that "force is controlled by controlling position". Using programmable stiffness and damping matrices in the impedance model, a compromise is reached between contact force and position accuracy as a result of unexpected interactions with the environment.

In the following, we will first introduce the convenient format of the manipulator dynamic model in task space coordinates. The imposition of a desired impedance at the end-effector level will then be obtained by an inverse dynamics scheme designed in the proper task space coordinates.

4.1. IMPEDANCE CONTROL

Simplified control laws are then recovered from this general scheme.

4.1.1 Task space dynamic model

When additional external forces act from the environment on the manipulator end effector, the dynamic model (in the absence of friction forces) becomes

$$H(q)\ddot{q} + C(q,\dot{q})\dot{q} + g(q) = u + J^T(q)f, \qquad (4.1)$$

with the usual notation for the dynamic terms on the left-hand side, and where

$$f = \begin{pmatrix} \gamma \\ \mu \end{pmatrix} \qquad (4.2)$$

is the $(m \times 1)$ vector of generalized forces expressed in the base frame; respectively, linear forces γ and angular moments μ. Notice that the forces $J^T f$ on the right-hand side of (4.1) have been taken with the positive sign, since they are exerted from the environment on the end effector.

In order to focus on the main aspects of the control problem, we will consider only the case $m = n$ and, without loss of generality, $n = 6$. Moreover, in the following developments it is assumed that *no kinematic singularities* are encountered. The square matrix $J(q)$ in (4.1) is the *geometric Jacobian* of the manipulator relating end-effector velocity to joint velocity

$$v = \begin{pmatrix} \dot{p} \\ \omega \end{pmatrix} = J(q)\dot{q}. \qquad (4.3)$$

Since the interaction between the manipulator and the environment occurs at the task level, it is useful to rewrite the dynamic model directly in the task space. With reference to the *analytical Jacobian*, the differential kinematics equation can be written in the form

$$\dot{x} = J_a(q)\dot{q} = T_a^{-1}(\phi_e)J(q)\dot{q}, \qquad (4.4)$$

where we have assumed that no representation singularities of ϕ_e occur.

The two sets of generalized forces f and f_a performing work on v and $\dot{x}$, respectively, are related by the virtual work principle, i.e.,

$$J^T(q)f = J^T(q)T_a^{-T}(\phi_e)f_a = J_a^T(q)f_a. \qquad (4.5)$$

Then, the model (4.1) can be rewritten as

$$H(q)\ddot{q} + C(q,\dot{q})\dot{q} + g(q) = u + J_a^T(q)f_a, \qquad (4.6)$$

with an alternate format for the external force term. By further differentiation of (4.4)

$$\ddot{x} = J_a(q)\ddot{q} + \dot{J}_a(q)\dot{q} \tag{4.7}$$

and substitution into (4.6), it is easy to see that the dynamic model in the task space becomes

$$H_x(q)\ddot{x} + C_x(q,\dot{q})\dot{x} + g_x(q) = J_a^{-T}(q)u + f_a, \tag{4.8}$$

where

$$\begin{aligned} H_x(q) &= J_a^{-T}(q)H(q)J_a^{-1}(q) \\ C_x(q,\dot{q}) &= J_a^{-T}(q)C(q,\dot{q})J_a^{-1}(q) - H_x(q)\dot{J}_a(q)J_a^{-1}(q) \\ g_x(q) &= J_a^{-T}(q)g(q). \end{aligned} \tag{4.9}$$

Indeed, similar relationships are obtained if we desire that the vector f appears on the right-hand side of (4.8) —just drop the subscripts a.

Note that the functional dependence of nonlinear terms in (4.8) is still on q, $\dot{q}$, and not on the new state variables x, $\dot{x}$. This substitution is not essential and could be easily performed by using inverse kinematics relationship. From the computational point of view, e.g., for simulation purposes, it is more advantageous to keep the explicit dependence on joint variables.

As for the joint space dynamic model, analogous structural properties hold for (4.8). The following ones are of particular interest hereafter.

Property 4.1 The inertia matrix H_x is a symmetric positive definite matrix, provided that J_a is full-rank, which verifies

$$\lambda_{mx}I \leq H_x(q) \leq \lambda_{Mx}I \tag{4.10}$$

where λ_{mx} ($\lambda_{Mx} < \infty$) denotes the strictly positive minimum (maximum) eigenvalue of H_x for all configurations q.

□

Property 4.2 The matrix $\dot{H}_x - 2C_x$ obtained via (4.9) is skew-symmetric, provided that the matrix $\dot{H} - 2C$ is skew-symmetric.

□

Property 4.3 The matrix C_x verifies

$$\|C_x\| \leq k_{Cx}\|\dot{x}\| \tag{4.11}$$

for some bounded constant $k_{Cx} > 0$.

□

4.1.2 Inverse dynamics control

To achieve a desired dynamic characteristic for the interaction between the manipulator and the environment at the contact, the design of the control input u in (4.8) (or (4.6)) is carried out in two steps.

The first step decouples and linearizes the closed-loop dynamics in task space coordinates so as to obtain

$$\ddot{x} = u_0, \qquad (4.12)$$

where u_0 is an external auxiliary input still available for control. Based on (4.8), this is achieved by choosing

$$u = J_a^T(q)\big(H_x(q)u_0 + C_x(q,\dot{q})\dot{x} + g_x(q) - f_a\big) \qquad (4.13)$$

or, in terms of the original model components,

$$u = H(q)J_a^{-1}(q)\big(u_0 - \dot{J}_a(q)\dot{q}\big) + C(q,\dot{q})\dot{q} + g(q) - J_a^T(q)f_a. \qquad (4.14)$$

This *inverse dynamics control* law is formally equivalent to the direct task space control law presented in the previous chapter, with the addition of *force feedback*.

In the second step, the desired impedance model that dynamically balances contact forces f_a at the manipulator end effector is chosen as a linear second-order mechanical system described by

$$H_m(\ddot{x} - \ddot{x}_d) + D_m(\dot{x} - \dot{x}_d) + K_m(x - x_d) = f_a, \qquad (4.15)$$

where H_m is the *apparent inertia matrix*, while D_m and K_m are the desired *damping and stiffness matrices*, respectively. The vector $x_d(t)$ specifies a reference trajectory which can be exactly executed only if $f_a = 0$, i.e., during free motion. When the manipulator is in contact, the automatic balance of dynamic forces will produce a different motion behaviour. If (4.15) has to represent a *physical impedance*, positive definite H_m and K_m matrices and a positive semi-definite D_m matrix should be chosen. All matrices are symmetric —typically diagonal. Notice also that the control objective of imposing the dynamic behaviour (4.15) to the original robotic system (4.8) can be recast in the general framework of *model matching* problems for nonlinear systems.

The desired mechanical impedance is then obtained by choosing u_0 in (4.14) as

$$u_0 = \ddot{x}_d + H_m^{-1}\big(D_m(\dot{x}_d - \dot{x}) + K_m(x_d - x) + f_a\big), \qquad (4.16)$$

so that the overall *impedance control* law becomes

$$u = J_a^T(q)\Big(H_x(q)\ddot{x}_d + C_x(q,\dot{q})\dot{x} + g_x(q)$$
$$+ H_x(q)H_m^{-1}\big(D_m(\dot{x}_d - \dot{x}) + K_m(x_d - x)\big)$$
$$+ \big(H_x(q)H_m^{-1} - I\big) f_a\Big). \qquad (4.17)$$

The implementation of (4.17) requires feedback from the manipulator state $(q, \dot{q})$ and measure of the contact force f_a. However, no explicit force loop is imposed in this control law.

Some further considerations should be made about the choice of an impedance model. Although any choice for H_m, D_m, K_m that complies with physical characteristics is feasible, some limitations arise in practice. In particular, if the location where contact occurs is not exactly known, the elements of the usually diagonal matrices H_m and K_m should be chosen so as to avoid excessive impact forces.

As a suggestive representation, let us imagine we are searching for the electric switch panel on the wall of a dark room. During this guarded motion task, the human arm is "stiff" and moves fast only where no contact is expected (e.g., sideways) while it is rather "compliant" and slow in those directions where there may be an impact, based on the approximate knowledge of the distance from the wall. Moreover, the harder is the environment surface according to our prediction, the smoother and more careful will be the arm motion in the direction of approach. As a result, in this task we are neither controlling the contact force nor accurately positioning our dexterous end effector; we rather assign a selective dynamic behaviour to the arm in preview of the specific environment.

A robot manipulator under impedance control will mimic the human performance in the presence of environment uncertainty by choosing large inertial components H_{mi} and/or low stiffness coefficients K_{mi} along the anticipated directions of contact; vice-versa, large stiffness coefficients K_{mj} are imposed along those directions which are assumed to be free, thus closely realizing the associated motion command $x_{dj}(t)$. Instead, the choice of damping D_m is related only to the desired transient characteristics.

A relevant simplification occurs in (4.17) when a *nonlinear impedance* is prescribed. Replacing formally

$$H_m = H_x(q) \qquad (4.18)$$

in (4.15) leads to

$$u = J_a^T(q)\left(H_x(q)\ddot{x}_d + C_x(q,\dot{q})\dot{x} + g_x(q) + D_m(\dot{x}_d - \dot{x}) + K_m(x_d - x)\right)$$
$$(4.19)$$

4.1. IMPEDANCE CONTROL

or, again in the original coordinates,

$$u = H(q)J_a^{-1}(q)(\ddot{x}_d - \dot{J}_a(q)\dot{q}) + C(q,\dot{q})\dot{q} + g(q)$$
$$+ J_a^T(q)(D_m(\dot{x}_d - \dot{x}) + K_m(x_d - x)). \quad (4.20)$$

Remarkably, *no force feedback* is required in this case. In this respect, (4.20) can be seen as a pure motion controller, yet intended to keep limited the end-effector forces arising from contacts with the environment. The price to pay is that the effective inertia of the manipulator will be the natural one in the *task space*.

4.1.3 PD control

Provided that *quasi-static* assumptions are made ($\dot{x}_d = 0$ and $\dot{q} \approx 0$ in the nonlinear dynamic terms), the following controller is obtained from (4.20):

$$u = J_a^T(q)(K_m(x_d - x) - D_m\dot{x}) + g(q). \quad (4.21)$$

It can be recognized that this is nothing but the *task space PD control* law, with added gravity compensation, that was considered in the previous chapter. In (4.21), the equivalent task space elastic and viscous forces respectively due to the position error $x_d - x$ and to the motion $\dot{x}$, are transformed into joint space torques through the usual Jacobian transpose.

When a *constant* reference value x_d is considered in the absence of contact forces ($f_a = 0$), it has been shown in the previous chapter that this control law enforces asymptotic stability of x_d, provided that no kinematic singularities are encountered. In alternative, and with the aim of including in the analysis also contact forces, the following Lyapunov candidate could have been used

$$V_1 = \frac{1}{2}\dot{x}^T H_x(q)\dot{x} + \frac{1}{2}(x_d - x)^T K_m(x_d - x), \quad (4.22)$$

fully defined in terms of task space components. The time derivative of (4.22) along the trajectories of the closed-loop system (4.8) and (4.21) is

$$\dot{V}_1 = \dot{x}^T H_x(q)\ddot{x} + \frac{1}{2}\dot{x}^T \dot{H}_x(q)\dot{x} - \dot{x}^T K_m(x_d - x)$$
$$= \dot{x}^T \left(-C_x(q,\dot{q})\dot{x} - g_x(q) + J_a^{-T}(q)g(q) + K_m(x_d - x) - D_m\dot{x}\right)$$
$$+ \frac{1}{2}\dot{x}^T \dot{H}_x(q)\dot{x} - \dot{x}^T K_m(x_d - x)$$
$$= -\dot{x}^T D_m\dot{x} \leq 0, \quad (4.23)$$

where Property 4.2 and $g_x = J_a^{-T} g$ have been used. Asymptotic stability follows from La Salle's invariant set theorem, provided that the matrix $J_a^T(q)$ remains always of full rank.

When a contact force $f_a \neq 0$ arises during motion, then asymptotic stability of x_d is no longer ensured. In fact, a steady-state compromise between environment deformation and positional compliance of the controller, governed by the (inverse of) matrix K_m, will be reached. The stability analysis can be carried out assuming that the environment deformation is modelled as a generalized spring with symmetric *stiffness matrix* K_e and rest location x_e.

By a suitable selection of reference frames, a "simple" environment will generate reaction forces and torques if $x > x_e$, componentwise. As a result, when the manipulator is in full contact with the environment, a generalized force of the form

$$f_a = K_e(x_e - x) \qquad (4.24)$$

is felt at the end-effector level. The (unique) equilibrium point x_E of the manipulator under the environment reaction force (4.24) and the control law (4.21) is found through the steady-state balance condition

$$K_m(x_d - x) + K_e(x_e - x) = 0, \qquad (4.25)$$

leading to

$$x_E = (K_m + K_e)^{-1}(K_m x_d + K_e x_e) \qquad (4.26)$$

and to the equilibrium force

$$f_E = K_e(x_e - x_E). \qquad (4.27)$$

Lemma 4.1 *The equilibrium point (4.26) of the closed-loop system (4.8) and (4.21) is globally asymptotically stable.*

⋄ ⋄ ⋄

Proof. Consider the modified Lyapunov candidate of the form

$$V_2 = V_1 + \frac{1}{2}(x_e - x)^T K_e(x_e - x) - V_2(x_E, 0) \qquad (4.28)$$

where

$$V_2(x_E, 0) = \frac{1}{2}(x_d - x_E)^T K_m(x_d - x_E) + \frac{1}{2}(x_e - x_E)^T K_e(x_e - x_E) \geq 0. \qquad (4.29)$$

Proceeding as with (4.22) and taking into account (4.24) gives

$$\dot{V}_2 = -\dot{x}^T D_m \dot{x} \leq 0. \qquad (4.30)$$

4.1. IMPEDANCE CONTROL

From the structure of $V_2 \geq 0$, this implies that $\dot{x} \to 0$ as $t \to \infty$, and that both terms $(x_e - x)$ and $(x_d - x)$ are bounded. As a consequence, the location x_E is the asymptotically stable equilibrium point of the closed-loop system.

$\diamond$

Remarks

- It follows from (4.26) that when the environment is very stiff along a component i, $k_{ei} \gg k_{mi}$ and the equilibrium is $x_{Ei} \approx x_{ei}$. Vice-versa, if $k_{mj} \gg k_{ej}$ then $x_{Ej} \approx x_{dj}$, with a larger environment deformation.

- The control law (4.21) features a damping term in the task space. The presence of such a damping is crucial as revealed by both (4.23) and (4.30). Therefore, in case of redundant manipulators and/or occurrence of kinematic singularities, it may be advisable to operate the damping action in the joint space, i.e., by choosing the controller as

$$u = J_a^T(q) K_m(x_d - x) - D_{mq}\dot{q} + g(q). \tag{4.31}$$

with obvious meaning of D_{mq}.

Stiffness control

A simplified control can be achieved if we assume that small deformations occur

$$(x_d - x) \approx J_a(q)(q_d - q). \tag{4.32}$$

If gravity terms are mechanically balanced, that is to say $g(q) = 0$, and no additional damping is included ($D_m = 0$), then (4.21) reduces to the so-called *stiffness control*:

$$u = J_a^T(q) K_m J_a(q)(q_d - q) = K_x(q)(q_d - q). \tag{4.33}$$

Here, a configuration dependent joint torque is generated in response to small joint position errors so as to correspond to a constant task space stiffness matrix K_m. This simple control scheme is the *active* counterpart of a passive compliant end-effector device (such as the Remote Center of Compliance) that possesses directional-dependent but only fixed accommodation characteristics.

4.2 Parallel control

The above impedance control schemes perform only indirect force control, through a closed-loop position controller, without explicit closure of a force feedback loop. In other words, it is not possible to specify a desired amount of contact force with an impedance controller, but only a satisfactory dynamic behaviour between end-effector force and displacement at the contact.

An effective strategy to embed the possibility of *force regulation* is given by the so-called *parallel control*. The key concept is to provide an impedance controller with the ability of controlling both position and force. This is obtained by closing an outer force control loop around the inner position control loop, which is typically available in a robot manipulator. By a suitable design of the force control action (typically an integral term), it is possible to regulate the contact force to a desired value. In order to provide motion control along the feasible task space directions, also a desired position is input to the inner loop. The result is two control actions, working in parallel; namely, a force control action and a position control action. In order to ensure force control along the task space directions where environment reaction forces exist, the force action is designed so as to dominate the position action.

Following the previous analysis of impedance control, the environment can be modelled as in (4.24). We consider here the case of $m = n = 3$ and contact with a *planar* surface. The choice of a planar surface is motivated by noticing that it is locally a good approximation to surfaces of regular curvature. Hence, K_e is a (3×3) constant symmetric stiffness matrix of rank 1, i.e., the contact force f_a is directed along the normal to the plane. Let

$$R_e = (\,n_e \quad s_e \quad a_e\,) \tag{4.34}$$

be the rotation matrix describing the orientation of a contact frame attached to the plane with respect to some base frame, where a_e is the unit vector along the normal and n_e, s_e lie in the plane. Then, the stiffness matrix can be written as

$$K_e = R_e \,\mathrm{diag}\{0, 0, k_e\}\, R_e^T = k_e \, a_e a_e^T \tag{4.35}$$

where $k_e > 0$ is the stiffness coefficient.

The elastic contact model (4.24) and (4.35) suggests that a null force error can be obtained only if the desired contact force f_{ad} is aligned with a_e. On the other hand, null position errors can be obtained only on the contact plane, while the component of position x along a_e has to accommodate the force requirement specified by f_{ad}; thus, the desired position x_d can

4.2. PARALLEL CONTROL

be freely reached only in the null space of K_e, i.e., where the environment provides no reaction forces.

4.2.1 Inverse dynamics control

With reference to the task space dynamic model (4.8), the design of the control input u can be carried out as for the above inverse dynamics controller in (4.14), where force measurements of f_a are assumed to be available. According to the parallel control approach, the new control input u_0 can be designed as the sum of a position control action and a force control action; namely, as

$$u_0 = u_{0x} + u_{0f} \tag{4.36}$$

with

$$u_{0x} = \ddot{x}_d + H_m^{-1}\left(D_m(\dot{x}_d - \dot{x}) + K_m(x_d - x)\right) \tag{4.37}$$

$$u_{0f} = -H_m^{-1}\left(K_F(f_{ad} - f_a) + K_I \int_0^t (f_{ad} - f_a)d\tau\right) \tag{4.38}$$

where H_m, D_m, K_m assume the same meaning as for the above impedance controller, and K_F, K_I are suitable force feedback matrix gains characterizing a PI action on the force error. It is worth noticing that by setting $K_F = 0$ and $K_I = 0$ the original impedance control law (4.16) is recovered.

Substituting (4.36) with (4.37) and (4.38) in (4.12) yields

$$H_m(\ddot{x} - \ddot{x}_d) + D_m(\dot{x} - \dot{x}_d) + K_m(x - x_d) = K_F(f_a - f_{ad}) + K_I \int_0^t (f_a - f_{ad})d\tau \tag{4.39}$$

which reveals that, thanks to the *integral* action, the force error $(f_{ad} - f_a)$ is allowed to prevail over the position error $(x_d - x)$ at steady state.

Assuming that the desired force f_{ad} is aligned with a_e and x_d has a constant component along a_e, the equilibrium trajectory of system (4.39) is characterized by

$$x_E = (I - a_e a_e^T)x_d + a_e a_e^T(x_e - k_e^{-1} f_{ad}) \tag{4.40}$$

$$f_E = k_e a_e a_e^T(x_e - x_E) = f_{ad}. \tag{4.41}$$

Stability of the closed-loop system can be developed according to classical linear systems theory. Choosing diagonal matrices in (4.39) as $H_m = h_m I$, $D_m = d_m I$, $K_m = k_m I$, $K_F = k_F I$, $K_I = k_I I$, and plugging (4.24) in (4.39) gives

$$h_m \ddot{x} + d_m \dot{x} + (k_m I + k_F k_e a_e a_e^T)x + k_I k_e \, a_e a_e^T \int_0^t x \, d\tau \tag{4.42}$$

$$= h_m \ddot{x}_d + d_m \dot{x}_d + k_m x_d - k_F(f_{ad} - k_e\, a_e a_e^T x_e) - k_I \int_0^t (f_{ad} - k_e\, a_e a_e^T x_e) d\tau$$

which represents a third-order linear system, whose stability can be analyzed by referring to the unforced system as long as the input is bounded. Therefore, setting to zero the right-hand side of (4.42), accounting for (4.34), and projecting the position vector on the contact frame as

$$R_e^T x = \begin{pmatrix} x_n \\ x_s \\ x_a \end{pmatrix} \qquad (4.43)$$

leads to the system of three scalar decoupled equations

$$h_m \ddot{x}_n + d_m \dot{x}_n + k_m x_n = 0 \qquad (4.44)$$
$$h_m \ddot{x}_s + d_m \dot{x}_s + k_m x_s = 0 \qquad (4.45)$$
$$h_m \ddot{x}_a + d_m \dot{x}_a + (k_m + k_F k_e) x_a + k_I k_e \int_0^t x_a d\tau = 0, \qquad (4.46)$$

revealing that a stable behaviour is ensured by a proper choice of the gains k_m, d_m, k_F, k_I for the third equation.

Remarks

- The parallel inverse dynamics control scheme ensures tracking of the desired position along the task space directions on the plane together with regulation of the desired force along the task space direction normal to the plane, but the control law is based on force and position measurements along all task space directions notwithstanding.

- The scheme requires that the desired force is correctly planned. If the end effector is about to make contact with the environment and no information about its geometry is available, i.e., the direction of a_e is unknown, it is convenient to set $f_{ad} = 0$ which is anyhow in the range space of any matrix K_e. Then, once contact is established and a rough estimate of a_e is available from the force measurements, we can switch to the non-zero value of desired force.

- If the desired force set point f_{ad} is not aligned with a_e, the equilibrium trajectory (4.40) and (4.41) is modified into

$$\bar{x}_E = x_E + v_E t \qquad (4.47)$$
$$\bar{f}_E = a_e a_e^T f_{ad} \qquad (4.48)$$

4.2. PARALLEL CONTROL

where x_E is as in (4.40) and

$$v_E = \frac{k_I}{k_m}(I - a_e a_e^T)f_{ad}, \tag{4.49}$$

showing that the misalignment of f_{ad} with a_e causes a drift motion due to the presence of the integral action ($k_I \neq 0$); notice, however, that choosing $k_m \gg k_I$ attenuates the magnitude of such a drift.

- In the case of a curved surface, a linear contact force model as in (4.24) still holds but K_e becomes a function of x_e which is in turn a function of x. The problem becomes more involved and we have to resort to hybrid force/motion control, as discussed in Section 4.3.

4.2.2 PID control

The parallel control scheme presented above requires complete knowledge of the manipulator dynamic model in order to ensure tracking of the desired end-effector position on the environment surface. Nevertheless, a computationally lighter control scheme can be devised when a *force and position regulation* task is considered. Such a scheme can be regarded as an extension of the PD controller based on (4.21).

Consider the *constant* set points x_d and f_{ad}, where f_{ad} is aligned with a_e, and the *PID control* law

$$u = J_a^T(q)\left(k_m(x_d-x) - d_m\dot{x} - (f_{ad} + k_F(f_{ad}-f_a) + k_I\int_0^t (f_{ad}-f_a)d\tau)\right) \\ + g(q) \tag{4.50}$$

where $k_m, d_m, k_F, k_I > 0$. This controller corresponds to a position PD action + gravity compensation + desired force feedforward + a force PI action. In the case of perfect gravity compensation, the equilibrium for the closed-loop system is described by the same position in (4.40) and force in (4.41).

In order to study stability of system (4.8) under control (4.50) with the environment model (4.24) and (4.35), it is convenient to introduce a suitable state vector. Consider the end-effector deviation from the equilibrium position

$$\epsilon = x_e - x = x_d - x + d\,a_e \tag{4.51}$$

where

$$d = a_e a_e^T \left(k_e(x_e - x_d) - k_e^{-1} f_{ad}\right) \tag{4.52}$$

is a constant quantity taking into account the effects of the environment contact force and of the desired force set point along the normal to the plane.

Differentiating (4.51) with respect to time gives

$$\dot{\epsilon} = -\dot{x}. \tag{4.53}$$

From (4.51) and (4.52), the position and force errors can be respectively expressed as

$$x_d - x = \epsilon - da_e \tag{4.54}$$
$$f_{ad} - f_a = -k_e\, a_e a_e^T \epsilon. \tag{4.55}$$

Substituting (4.54) and (4.55) in (4.50) and using (4.53) leads to the closed-loop dynamics

$$H_x \ddot{\epsilon} + (C_x + d_m I)\dot{\epsilon} + (k_m I + k'_F k_e\, a_e a_e^T)\epsilon + k_I k_e w a_e = 0, \tag{4.56}$$

where $k'_F = 1 + k_F$ and

$$w = -k_e^{-1} a_e^T \left(\int_0^t (f_{ad} - f_a)d\tau + k_m k_I^{-1} d\, a_e \right). \tag{4.57}$$

Differentiating (4.57) with respect to time and using (4.55) gives

$$\dot{w} = a_e^T \epsilon. \tag{4.58}$$

Equations (4.56) and (4.58) provide a description of the closed-loop system in terms of the (7×1) state vector

$$z = (\dot{\epsilon}^T \quad \epsilon^T \quad w)^T \tag{4.59}$$

which can be written in the standard compact homogeneous form

$$\dot{z} = Fz \tag{4.60}$$

with

$$F = \begin{pmatrix} -H_x^{-1}(C_x + d_m I) & -H_x^{-1}(k_m I + k'_F k_e a_e a_e^T) & -k_I k_e H_x^{-1} a_e \\ I & 0 & 0 \\ 0 & a_e^T & 0 \end{pmatrix}. \tag{4.61}$$

Theorem 4.1 *There exists a choice of gains k_m, d_m, k_F, k_I that makes the origin of the state space of system (4.60) and (4.61) locally asymptotically stable.*

◇ ◇ ◇

4.2. PARALLEL CONTROL

Proof. Consider the Lyapunov function candidate

$$V = \frac{1}{2} z^T P z \tag{4.62}$$

where

$$P = \begin{pmatrix} H_x & \rho H_x & 0 \\ \rho H_x & (k_m + \rho d_m)I + k'_F k_e a_e a_e^T & k_I k_e a_e \\ 0 & k_I k_e a_e^T & \rho k_I k_e \end{pmatrix} \tag{4.63}$$

with $\rho > 0$. Computing the time derivative of V along the trajectories of system (4.60) and (4.61) gives

$$\dot{V} = -\dot{\epsilon}^T (d_m I - \rho H_x)\dot{\epsilon} + \rho \epsilon^T C_x^T \dot{\epsilon} - \epsilon^T \left(\rho k_m I + (\rho k'_F - k_I) k_e \, a_e a_e^T\right)\epsilon \tag{4.64}$$

where Property 4.2 has been conveniently exploited.

Consider the region of the state space

$$Z = \{ z : \|\epsilon\| < \Phi \}. \tag{4.65}$$

The term $\rho \epsilon^T C_x^T \dot{\epsilon}$ in (4.64) can be upper bounded in the region (4.65) as

$$\rho \epsilon^T C_x^T \dot{\epsilon} \leq \rho \Phi k_{Cx} \|\dot{\epsilon}\|^2 \tag{4.66}$$

where Property 4.3 has been used.

From (4.66), the function (4.64) can be upper bounded as

$$\dot{V} \leq -(d_m - \rho \lambda_{Mx} - \rho \Phi k_{Cx}) \|\dot{\epsilon}\|^2 - \rho k_m \|\epsilon\|^2 - k_e (\rho k'_F - k_I)(a_e^T \epsilon)^2 \tag{4.67}$$

where Property 4.1 has been used.

On the other hand, the function candidate (4.62) and (4.63) can be lower bounded as

$$V \geq \frac{1}{2} (\dot{\epsilon}^T H_x \quad \epsilon^T) \begin{pmatrix} (1/\lambda_{Mx})I & \rho I \\ \rho I & (k_m + \rho d_m)I \end{pmatrix} \begin{pmatrix} H_x \dot{\epsilon} \\ \epsilon \end{pmatrix}$$

$$+ \frac{k_e}{2} (a_e^T \epsilon \quad w) \begin{pmatrix} k'_F & k_I \\ k_I & \rho k_I \end{pmatrix} \begin{pmatrix} a_e^T \epsilon \\ w \end{pmatrix}. \tag{4.68}$$

From (4.68) the function V is positive definite provided that

$$k_m + \rho d_m > \rho^2 \lambda_{Mx} \tag{4.69}$$

$$\rho(1 + k_F) > k_I, \tag{4.70}$$

whereas from (4.67) the function $\dot{V}$ is negative semi-definite provided that

$$d_m > \rho(\lambda_{Mx} + \Phi k_{Cx}) \tag{4.71}$$

and (4.70) again holds. Observing that condition (4.71) implies (4.69), it can be recognized that a choice of d_m, k_F and k_I exists such that conditions (4.70) and (4.71) are satisfied.

Since V is only negative semi-definite, the inequality $\dot{V} \leq 0$ must be further analyzed to prove asymptotic stability. In particular, $\dot{V} = 0$ implies $\dot{\epsilon} = 0$ and $\epsilon = 0$. From (4.60) and (4.61) local asymptotic stability around $z = 0$ follows from La Salle's invariant set theorem.

◇

Remarks

- The choice of the Lyapunov function (4.62) can be motivated by an energy-based argument. The diagonal terms in (4.63) express essentially the manipulator kinetic energy, the potential energy associated with the deviation from the equilibrium end-effector position, and the potential energy stored along the normal to the plane, respectively.

- The free parameter ρ is not used in the control law (4.50) and allows then an opportune choice of the gains d_m, k_F and k_I. Notice that k_m is not involved in the conditions (4.70) and (4.71) and thus is available to meet further design requirements during the non-contact phase of the task. Also, by increasing d_m a larger value of Φ can be tolerated which in turn determines the local nature of the stability proof.

4.3 Hybrid force/motion control

When the environment is quite rigid and the manipulator has to continuously keep its end effector in contact, a *constrained approach* may be more convenient to describe the system dynamics and to set up the control problem. This approach considers the generalized surface of the environment as an infinitely stiff, bilateral constraint with no friction.

In this ideal case, an algebraic vector equation restricts the feasible locations (positions and orientations) of the manipulator end effector and induces kinematic constraints on the robot manipulator motion, effectively reducing the number of generalized coordinates needed to describe the manipulator dynamics. In a dual fashion, this approach explicitly provides also the forces and torques occurring at the contact. In fact, these are obtained from the existing orthogonality relations between *admissible* generalized velocities v and generalized *reaction* forces f in the task space (Cartesian space).

4.3. HYBRID FORCE/MOTION CONTROL

The constrained formulation is best suited for cases where a direct control of arbitrary forces exerted against the environment is needed together with motion control along a reduced number of task space directions. Using the whole information about the environment geometry, *hybrid force/motion control* is then possible. When unavoidable modelling errors come into play, the constrained approach provides also a convenient setting for handling force and displacement error signals at the joint or at the end-effector level.

We start with the derivation of the constrained manipulator dynamic equations. These can be given two different formats: one of *full* dimension, with filtering of the applied torques so as to maintain consistency with the constraint; another of *reduced* dimension, eliminating as many variables as constraints.

4.3.1 Constrained dynamics

In the following, we assume that the robot manipulator is subject to a *holonomic* constraint expressed directly in the *joint space*

$$\phi(q) = 0, \qquad \phi : R^n \to R^m, \quad m < n. \tag{4.72}$$

This nonlinear equality constraint is the equivalent representation of a Cartesian rigid and frictionless environment surface on which the manipulator end effector "lies". We consider only the time-invariant (*scleronomic*) case (4.72), but the extension to *rheonomic* constraints, $\phi(q,t) = 0$, is straightforward.

It is assumed that the vector constraint (4.72) is twice differentiable and that its m components are linearly independent. Differentiation of (4.72) yields then

$$\frac{\partial \phi}{\partial q}\dot{q} = J_\phi(q)\dot{q} = 0, \tag{4.73}$$

and

$$J_\phi(q)\ddot{q} + \dot{J}_\phi(q)\dot{q} = 0. \tag{4.74}$$

The assumed regularity condition implies that the rank of $J_\phi(q)$ will be equal to the number m of its rows, globally for q or at least locally in a neighbourhood of the operating point.

The manipulator dynamics subject to (4.72) is written as

$$H(q)\ddot{q} + C(q,\dot{q})\dot{q} + g(q) = u + u_f, \tag{4.75}$$

where the $(n \times 1)$ vector of generalized forces u_f arising from the presence of the constraint can be expressed in terms of an $(m \times 1)$ vector of multipliers

λ as
$$u_f = J_\phi^T(q)\lambda. \qquad (4.76)$$
This characterization of the constraining forces associated with (4.72) follows from the virtual work principle.

The force multipliers λ can be eliminated by solving (4.75) for $\ddot{q}$ and substituting it into (4.74). We obtain

$$\lambda = \left(J_\phi(q)H^{-1}(q)J_\phi^T(q)\right)^{-1} \cdot$$
$$\left(J_\phi(q)H^{-1}(q)(C(q,\dot{q})\dot{q} + g(q) - u) - \dot{J}_\phi^T(q)\dot{q}\right), \qquad (4.77)$$

from which it follows that the value of the multipliers instantaneously depends also on the applied input torque u. Substituting (4.76) and (4.77) into (4.75) yields a (*full*) n-*dimensional* set of second-order differential equations whose solutions, when initialized on the constraint, automatically satisfy (4.72) for all times, i.e.,

$$H(q)\ddot{q} + J_\phi^T(q)\left(J_\phi(q)H^{-1}(q)J_\phi^T(q)\right)^{-1}\dot{J}_\phi(q)\dot{q} =$$
$$P(q)(u - C(q,\dot{q})\dot{q} - g(q)), \qquad (4.78)$$

where

$$P(q) = I - J_\phi^T(q)\left(J_\phi(q)H^{-1}(q)J_\phi^T(q)\right)^{-1}J_\phi(q)H^{-1}(q). \qquad (4.79)$$

Since $PJ_\phi^T = 0$ and $P^2 = P$ (idempotent), the $(n \times n)$ matrix $P(q)$ is a projection operator that filters out all joint torques lying in the range of the transpose of the constraint Jacobian. These correspond to generalized forces that tend to violate the imposed joint space constraint.

As opposed to the above description, a reduction procedure can be set up to obtain a more compact representation of the robotic system. Using the implicit function theorem, we can always partition the joint vector q in two vectors; namely, an $(m \times 1)$ vector q_1 and an $((n-m) \times 1)$ vector q_2 such that

$$\text{rank}\,\frac{\partial \phi}{\partial q_1} = m. \qquad (4.80)$$

As a consequence, the constraint (4.72) can be expressed in terms of the variables q_2 only:

$$q_1 = \psi(q_2) \quad \Longrightarrow \quad \phi(\psi(q_2), q_2) = 0. \qquad (4.81)$$

The proper use of (4.81) eliminates from the n-dimensional constrained system the presence of the m variables q_1 together with all the constraint

4.3. HYBRID FORCE/MOTION CONTROL

equations, leading to a *reduced $(n-m)$-dimensional* unconstrained dynamical system. For this, it is convenient to apply the change of coordinates

$$\theta = \begin{pmatrix} \theta_1 \\ \theta_2 \end{pmatrix} = \begin{pmatrix} q_1 - \psi(q_2) \\ q_2 \end{pmatrix} = \Theta(q), \quad (4.82)$$

with inverse transformation

$$q = \begin{pmatrix} \theta_1 + \psi(\theta_2) \\ \theta_2 \end{pmatrix} = Q(\theta). \quad (4.83)$$

In the new coordinates, satisfaction of the constraint (4.72) implies $\theta_1 = 0$. We will also make use of the differential relation

$$\dot{q} = \begin{pmatrix} I & \frac{\partial \psi}{\partial \theta_2} \\ 0 & I \end{pmatrix} \dot{\theta} = T(\theta_2)\dot{\theta}. \quad (4.84)$$

Moreover, the input torques u_θ performing work on the new coordinates satisfy

$$u_\theta^T \dot{\theta} = u^T \dot{q} = u^T T(\theta_2)\dot{\theta}, \quad (4.85)$$

from which

$$u_\theta = T^T(\theta_2)u. \quad (4.86)$$

Differentiation of (4.84) gives

$$\ddot{q} = T(\theta_2)\ddot{\theta} + \dot{T}(\theta_2)\dot{\theta}, \quad (4.87)$$

and its substitution into (4.75) premultiplied by $T^T(\theta_2)$ yields

$$T^T(\theta_2)\Big(H(Q(\theta))T(\theta_2)\ddot{\theta} + H(Q(\theta))\dot{T}(\theta_2)\dot{\theta} \quad (4.88)$$
$$+ C(Q(\theta), T(\theta_2)\dot{\theta})T(\theta_2)\dot{\theta} + g(Q(\theta))\Big) = T^T(\theta_2)(u + u_f).$$

Defining

$$\hat{H}(\theta) = T^T(\theta_2)H(Q(\theta))T(\theta_2)$$
$$\hat{n}(\theta, \dot{\theta}) = T^T(\theta_2)H(Q(\theta))\dot{T}(\theta_2)\dot{\theta} \quad (4.89)$$
$$+ T^T(\theta_2)C(Q(\theta), T(\theta_2)\dot{\theta})T(\theta_2)\dot{\theta} + T^T(\theta_2)g(Q(\theta))$$

provides the more compact form

$$\hat{H}(\theta)\ddot{\theta} + \hat{n}(\theta, \dot{\theta}) = T^T(\theta_2)u + T^T(\theta_2)J_\phi^T(Q(\theta))\lambda, \quad (4.90)$$

which is again the full manipulator dynamics, expressed in the θ coordinates.

Analyzing independently the first m and the last $n - m$ second-order differential equations in (4.90) will be useful for the subsequent control developments. Introduce the simple matrices

$$\begin{pmatrix} E_1 \\ E_2 \end{pmatrix} = \begin{pmatrix} I & 0 \\ 0 & I \end{pmatrix} = (E_1^T \quad E_2^T), \tag{4.91}$$

where the identity and null block matrices have dimensions so that E_1 is $(m \times n)$ and E_2 is $((n-m) \times n)$. Then, evaluate the dynamic terms in (4.90) on the constraint $\phi(q) = 0$, i.e., for $\theta_1(t) \equiv 0$ (thus, $\dot\theta_1 = \ddot\theta_1 = 0$). With a slight abuse of notation, we will write $\hat{H}(\theta_2)$ in place of $\hat{H}(0, \theta_2)$ and similarly for other restricted terms. This leads to

$$E_1 \hat{H}(\theta_2) E_2^T \ddot\theta_2 + E_1 \hat{n}(\theta_2, \dot\theta_2) = E_1 T^T(\theta_2) u + E_1 T^T(\theta_2) J_\phi^T(\theta_2) \lambda \tag{4.92}$$

and

$$E_2 \hat{H}(\theta_2) E_2^T \ddot\theta_2 + E_2 \hat{n}(\theta_2, \dot\theta_2) = E_2 T^T(\theta_2) u. \tag{4.93}$$

We remark that the term $E_2 T^T(\theta_2) J_\phi^T(\theta_2) \lambda$ was dropped from (4.93), being identically zero. To show this, differentiate $\phi(q) = 0$ with respect to time and use (4.81):

$$\dot\phi(\psi(q_2), q_2) = \left(\frac{\partial \phi}{\partial q_1} \frac{\partial \psi}{\partial q_2} + \frac{\partial \phi}{\partial q_2} \right) \dot q_2 \tag{4.94}$$

$$= \begin{pmatrix} \frac{\partial \phi}{\partial q_1} & \frac{\partial \phi}{\partial q_2} \end{pmatrix} \begin{pmatrix} \frac{\partial \psi}{\partial q_2} \\ I \end{pmatrix} \dot q_2 = J_\phi(q) T(q_2) E_2^T \dot q_2 = 0.$$

Since (4.94) holds for all $\dot q_2$, it follows that $E_2 T^T J_\phi^T = 0$.

The first set of dynamic equations (4.92) can be discarded, since it provides only the actual value of the multipliers λ. In fact, solving for $\ddot\theta_2$ from (4.93) and substituting in (4.92) yields

$$E_1 \hat{H}(\theta_2) E_2^T (E_2 \hat{H}(\theta_2) E_2^T)^{-1} E_2 (T^T(\theta_2) u - \hat{n}(\theta_2, \dot\theta_2)) + E_1 \hat{n}(\theta_2, \dot\theta_2)$$
$$= E_1 T^T(\theta_2) u + E_1 T^T(\theta_2) J_\phi^T(\theta_2) \lambda, \tag{4.95}$$

from which the force multipliers are obtained as

$$\lambda = \left(E_1 T^T(\theta_2) J_\phi^T(\theta_2) \right)^{-1} \cdot$$
$$\left(E_1 - \hat{H}_{12}(\theta_2) \hat{H}_{22}^{-1}(\theta_2) E_2 \right) \left(\hat{n}(\theta_2, \dot\theta_2) - T^T(\theta_2) u \right), \tag{4.96}$$

4.3. HYBRID FORCE/MOTION CONTROL

being the matrix $E_1 T^T J_\phi^T = (\partial \phi / \partial q_1)^T$ square and nonsingular by assumption, and where block matrix notation has been used. The above expression for λ is equivalent to (4.77), but it is based on the reduced form of the dynamic equations. The overall dynamic behaviour will be driven only by (4.93) which, incidentally, can be rewritten simply as

$$\hat{H}_{22}(\theta_2)\ddot{\theta}_2 + \hat{n}_2(\theta_2, \dot{\theta}_2) = u_{\theta_2}. \tag{4.97}$$

This has to be completed by $\theta_1(t) \equiv 0$.

For illustration, the reduced equations of motion can be easily derived in the simple case of

$$\phi(q) = q_1 = 0, \tag{4.98}$$

a *linear* joint space constraint already set in explicit form (here, $\theta = q$). Since $J_\phi = (I \ 0)$, $T = I$, and then $\hat{H} = H$ and $\hat{n} = n = C\dot{q} + g$, eqs. (4.92) and (4.93) become

$$\begin{aligned} H_{12}(\theta_2)\ddot{\theta}_2 + n_1(\theta_2, \dot{\theta}_2) &= u_1 + \lambda \\ H_{22}(\theta_2)\ddot{\theta}_2 + n_2(\theta_2, \dot{\theta}_2) &= u_2 \end{aligned} \tag{4.99}$$

while the multipliers are given by

$$\lambda = n_1(\theta_2, \dot{\theta}_2) - u_1 - H_{12}(\theta_2) H_{22}^{-1}(\theta_2) \left(n_2(\theta_2, \dot{\theta}_2) - u_2 \right). \tag{4.100}$$

4.3.2 Inverse dynamics control

The control problem requires the asymptotic tracking of the following joint *motion* and contact *force* trajectories:

$$q = q_d(t) \qquad u_f = u_{fd}(t) \qquad t \geq 0, \tag{4.101}$$

on the assumption that these specifications are *consistent* with the model of the constraining surface. This implies that the $2n$ time functions in (4.101) are such that:

- $\phi(q_d(t)) = 0$ for all $t \geq 0$;
- for all $t \geq 0$, there exists an $(m \times 1)$ vector $\lambda_d(t)$ satisfying

$$u_{fd}(t) = J_\phi^T(q_d(t))\lambda_d(t). \tag{4.102}$$

Notice that the vector λ_d is a parameterization of the desired contact force in the constraint coordinates θ. In the full-rank hypothesis for J_ϕ, the unique value satisfying (4.102) is obtained through pseudoinversion

$$\lambda_d = (J_\phi(q_d) J_\phi^T(q_d))^{-1} J_\phi(q_d) u_{fd}, \tag{4.103}$$

for any given consistent choice of u_{fd}.

An *inverse dynamics* approach will be followed for the design of a control law that realizes the assigned task of motion in contact with the constraining surface with the prescribed exchange of forces. In doing so, the most convenient strategy will be to reconsider the task as specified by the $n - m$ motion assignments $\theta_{2d}(t)$ and by the m force multipliers $\lambda_d(t)$. In this way, a set of n *independent* time evolutions are assigned which can be realized by the suitable choice of the n control inputs u. Note also that from these assignments and from $\theta_{1d} = 0$, the whole motion vector q_d can be recovered using (4.83) while the n-dimensional joint space force is obtained through $u_{fd} = J_\phi^T(q_d)\lambda_d$. Vice-versa, (4.82) and (4.103) are used for moving the task specification from q_d and u_{fd} to θ_{2d} and λ_d.

The reduced form (4.92) and (4.93) of the system dynamic equations will be used for deriving an inversion control law. In particular, the control u will be found from the following m equations

$$E_1 T^T(\theta_2) u = E_1 \hat{n}(\theta_2, \dot{\theta}_2)$$
$$+ E_1 \hat{H}(\theta_2) E_2^T \left(\ddot{\theta}_{2d} + K_D(\dot{\theta}_{2d} - \dot{\theta}_2) + K_P(\theta_{2d} - \theta_2) \right) \quad (4.104)$$
$$- E_1 T^T(\theta_2) J_\phi^T(\theta_2) \left(\lambda_d + K_\lambda(\lambda_d - \lambda) + K_I \int_0^t (\lambda_d - \lambda) d\tau \right),$$

and the $n - m$ equations

$$E_2 T^T(\theta_2) u = E_2 \hat{n}(\theta_2, \dot{\theta}_2)$$
$$+ E_2 \hat{H}(\theta_2) E_2^T \left(\ddot{\theta}_{2d} + K_D(\dot{\theta}_{2d} - \dot{\theta}_2) + K_P(\theta_{2d} - \theta_2) \right). (4.105)$$

We note that a PD action on the motion error has been added to the feedforward acceleration $\ddot{\theta}_{2d}$, while a PI action was chosen on the force multiplier error, apart from the feedforward of λ_d.

Combining the above n equations and solving for u gives finally

$$u = T^{-T}(\theta_2) \hat{H}(\theta_2) E_2^T \left(\ddot{\theta}_{2d} + K_D(\dot{\theta}_{2d} - \dot{\theta}_2) + K_P(\theta_{2d} - \theta_2) \right)$$
$$+ T^{-T}(\theta_2) \hat{n}(\theta_2, \dot{\theta}_2) - T^{-T}(\theta_2) E_1^T E_1 T^T(\theta_2) J_\phi^T(\theta_2) \cdot$$
$$\left(\lambda_d + K_\lambda(\lambda_d - \lambda) + K_I \int_0^t (\lambda_d - \lambda) d\tau \right). \quad (4.106)$$

The analysis of the characteristics of the closed-loop system is made in terms of the following errors:

$$e_1 = \theta_1 = q_1 - \psi(q_2)$$
$$e_2 = \theta_2 - \theta_{2d} = q_2 - q_{2d}$$
$$e_\lambda = \lambda - \lambda_d. \quad (4.107)$$

4.3. HYBRID FORCE/MOTION CONTROL

The closed-loop equations obtained from (4.92), (4.93), and (4.106) are

$$E_1 \hat{H}(\theta_2) E_2^T (\ddot{e}_2 + K_D \dot{e}_2 + K_P e_2) =$$
$$E_1 T^T(\theta_2) J_\phi^T(\theta_2) \left(e_\lambda + K_\lambda e_\lambda + K_I \int_0^t e_\lambda \, d\tau \right), \quad (4.108)$$

and

$$E_2 \hat{H}(\theta_2) E_2^T (\ddot{e}_2 + K_D \dot{e}_2 + K_P e_2) = 0, \quad (4.109)$$

together with $e_1(t) \equiv 0$. From (4.109), since the block $\hat{H}_{22} = E_2 \hat{H} E_2^T$ on the diagonal of the inertia matrix is always invertible, it follows that

$$e_2(t) \to 0 \quad \Longrightarrow \quad q_2(t) \to q_{2d}(t) \quad (4.110)$$

as $t \to \infty$, provided that $K_P > 0$ and $K_D > 0$. Then we have also

$$e_1(t) \equiv 0 \quad \Longrightarrow \quad q_1(t) \to q_{1d}(t) = \psi(q_{2d}(t)). \quad (4.111)$$

Therefore, the study of eq. (4.108) reduces to the analysis of the asymptotic stability of the linear equation

$$e_\lambda + K_\lambda e_\lambda + K_I \int_0^t e_\lambda(\tau) d\tau = 0, \quad (4.112)$$

and then

$$e_\lambda(t) \to 0 \quad \Longrightarrow \quad \lambda(t) \to \lambda_d(t) \quad (4.113)$$

as $t \to \infty$, provided that $K_\lambda \geq 0$ and $K_I \geq 0$. Note that both feedback gains in the force loop can be set to zero, without loosing the asymptotic convergence to zero of the force error. In that case, the control law (4.106) would require *no force feedback* but only the feedforward of the desired force multipliers. This analysis, however, holds only in nominal conditions, i.e., when the environment surface is perfectly known and the only disturbance is a mismatch in the initial condition of motion *on the constraining surface*. When other uncertainties and disturbances are present, the benefits introduced by the force feedback terms become self-evident. In particular, the integral action is able to compensate for a constant bias disturbance.

The inversion controller (4.106) can also be rewritten in terms of the original variables q, using

$$\theta_d = \Theta(q_d) \quad (4.114)$$
$$\dot{\theta}_d = T^{-1}(q_{2d}) \dot{q}_d \quad (4.115)$$
$$\ddot{\theta}_d = T^{-1}(q_{2d}) \left(\ddot{q}_d - \dot{T}(q_{2d}) T^{-1}(q_{2d}) \dot{q}_d \right), \quad (4.116)$$

where $q_d(t)$ is a specification consistent with the constraint $\phi(q) = 0$. After some manipulation, noting also that $\ddot{\theta}_{2d} = E_2 \ddot{\theta}_d$, we finally get

$$u = H(q)T(q)\Big(T^{-1}(q_d)\ddot{q}_d - T^{-1}(q_d)\dot{T}(q_d)T^{-1}(q_d)\dot{q}_d$$
$$+ E_2^T K_P E_2(\Theta(q_d) - \Theta(q)) + E_2^T K_D E_2(T^{-1}(q_d)\dot{q}_d - T^{-1}(q)\dot{q})\Big)$$
$$+ H(q)\dot{T}(q)T^{-1}(q)\dot{q} + C(q,\dot{q})\dot{q} + g(q)$$
$$- J_\phi^T(q)\left(\lambda_d + K_\lambda(\lambda_d - \lambda) + K_I \int_0^t (\lambda_d(\tau) - \lambda(\tau))d\tau\right). \quad (4.117)$$

In the form (4.117), it is apparent that this control law performs a cancellation of the manipulator dynamic nonlinearities as well as of the inertial couplings due to constraint curvature ($\dot{T} \neq 0$). Furthermore, a desired linear error dynamics is imposed via the matrix gains K_P, K_D, K_λ and K_I, which are typically chosen diagonal so as to preserve the independent specification of desired force multipliers and of free admissible motion components. We recognize that these considerations are the spirit of an exact linearization and decoupling approach based on nonlinear state feedback.

A slightly different inversion controller can also be derived if we choose to react through u to force errors defined directly in terms of vector u_f in (4.75), i.e., to

$$e_f = J_\phi^T(q)(\lambda - \lambda_d)$$
$$= u_f - J_\phi^T(q)\left(J_\phi(q_d)J_\phi^T(q_d)\right)^{-1} J_\phi(q_d)u_{fd}, \quad (4.118)$$

where (4.76) and (4.103) have been used. The control law u will now be obtained by imposing the following alternate set of equations in place of (4.104) and (4.105):

$$E_1 T^T(\theta_2)u = E_1 \hat{n}(\theta_2, \dot{\theta}_2)$$
$$+ E_1 \hat{H}(\theta_2) E_2^T \left(\ddot{\theta}_{2d} + K_D(\dot{\theta}_{2d} - \dot{\theta}_2) + K_P(\theta_{2d} - \theta_2)\right) \quad (4.119)$$
$$- E_1 T^T(\theta_2) J_\phi^T(\theta_2)\lambda_d - E_1 E_1^T K_F E_1 T^T(\theta_2) J_\phi^T(\theta_2)(\lambda_d - \lambda)$$

and

$$E_2 T^T(\theta_2)u = E_2 \hat{n}(\theta_2, \dot{\theta}_2)$$
$$+ E_2 \hat{H}(\theta_2) E_2^T \left(\ddot{\theta}_{2d} + K_D(\dot{\theta}_{2d} - \dot{\theta}_2) + K_P(\theta_{2d} - \theta_2)\right) \quad (4.120)$$
$$- E_2 T^T(\theta_2) J_\phi^T(\theta_2)\lambda_d - E_2 E_1^T K_F E_1 T^T(\theta_2) J_\phi^T(\theta_2)(\lambda_d - \lambda).$$

We point out that the last two terms in (4.120) are zero because (4.94) holds and $E_2 E_1^T = 0$. Similarly, $E_1 E_1^T = I$ in the last term of (4.119) so

4.3. HYBRID FORCE/MOTION CONTROL

that it could be dropped. However, these terms are kept for the sake of symmetry between (4.119) and (4.120).

Combining the above two equations and solving for u gives now

$$u = T^{-T}(\theta_2)\hat{H}(\theta_2)E_2^T\left(\ddot{\theta}_{2d} + K_D(\dot{\theta}_{2d} - \dot{\theta}_2) + K_P(\theta_{2d} - \theta_2)\right)$$
$$+ T^{-T}(\theta_2)\hat{n}(\theta_2, \dot{\theta}_2) - J_\phi^T(\theta_2)\lambda_d$$
$$- T^{-T}(\theta_2)E_1^T K_F E_1 T^T(\theta_2) J_\phi^T(\theta_2)(\lambda_d - \lambda). \qquad (4.121)$$

In this case, the controller contains only a proportional action on the force error. It can be easily shown that $K_F \geq 0$ guarantees asymptotic stability of the force error ($e_f \to 0$). The inclusion of a robustifying integral term is more problematic in this context, due to the time-varying nature of the direction of the force error e_f —coming from the presence of $J_\phi^T(q)$ in (4.118).

As before, also the control law (4.121) can be rewritten in terms of the original coordinates q. This leads to

$$u = H(q)T(q)\left(T^{-1}(q_d)\ddot{q}_d - T^{-1}(q_d)\dot{T}(q_d)T^{-1}(q_d)\dot{q}_d \right.$$
$$\left. + E_2^T K_P E_2(\Theta(q_d) - \Theta(q)) + E_2^T K_D E_2(T^{-1}(q_d)\dot{q}_d - T^{-1}(q)\dot{q})\right)$$
$$+ H(q)\dot{T}(q)T^{-1}(q)\dot{q} + C(q, \dot{q})\dot{q} + g(q) \qquad (4.122)$$
$$- J_\phi^T(q)(J_\phi(q_d)J_\phi^T(q_d))^{-1}J_\phi(q_d)u_{fd}$$
$$- T^{-T}(q)E_1^T K_F E_1 T^T(q) \left(J_\phi^T(q)(J_\phi(q_d)J_\phi^T(q_d))^{-1}J_\phi(q_d)u_{fd} - u_f\right).$$

Remarks

- In the control law (4.122), the matrices

$$\Sigma_F = E_1^T K_F E_1$$
$$\Sigma_P = E_2^T K_P E_2 \qquad (4.123)$$
$$\Sigma_D = E_2^T K_D E_2,$$

represent filtering matrices on the force, position, and velocity errors, respectively. This allows us to *selectively* react to error components depending on their specific directions. In this way it is guaranteed that no superposition of contrasting control actions results; there is one and only one effective linear controller for each force or complementary motion component, leading to m force loops and $n - m$ motion (position/velocity) loops.

- In practice, the whole approach starts from a p-dimensional ($p < n$) constraint task vector $r(x) = 0$ imposed on the end-effector location. We have then

$$r(x) = r(k(q)) = \phi(q) = 0, \qquad (4.124)$$

while the developments remain the same (with $m = p$). However, in this form it becomes evident that both the characteristics of the constraint task in $J_r(x) = \partial r/\partial x$ and the analytical Jacobian $J_a(q) = \partial k/\partial q$ are crucial, being $J_\phi(q) = J_r(k(q))J_a(q)$.

- When the kinematic constraint is time-varying, i.e.,

$$r(x,t) = 0 \qquad r : R^m \times R^+ \to R^p \quad p < n, \qquad (4.125)$$

further terms will appear in its first time derivative

$$\frac{\partial r}{\partial x}\dot{x} + \frac{\partial r}{\partial t} = J_r(x,t)\dot{x} + v_r(x,t) = 0 \qquad (4.126)$$

as well as in the second time derivative

$$J_r(x,t)\ddot{x} + \dot{J}_r(x,t)\dot{x} + \dot{v}_r(x,t) = J_r(x,t)\ddot{x} + a_r(x,\dot{x},t) = 0 \quad (4.127)$$

which takes the place of (4.74). By substituting

$$\begin{aligned} J_\phi(q) &\longmapsto J_r(k(q),t)J_a(q) \\ \dot{J}_\phi(q)\dot{q} &\longmapsto J_r(k(q),t)\dot{J}_a(q)\dot{q} + a_r(k(q), J_a(q)\dot{q}, t) \end{aligned} \qquad (4.128)$$

in all subsequent equations, the rest follows accordingly.

4.3.3 Hybrid task specification and control

In the constrained approach we have assumed that the robot manipulator is always in contact with the environment, as represented by the vector equation (4.72). In practice, uncertainty in the geometric knowledge of the object surfaces and the nonideal characteristics (finite stiffness, friction) of the constraint, together with the presence of various disturbances, introduce small but systematic position and force errors. These errors can be handled by a relaxation of the constrained framework. For this reason, it is convenient to restate the main point of this approach: due to the presence of a constraint expressed in suitable coordinates, it is possible to characterize generalized directions of admissible velocity and of reaction forces. At this point, this idea can be used even *without* the explicit introduction of a constraint in the form (4.72).

4.3. HYBRID FORCE/MOTION CONTROL

For most tasks there exists a convenient space where the vectors of admissible velocity and of contact reaction force can be used to form an *orthogonal* and *complete* basis set. In particular, we can define a time-varying *task frame* with the following property; at every instant, in each coordinate direction either a force or a velocity can be independently assigned, the other quantity being naturally imposed by the task geometry. We notice that this orthogonality can be stated only between force and velocity (or displacement) generalized vectors, not between force and position, justifying the current terminology of *hybrid force/velocity control*. The key point in the control scheme is to use the task frame concept also in off-nominal conditions for "filtering out" errors that are not consistent with the model, thus reacting only to force or motion error components along "expected" directions.

The steps for deriving a hybrid controller follow closely the previous derivations, taking also into account the above final remarks. Relevant dynamic quantities (velocities and forces) are first transferred in the suitable *task space*, through kinematic transformations. They are next compared with the current desired values, generating directional errors and simply eliminating those that do not agree with the task model (e.g., considering only measured forces that are normal to an assumed frictionless plane on which the manipulator end effector is moving). This filtering action plays a role similar to that of the Σ matrices in (4.123). In response to these errors, commands are generated by a set of m scalar force controllers and $n - m$ scalar velocity controllers, each independently designed for one task coordinate. In particular, in the design of the force controllers we may take advantage of an assumed compliant contact model, similar to (4.24). Finally, these control commands, which are dimensionally homogeneous to accelerations, are brought back to the joint space and are used to produce the applied input torques —possibly, based on the manipulator dynamics if an inversion controller is desired.

The resulting hybrid force/velocity control law can be easily derived from the previously presented ones, e.g., from (4.122). The only difference is that all dynamic quantities are evaluated *at the current state*, not necessarily satisfying a constraining equation.

Contact with dynamic environments

As opposed to "completely static" (constrained) or "purely dynamic" (impedance) interactions, there is a whole class of tasks which are accurately modelled only by considering a more general dynamic behaviour for the environment and a more flexible description of the robot manipulator contact. In this respect, the manipulator end effector may exert *dynamic* forces, i.e., forces not compensated by a constraint reaction and producing active work on the environment geometry, while still being subject to further kinematic constraints. A paradigmatic example is the task of a manipulator turning a crank, when crank dynamics is relevant. Also, the dynamic world seen by a manipulator may be another manipulator itself, as in the case of cooperative robotic tasks. An effective modelling technique should be able to handle mixed situations in which the end effector is kinematically constrained and/or dynamically coupled with the external world. Finally, purely *compliant* interactions should also be considered, typically associated with the potential energy of contact deformation.

In these situations, the hybrid control cannot be applied as such. Orthogonality can be stated only between two reduced sets of task directions, one in which it is possible to control only motion parameters and one in which only reaction forces can be controlled. Instead, along dynamic directions *either* a force parameter *or* a motion parameter can be directly controlled by the hybrid scheme, the other quantity resulting from the dynamic balance of the interaction between the manipulator and environment.

4.4 Further reading

Early work on force control can be found in [52, 53], typically applied to assembly tasks. Stiffness control was proposed in [42] and is conceptually equivalent to compliance control [39]. The original idea of a mechanical impedance model used for controlling the interaction between manipulator and environment is due to [24], and a similar formulation is given in [28]. Achieving the expected impedance may be difficult due to uncertainties on the manipulator dynamic model. Linear [12] and nonlinear [29, 33] adaptive impedance control algorithms have been proposed to overcome this problem, while a robust scheme can be found in [34]. An approach based on impedance control where the reference position trajectory is adjusted in order to fulfill a force control objective has been recently proposed in [8]. A survey of all the above schemes in quasi-static operation can be found in [54].

The parallel approach to force/position control of robot manipulators was introduced in [10] using an inverse dynamics controller. As an alternative to the inverse dynamics scheme, a passivity-based approach to force regulation and motion control has been proposed in [44], which naturally allows derivation of an adaptive control scheme [46]. The PID parallel controller for force and position regulation was developed in [11], where exponential stability is proved in addition to asymptotic stability. More recently, the regulator has been extended in an output feedback setting to

4.4. FURTHER READING

avoid velocity measurements [45]. The case of imperfect gravity compensation is treated in [43] where an adaptive scheme is proposed to resume the original equilibrium. On the other hand, the original idea of closing an outer force control loop around an inner position control loop dates back to [18]. The use of an integral action for removing steady-state force errors was also proposed in [51], and robustness with respect to force measurement delays was investigated in [50, 55].

The original hybrid position/force control concept was introduced in [41], based on the natural and artificial constraint task formulation of [35]. The explicit inclusion of the manipulator dynamic model was presented in [30], and the problem of correct definition of a time-varying task frame was pointed out in [15]. The constrained task formulation with inverse dynamics controllers is treated in [57, 58] using a Cartesian constraint as well as in [36] using a joint space constraint. A projection technique of Cartesian constraints to the joint space was presented in [5]. A comparison of the benefits of adopting a reduced vs. a complete set of dynamic model equations can be found in [26]. The constrained approach was also used in [37] with a controller based on linearized equations, and in [59] with a dynamic state feedback controller. The hybrid task specification in connection with an impedance controller on the constrained directions was proposed in [3], as well as in [16] where a sliding mode controller is used for robustness purposes. On-line identification of the task frame orientation with hybrid control experiments was presented in [22]. Theoretical issues on orthogonality of generalized force and velocity directions are discussed in [31, 19]. Transposition of model-based schemes from unconstrained motion control to constrained cases was accomplished for adaptive control in [47, 32, 4], for robust control in [48], and for learning control in [1], respectively. On the other hand, adaptation with respect to the compliance characteristics of the environment has been proposed in [9, 56]. Modelling of interaction between a manipulator and a dynamic environment was presented in [14], while hybrid control schemes based on these models can be found in [13].

It has been generally recognized that force control may cause unstable behaviour during contact with environment. Dynamic models for explaining this phenomenon were introduced in [20, 21]. Experimental investigations can be found in [2] and [49]. Emphasis on the problems with a stiff environment is given in [27, 23], while stability of impedance control was analyzed in [25]. Moreover, control schemes are usually derived on the assumption that the manipulator end effector is in contact with the environment and this contact is not lost. Impact phenomena may occur which deserve careful consideration, and there is a need for global analysis of control schemes including the transition from non-contact into contact and vice-versa [38, 6].

The automatic specification of robotic tasks that involve multiple and time-varying contacts of the manipulator end-effector with the environment is a challenging issue. General formalisms addressing this problem were proposed in [17, 40, 7].

We conclude by pointing out that the task of controlling the end-effector velocity of a single manipulator and its exchanged forces with the environment is the simplest of a series of robotic problems where similar issues arise. Among them we wish to mention *telemanipulation*, where the use of force feedback enhances the human operator capabilities of remotely handling objects with a slave manipulator, and *cooperative robot systems*, where two or more manipulators (viz., the fingers of a dexterous robot hand) should monitor the exchanged forces so as to avoid "squeezing" of the commonly held object.

References

[1] M. Aicardi, G. Cannata, and G. Casalino, "Hybrid learning control for constrained manipulators," *Advanced Robotics*, vol. 6, pp. 69–94, 1992.

[2] C.H. An and J.M. Hollerbach, "The role of dynamic models in Cartesian force control of manipulators," *Int. J. of Robotics Research*, vol. 8, no. 4, pp. 51–72, 1989.

[3] R.J. Anderson and M.W. Spong, "Hybrid impedance control of robotic manipulators," *IEEE J. of Robotics and Automation*, vol. 4, pp. 549–556, 1988.

[4] S. Arimoto, Y.H. Liu, and T. Naniwa, "Model-based adaptive hybrid control for geometrically constrained robots," *Proc. 1993 IEEE Int. Conf. on Robotics and Automation*, Atlanta, GA, pp. 618–623, 1993.

[5] S. Arimoto, T. Naniwa, and T. Tsubouchi, "Principle of orthogonalization for hybrid control of robot manipulators," in *Robotics, Mechatronics and Manufacturing Systems*, T. Takamori and K. Tsuchiya (Eds.), Elsevier, Amsterdam, NL, pp. 295–302, 1993.

[6] B. Brogliato and P. Orhant, "On the transition phase in robotics: Impact models, dynamics and control," *Proc. 1994 IEEE Int. Conf. on Robotics and Automation*, San Diego, CA, pp. 346–351, 1994.

[7] H. Bruyninckx, S. Dumey, S. Dutré, and J. De Schutter, "Kinematic models for model-based compliant motion in the presence of uncertainty," *Int. J. of Robotics Research*, vol. 14, pp. 465–482, 1995.

[8] C. Canudas de Wit and B. Brogliato, "Direct adaptive impedance control," *Postpr. 4th IFAC Symp. on Robot Control*, Capri, I, pp. 345–350, 1994.

[9] R. Carelli, R. Kelly, and R. Ortega, "Adaptive force control of robot manipulators," *Int. J. of Control*, vol. 52, pp. 37–54, 1990.

[10] S. Chiaverini and L. Sciavicco, "The parallel approach to force/position control of robotic manipulators," *IEEE Trans. on Robotics and Automation*, vol. 9, pp. 361–373, 1993.

[11] S. Chiaverini, B. Siciliano, and L. Villani, "Force/position regulation of compliant robot manipulators," *IEEE Trans. on Automatic Control*, vol. 39, pp. 647–652, 1994.

[12] R. Colbaugh, H. Seraji, and K. Glass, "Direct adaptive impedance control of robot manipulators," *J. of Robotic Systems*, vol. 10, pp. 217–248, 1993.

[13] A. De Luca and C. Manes, "Hybrid force-position control for robots in contact with dynamic environments," *Proc. 3rd IFAC Symp. on Robot Control*, Vienna, A, pp. 177–182, 1991.

[14] A. De Luca and C. Manes, "Modeling robots in contact with a dynamic environment," *IEEE Trans. on Robotics and Automation*, vol. 10, pp. 542–548, 1994.

[15] A. De Luca, C. Manes, and F. Nicolò, "A task space decoupling approach to hybrid control of manipulators," *Proc. 2nd IFAC Symp. on Robot Control*, Karlsruhe, D, pp. 157–162, 1988.

[16] A. De Luca, C. Manes, and G. Ulivi, "Robust hybrid dynamic control of robot arms," *Proc. 28th IEEE Conf. on Decision and Control*, Tampa, FL, pp. 2641–2646, 1989.

[17] J. De Schutter and H. Van Brussel, "Compliant robot motion I. A formalism for specifying compliant motion tasks," *Int. J. of Robotics Research*, vol. 7, no. 4, pp. 3–17, 1988.

[18] J. De Schutter and H. Van Brussel, "Compliant robot motion II. A control approach based on external control loops," *Int. J. of Robotics Research*, vol. 7, no. 4, pp. 18–33, 1988.

[19] J. Duffy, "The fallacy of modern hybrid control theory that is based on 'orthogonal complements' of twist and wrench spaces," *J. of Robotic Systems*, vol. 7, pp. 139–144, 1990.

[20] S.D. Eppinger and W.P. Seering, "Introduction to dynamic models for robot force control," *IEEE Control Systems Mag.*, vol. 7, no. 2, pp. 48–52, 1987.

[21] S.D. Eppinger and W.P. Seering, "Understanding bandwidth limitations on robot force control," *Proc. 1987 IEEE Int. Conf. on Robotics and Automation*, Raleigh, NC, pp. 904–909, 1987.

[22] A. Fedele, A. Fioretti, C. Manes, and G. Ulivi, "On-line processing of position and force measures for contour identification and robot control," *Proc. 1993 IEEE Int. Conf. on Robotics and Automation*, Atlanta, GA, vol. 1, pp. 369–374, 1993.

[23] G. Ferretti, G. Magnani, and P. Rocco, "On the stability of integral force control in case of contact with stiff surfaces," *ASME J. of Dynamic Systems, Measurement, and Control*, vol. 117, pp. 547–553, 1995.

[24] N. Hogan, "Impedance control: An approach to manipulation: Parts I—III," *ASME J. of Dynamic Systems, Measurement, and Control*, vol. 107, pp. 1–24, 1985.

[25] N. Hogan, "On the stability of manipulators performing contact tasks," *IEEE J. of Robotics and Automation*, vol. 4, pp. 677–686, 1988.

[26] R.K. Kankaanranta and H.N. Koivo, "Dynamics and simulation of compliant motion of a manipulator," *IEEE J. of Robotics and Automation*, vol. 4, pp. 163–173, 1988.

[27] H. Kazerooni, "Contact instability of the direct drive robot when constrained by a rigid environment," *IEEE Trans. on Automatic Control*, vol. 35, pp. 710–714, 1990.

[28] H. Kazerooni, T.B. Sheridan, and P.K. Houpt, "Robust compliant motion for manipulators, Part I: The fundamental concepts of compliant motion," *IEEE J. of Robotics and Automation*, vol. 2, pp. 83–92, 1986.

[29] R. Kelly, R. Carelli, M. Amestegui, and R. Ortega, "Adaptive impedance control of robot manipulators," *IASTED Int. J. of Robotics and Automation*, vol. 4, no. 3, pp. 134–141, 1989.

[30] O. Khatib, "A unified approach to motion and force control of robot manipulators: The operational space formulation," *IEEE J. of Robotics and Automation*, vol. 3, pp. 43–53, 1987.

REFERENCES

[31] H. Lipkin and J. Duffy, "Hybrid twist and wrench control for a robotic manipulator," *ASME J. of Mechanism, Transmissions, and Automation in Design*, vol. 110, pp. 138–144, 1988.

[32] R. Lozano and B. Brogliato, "Adaptive hybrid force-position control for redundant manipulators," *IEEE Trans. on Automatic Control*, vol. 37, pp. 1501–1505, 1992.

[33] W.-S. Lu and Q.-H. Meng, "Impedance control with adaptation for robotic manipulators," *IEEE Trans. on Robotics and Automation*, vol. 7, pp. 408–415, 1991.

[34] Z. Lu and A.A. Goldenberg, "Robust impedance control and force regulation: Theory and experiments," *Int. J. of Robotics Research*, vol. 14, pp. 225–254, 1995.

[35] M.T. Mason, "Compliance and force control for computer controlled manipulators," *IEEE Trans. on Systems, Man, and Cybernetics*, vol. 11, pp. 418–432, 1981.

[36] N.H. McClamroch and D. Wang, "Feedback stabilization and tracking in constrained robots," *IEEE Trans. on Automatic Control*, vol. 33, pp. 419–426, 1988.

[37] J.K. Mills and A.A. Goldenberg, "Force and position control of manipulators during constrained motion tasks," *IEEE Trans. on Robotics and Automation*, vol. 5, pp. 30–46, 1989.

[38] J.K. Mills and D.M. Lokhorst, "Control of robotic manipulators during general task execution: A discontinuous control approach," *Int. J. of Robotics Research*, vol. 12, pp. 146–163, 1993.

[39] R. Paul and B. Shimano, "Compliance and control," *Proc. 1976 Joint Automatic Control Conf.*, San Francisco, CA, pp. 694–699, 1976.

[40] M.A. Peshkin, "Programmed compliance for error corrective assembly," *IEEE Trans. on Robotics and Automation*, vol. 6, pp. 473–482, 1990.

[41] M.H. Raibert and J.J. Craig, "Hybrid position/force control of manipulators," *ASME J. of Dynamic Systems, Measurement, and Control*, vol. 102, pp. 126–133, 1981.

[42] J.K. Salisbury, "Active stiffness control of a manipulator in Cartesian coordinates," *Proc. 19th IEEE Conf. on Decision and Control*, Albuquerque, NM, pp. 95–100, 1980.

[43] B. Siciliano and L. Villani, "An adaptive force/position regulator for robot manipulators," *Int. J. of Adaptive Control and Signal Processing*, vol. 7, pp. 389–403, 1993.

[44] B. Siciliano and L. Villani, "A passivity-based approach to force regulation and motion control of robot manipulators," *Automatica*, vol. 32, pp. 443–447, 1996.

[45] B. Siciliano and L. Villani, "A force/position regulator for robot manipulators without velocity measurements," *1996 IEEE Int. Conf. on Robotics and Automation*, Minneapolis, MN, pp. 2567–2572, 1996.

[46] B. Siciliano and L. Villani, "Adaptive compliant control of robot manipulators," *Control Engineering Practice*, vol. 4, pp. 705–712, 1996.

[47] J.-J.-E. Slotine and W. Li, "Adaptive strategies in constrained manipulation," *Proc. 1987 IEEE Int. Conf. on Robotics and Automation*, Raleigh, NC, pp. 595–601, 1987.

[48] C.-Y. Su, T.-P. Leung, and Q.-J. Zhou, "Force/motion control of constrained robots using sliding mode," *IEEE Trans. on Automatic Control*, vol. 37, pp. 668–672, 1992.

[49] R. Volpe and P. Khosla, "A theoretical and experimental investigation of impact control for manipulators," *Int. J. of Robotics Research*, vol. 12, pp. 351–365, 1993.

[50] R. Volpe and P. Khosla, "A theoretical and experimental investigation of explicit force control strategies for manipulators," *IEEE Trans. on Automatic Control*, vol. 38, pp. 1634–1650, 1993.

[51] J. Wen and S. Murphy, "Stability analysis of position and force control for robot arms," *IEEE Trans. on Automatic Control*, vol. 36, pp. 365–371, 1991.

[52] D.E. Whitney, "Force feedback control of manipulator fine motions," *ASME J. of Dynamic Systems, Measurement, and Control*, vol. 99, pp. 91–97, 1977.

[53] D.E. Whitney, "Quasi-static assembly of compliantly supported rigid parts," *ASME J. of Dynamic Systems, Measurement, and Control*, vol. 104, pp. 65–77, 1982.

[54] D.E. Whitney, "Historical perspective and state of the art in robot force control," *Int. J. of Robotics Research*, vol. 6, no. 1, pp. 3–14, 1987.

[55] L.S. Wilfinger, J.T. Wen, and S.H. Murphy, "Integral force control with robustness enhancement," *IEEE Control Systems Mag.*, vol. 14, no. 1, pp. 31–40, 1994.

[56] B. Yao and M. Tomizuka, "Adaptive control of robot manipulators in constrained motion — Controller design," *ASME J. of Dynamic Systems, Measurement, and Control*, vol. 117, pp. 320–328, 1995.

[57] T. Yoshikawa, "Dynamic hybrid position/force control of robot manipulators — Description of hand constraints and calculation of joint driving force," *IEEE J. of Robotics and Automation*, vol. 3, pp. 386–392, 1987.

[58] T. Yoshikawa, T. Sugie, and M. Tanaka, "Dynamic hybrid position/force control of robot manipulators — Controller design and experiment," *IEEE J. of Robotics and Automation*, vol. 4, pp. 699–705, 1988.

[59] X. Yun, "Dynamic state feedback control of constrained robot manipulators," *Proc. 27th IEEE Conf. on Decision and Control*, Austin, TX, pp. 622–626, 1988.

Part II

Flexible manipulators

Chapter 5

Elastic joints

This chapter deals with modelling and control of robot manipulators with joint flexibility. The presence of such a flexibility is a common aspect in many current industrial robots. When motion transmission elements such as harmonic drives, transmission belts and long shafts are used, a dynamic time-varying displacement is introduced between the position of the driving actuator and that of the driven link.

Most of the times, this intrinsic small deflection is regarded as a source of problems, especially when accurate trajectory tracking or high sensitivity to end-effector forces is mandatory. In fact, an oscillatory behaviour is usually observed when moving the links of a robot manipulator with nonnegligible joint flexibility. These vibrations are of small magnitude and occur at relatively high frequencies, but still within the bandwidth of interest for control.

On the other hand, there are cases when compliant elements (in our case, at the joints) may become useful in a robotic structure, e.g., as a protection against unexpected "hard" contacts during assembly tasks. Moreover, when using harmonic drives, the negative side effect of flexibility is balanced by the benefit of working with a compact, in-line component, with high reduction ratio and large power transmission capability.

From the modelling viewpoint, the above deformation can be characterized as being *concentrated* at the joints of the manipulator, and thus we often refer to this situation by the term *elastic joints* in lieu of flexible joints. This is a main feature to be recognized, because it will limit the complexity both of the model derivation and of the control synthesis. In particular, we emphasize the difference with lightweight manipulator links, where flexibility involves bodies of larger mass (as opposed to an elastic transmission shaft) undergoing deformations distributed over longer segments. In that case, flexibility cannot be reduced to an effect concentrated at the joint. As we will see, this has relevant consequences in the control analysis and design; the case of flexible joints should then be treated separately from that of flexible links.

We also remark that the assumption of perfect rigidity is an ideal one for all robot manipulators. However, the primary concern in deriving a mathematical model including any kind of flexibility is to evaluate quantitatively its relative effects, as superimposed to the rigid body motion. The additional modelling effort allows verifying whether a control law derived on the rigidity assumption (valid for rigid manipulators) will still work in practice, or should be modified and if so up to what extent. If high performance cannot be reached in this way, new specific control laws should be investigated, explicitly based on the more complete manipulator model.

When compared to the rigid case, the dynamic model of robot manipulators with elastic joints (but rigid links) requires *twice* the number of generalized coordinates to completely characterize the configuration of all rigid bodies (motors and links) constituting the manipulator. On the other hand, since actual joint deformations are quite small, elastic forces are typically limited to a domain of linearity.

The case of elastic joint manipulators is a first example in which the number of control inputs does not reach the number of mechanical degrees of freedom. Therefore, control tasks are supposed to be more difficult than the equivalent ones for rigid manipulators. In particular, the implementation of a full state feedback control law will require twice the number of sensors, measuring quantities that are *before and after* (or across) the elastic deformation.

Conversely, the strong couplings imposed by the elastic joints are helpful in obtaining a convenient behaviour for all variables. Also, both the elastic and input torques act on the same joint axes (they are *physically co-located*) and this induces nice control characteristics to the system.

The control goal is then to properly handle the vibrations induced by elasticity at the joints, so as to achieve fast positioning and accurate tracking at the manipulator end-effector level. In the following, in view of the assumed link rigidity, the task space control problem is not considered, but attention will be focused only on the motion of the links, i.e., in the *joint space*. In fact, by a proper choice of coordinates, the direct kinematics of a manipulator with elastic joints is exactly the same as that of a rigid manipulator, and no further problems arise in this respect. Notice that the position of a link is a variable already *beyond* the point where elasticity is introduced, and thus the control objective is definitely not a restricted one.

This chapter is organized in two parts, covering modelling issues and control problems.

First, the dynamic equations of general elastic joint robot manipulators are derived, their internal structure is highlighted, and possible simplifications are discussed. There are two different, and both common, modelling assumptions that lead to two kinds of dynamic models: a *complete* and a *reduced* one. In what follows, it will be clear that these two models do not share the same structural properties from the control point of view. Furthermore, when the manipulator elastic joints are relatively stiff, it will be possible to suitably rewrite the dynamic equations in a singularly perturbed form.

A series of control strategies are then investigated for the problems of set-point *regulation* and trajectory *tracking control*.

For point-to-point motion, *linear controllers* usually provide satisfactory performance. A single link driven through an elastic joint and moving on the horizontal plane is introduced as a paradigmatic case study for showing properties and difficulties encountered even in a linear setting. This simple example shows immediately what can be achieved with different sets of state measurements. More in general, the inclusion of gravity will be handled by the addition of a constant compensation term to a PD controller, with feedback only from the motor variables.

For the *reduced model* of multilink elastic joint manipulators, the trajectory tracking problem is solved using two nonlinear control methods: *feedback linearization* and *singular perturbation*. The former is a global exact method, while the latter exploits the feasibility of an incremental design, moving from the rigid case up to the desired accuracy order.

When the *complete model* is considered, use of *dynamic state feedback* for obtaining exact linearization and decoupling will be illustrated by means of an example. We also show that the key assumption in this result holds for the whole class of manipulators with elastic joints, thus guaranteeing that full linearization can be achieved in general. Finally, a *nonlinear regulation* approach will be presented; its implementation is quite simple, since the computational burden is considerably reduced.

Only nonadaptive control schemes based on full or partial state feedback will be discussed. Results for the unknown parameter case and on the use of state observers are quite recent and not yet completely settled down. They can be found in the list of references at the end of the chapter for further reading.

5.1 Modelling

We refer to a robot manipulator with elastic joints as an open kinematic chain having $n + 1$ rigid bodies, the base (link 0) and the n links, inter-

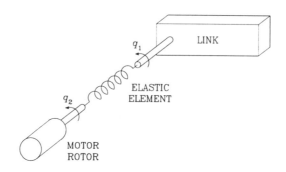

Figure 5.1: Schematic representation of an elastic joint.

connected by n joints undergoing elastic deformation. The manipulator is actuated by electrical drives which are assumed to be located at the joints. More specifically, we consider the standard situation in which motor i is mounted on link $i-1$ and moves link i. When reduction gears are present, they are modelled as being placed *before* the elastic element. All joints are considered to be elastic, though mixed situations may be encountered in practice due to the use of different transmission devices. The following quite general assumptions are made about the mechanical structure.

Assumption 5.1 Joint deformations are small, so that elastic effects are restrained to the domain of linearity.

□

Assumption 5.2 The elasticity in the joint is modelled as a spring, torsional for revolute joints and linear for prismatic joints; Fig. 5.1 shows an elastic revolute joint.

□

Assumption 5.3 The rotors of the actuators are modelled as uniform bodies having their centers of mass on the rotation axes.

□

We emphasize the relevance of Assumption 5.3 on the geometry of the rotors: it implies that both the inertia matrix and the gravity term in the dynamic model are independent of the actual internal position of the motors.

Following the usual Lagrange formulation, a set of generalized coordinates has to be introduced to characterize uniquely the system configuration. Since the manipulator chain is composed of $2n$ rigid bodies, $2n$ coordinates are needed. Let q_1 be the $(n \times 1)$ vector of link positions, and

5.1. MODELLING

q_2 represents the $(n \times 1)$ vector of actuator (rotor) positions, as reflected through the gear ratios. With this choice, the difference $q_{1i} - q_{2i}$ is joint i deformation. Moreover, the direct kinematics of the whole manipulator (and of each link end point) will be a function of the link variables q_1 only.

The *kinetic energy* of the manipulator structure is given as usual by

$$T = \frac{1}{2} \dot{q}^T H(q) \dot{q}, \tag{5.1}$$

where $q = (q_1^T \ q_2^T)^T$ and $H(q)$ is the $(2n \times 2n)$ inertia matrix, which is symmetric and positive definite for all q. Moreover, for revolute joints all elements of $H(q)$ are bounded. According to the previous assumptions, $H(q)$ has the following internal structure:

$$H(q) = H(q_1) = \begin{pmatrix} H_1(q_1) & H_2(q_1) \\ H_2^T(q_1) & H_3 \end{pmatrix}. \tag{5.2}$$

All blocks in (5.2) are $(n \times n)$ matrices: H_1 contains the inertial properties of the rigid links, H_2 accounts for the inertial couplings between each spinning actuator and the previous links, while H_3 is the constant diagonal matrix depending on the rotor inertias of the motors and on the gear ratios.

The *potential energy* is given by the sum of two terms. The first one is the gravitational term for both actuators and links; on the symmetric mass assumption for the rotors, it takes on the form

$$U_g = U_g(q_1). \tag{5.3}$$

The second one, arising from joint elasticity, can be written as

$$U_e = \frac{1}{2}(q_1 - q_2)^T K (q_1 - q_2) \tag{5.4}$$

in which $K = \mathrm{diag}\{k_1, \ldots, k_n\}$ is the joint *stiffness matrix*, $k_i > 0$ being the elastic constant of joint i. By defining the matrix

$$K_e = \begin{pmatrix} K & -K \\ -K & K \end{pmatrix}, \tag{5.5}$$

the elastic energy (5.4) can be rewritten as

$$U_e = \frac{1}{2} q^T K_e q. \tag{5.6}$$

The dynamic equations of motion are obtained from the Lagrangian function $L(q, \dot{q}) = T(q, \dot{q}) - U(q)$ as

$$\frac{d}{dt}\frac{\partial L}{\partial \dot{q}_i} - \frac{\partial L}{\partial q_i} = e_i \qquad i = 1, \ldots, 2n \tag{5.7}$$

where e_i is the generalized force performing work on q_i. Since only the motor coordinates q_2 are directly actuated, we collect all forces on the right-hand side of (5.7) in the $(2n \times 1)$ vector

$$e = \begin{pmatrix} 0 & \cdots & 0 & u_1 & \cdots & u_n \end{pmatrix}^T, \qquad (5.8)$$

where u_i denotes the external torque supplied by the motor at joint i. Link coordinates q_1 are indirectly actuated only through the elastic coupling.

Computing the derivatives needed in (5.7) leads to the set of $2n$ second-order nonlinear differential equations of the form

$$H(q_1)\ddot{q} + C(q,\dot{q})\dot{q} + K_e q + g(q_1) = e, \qquad (5.9)$$

in which the Coriolis and centrifugal terms are

$$C(q,\dot{q})\dot{q} = \dot{H}(q_1)\dot{q} - \frac{1}{2}\left(\frac{\partial}{\partial q}(\dot{q}^T H(q_1)\dot{q})\right)^T, \qquad (5.10)$$

and the gravity vector is

$$g(q_1) = \left(\frac{\partial U_g(q_1)}{\partial q}\right)^T = \begin{pmatrix} g_1(q_1) \\ 0 \end{pmatrix}, \qquad (5.11)$$

with $g_1 = (\partial U_g/\partial q_1)^T$. Eq. (5.9) is also said to be the *full model* of an elastic joint manipulator.

Viscous friction terms acting both on the link and on the motor sides of the elastic joints could be easily included in the dynamic model.

5.1.1 Dynamic model properties

Referring to the general dynamic model (5.9), the following useful properties can be derived, some of which are already present for the rigid manipulator model.

Property 5.1 The elements of $C(q,\dot{q})$ can always be defined so that the matrix $\dot{H} - 2C$ is *skew-symmetric*. In particular, one such feasible choice is provided by the Christoffel symbols, i.e.,

$$C_{ij}(q,\dot{q}) = \frac{1}{2}\left(\frac{\partial H_{ij}}{\partial q}\dot{q} + \sum_{k=1}^{2n}\left(\frac{\partial H_{ik}}{\partial q_j} - \frac{\partial H_{jk}}{\partial q_i}\right)\dot{q}_k\right), \qquad (5.12)$$

for $i,j = 1, \ldots, 2n$.

□

5.1. MODELLING

Property 5.2 If $C(q,\dot{q})$ is defined by (5.12), then it can be decomposed as

$$C(q,\dot{q}) = C_A(q_1,\dot{q}_2) + C_B(q_1,\dot{q}_1), \tag{5.13}$$

with

$$C_A(q_1,\dot{q}_2) = \begin{pmatrix} C_{A1}(q_1,\dot{q}_2) & 0 \\ 0 & 0 \end{pmatrix} \tag{5.14}$$

$$C_B(q_1,\dot{q}_1) = \begin{pmatrix} C_{B1}(q_1,\dot{q}_1) & C_{B2}(q_1,\dot{q}_1) \\ C_{B3}(q_1,\dot{q}_1) & 0 \end{pmatrix} \tag{5.15}$$

where the elements of the $(n \times n)$ matrices C_{A1}, C_{B1}, C_{B2}, C_{B3} are:

$$C_{A1ij}(q_1,\dot{q}_2) = \frac{1}{2}\left(\frac{\partial (H_2)_i}{\partial q_{1j}} - \frac{\partial (H_2)_j}{\partial q_{1i}}\right)\dot{q}_2 \tag{5.16}$$

$$C_{B1ij}(q_1,\dot{q}_1) = \frac{1}{2}\left(\frac{\partial H_{1ij}}{\partial q_1}\dot{q}_1 + \left(\frac{\partial (H_1)_i}{\partial q_{1j}} - \frac{\partial (H_1)_j}{\partial q_{1i}}\right)\dot{q}_1\right) \tag{5.17}$$

$$C_{B2ij}(q_1,\dot{q}_1) = \frac{1}{2}\left(\frac{\partial H_{2ij}}{\partial q_1}\dot{q}_1 - \frac{\partial (H_2^T)_j}{\partial q_{1i}}\dot{q}_1\right) \tag{5.18}$$

$$C_{B3ij}(q_1,\dot{q}_1) = \frac{1}{2}\left(\frac{\partial H_{2ji}}{\partial q_1}\dot{q}_1 + \frac{\partial (H_2^T)_i}{\partial q_{1j}}\dot{q}_1\right) \tag{5.19}$$

with $(H)_i$ denoting the i-th *row* of a matrix H. These expressions follow directly from the dependency of the inertia matrix (5.2) and from Property 5.1.

□

Property 5.3 Matrix $H_2(q_1)$ has the upper triangular structure

$$\begin{pmatrix} 0 & H_{212}(q_{11}) & H_{213}(q_{11},q_{12}) & \cdots & H_{21n}(q_{11},\ldots,q_{1,n-1}) \\ 0 & 0 & H_{223}(q_{12}) & \cdots & H_{22n}(q_{12},\ldots,q_{1,n-1}) \\ \vdots & \vdots & \vdots & \ddots & \vdots \\ 0 & 0 & 0 & \cdots & H_{2,n-1,n}(q_{1,n-1}) \\ 0 & 0 & 0 & \cdots & 0 \end{pmatrix}, \tag{5.20}$$

where the most general cascade dependence is shown for each single term. Indeed, the elements of H_2 can be obtained as

$$H_{2ij} = \frac{\partial^2 T}{\partial \dot{q}_{1j}\partial \dot{q}_{2i}}, \tag{5.21}$$

where the kinetic energy T is given by the sum of the kinetic energy of each link (including the stator of the successive motor) and of each motor rotor.

However, since the total kinetic energy of the links is a quadratic form of $\dot{q}_1$ only, by virtue of the chosen variable definition, contributions to H_2 may only come from that part of T which is due to the rotors. For rotor i, the kinetic energy is given by

$$T_{ri} = \frac{1}{2} m_{ri}{}^{r_i} v_{2i}^T {}^{r_i} v_{2i} + \frac{1}{2} {}^{r_i}\omega_{2i}^T {}^{r_i} I_{ri} {}^{r_i}\omega_{2i} \tag{5.22}$$

where ${}^{r_i}v_{2i}$ and ${}^{r_i}\omega_{2i}$ are respectively the linear and angular velocity of the rotor expressed in the frame r_i attached to the corresponding stator, while m_{ri} and ${}^{r_i}I_{ri}$ are respectively the mass and the inertia tensor of the rotor. Since the rotor center of mass lies on its axis of rotation, only the second term on the right-hand side of (5.22) will contribute to H_2. The angular velocity ${}^{r_i}\omega_{2i}$ can be calculated recursively as (for revolute joints)

$$\begin{aligned}{}^{r_i}\omega_{2i} &= {}^{r_i}R_{i-1}{}^{i-1}\omega_{1,i-1} + \dot{q}_{2i}{}^{r_i}a_{2i} \\ {}^{i}\omega_{1i} &= {}^{i}R_{i-1}(q_{1i})^{i-1}\omega_{1,i-1} + \dot{q}_{1i}{}^{i}a_{1i}\end{aligned} \tag{5.23}$$

where ${}^{i}\omega_{1i}$ is the angular velocity of link i in the frame i attached to the link itself, ${}^{r_i}R_{i-1}$ is the constant (3×3) rotation matrix from frame r_i attached to the rotor to frame $i-1$, ${}^{r_i}a_{2i} = (\,0\ \ 0\ \ 1\,)^T$, ${}^{i}R_{i-1}$ is the (3×3) rotation matrix from frame i to frame $i-1$, and ${}^{i}a_{1i} = (\,0\ \ 0\ \ 1\,)^T$. Eqs. (5.22) and (5.23) imply (5.20).

□

Property 5.4 A positive constant α exists such that

$$\left\| \frac{\partial g_1(q_1)}{\partial q_1} \right\| \leq \alpha \qquad \forall q_1. \tag{5.24}$$

This property follows from the fact that $g_1(q_1)$ is formed by trigonometric functions of the link variables q_{1i} in the case of revolute joints, and also by linear functions in q_{1i} if some prismatic joint is present. The previous inequality implies, by the mean value theorem, that

$$\|g_1(q_1) - g_1(q_1')\| \leq \alpha \|q_1 - q_1'\| \qquad \forall q_1, q_1'. \tag{5.25}$$

□

5.1.2 Reduced models

In many common manipulator kinematic arrangements, the block H_2 in the inertia matrix of the elastic joint model (5.9) is *constant*. For instance, this occurs in the case of a two-revolute-joint planar arm or of a three-revolute

5.1. MODELLING

joint anthropomorphic manipulator. This implies several simplifications in the dynamic model, due to vanishing of terms. In particular,

$$H_2 = \text{const} \quad \Longrightarrow \quad C_{A1} = C_{B2} = C_{B3} = 0 \tag{5.26}$$

so that Coriolis and centrifugal terms, which are always independent of q_2, become also independent of $\dot{q}_2$. As a result of (5.26), model (5.9) can be rewritten in partitioned form as

$$H_1(q_1)\ddot{q}_1 + H_2\ddot{q}_2 + C_1(q_1,\dot{q}_1)\dot{q}_1 + K(q_1 - q_2) + g_1(q_1) = 0$$
$$H_2^T\ddot{q}_1 + H_3\ddot{q}_2 + K(q_2 - q_1) = u, \tag{5.27}$$

where $C_1 = C_{B1}$ for compactness. Note that no velocity terms appear in the second set of n equations, the one associated with the motor variables.

For some special kinematic structures it is found that $H_2 = 0$ and further simplifications are induced; this is the case of a single elastic joint, and of a 2-revolute-joint polar arm, i.e., with orthogonal joint axes. As a consequence, no inertial couplings are present between the link and motor dynamics, i.e.,

$$H_1(q_1)\ddot{q}_1 + C_1(q_1,\dot{q}_1)\dot{q}_1 + K(q_1 - q_2) + g_1(q_1) = 0$$
$$H_3\ddot{q}_2 + K(q_2 - q_1) = u. \tag{5.28}$$

For general elastic joint manipulators, a *reduced model* of the form (5.28) can also be obtained by neglecting some contributions in the energy of the system. In particular, H_2 will be *forced to zero* if the angular part of the kinetic energy of each rotor is considered to be due only to its own rotation, i.e., $\omega_{2i} = \dot{q}_{2i}{}^{r_i}a_{2i}$ —compare with (5.23)— or

$$T_{ri} = \frac{1}{2}m_{ri}{}^{r_i}v_{2i}^{T}{}^{r_i}v_{2i} + \frac{1}{2}I_{mi}\dot{q}_{2i}^2 \tag{5.29}$$

with the positive scalar $I_{mi} = {}^{r_i}I_{rizz}$. When the gear reduction ratios are very large, this approximation is quite reasonable since the fast spinning of each rotor dominates the angular velocity of the previous carrying links.

We note that the full model (5.9) (or (5.27)) and the reduced model (5.28) display different characteristics with respect to certain control problems. As will be discussed later, while the reduced model is always feedback linearizable by *static* state feedback, the full model needs in general *dynamic* state feedback for achieving the same result.

5.1.3 Singularly perturbed model

A different modelling approach can be pursued, which is convenient for designing simplified control laws. When the joint stiffness is large, the system naturally exhibits a *two-time scale* dynamic behaviour in terms of rigid and elastic variables. This can be made explicit by properly transforming the $2n$ differential equations of motion in terms of the new generalized coordinates

$$\begin{pmatrix} q_1 \\ z \end{pmatrix} = \begin{pmatrix} I & 0 \\ -K & K \end{pmatrix} \begin{pmatrix} q_1 \\ q_2 \end{pmatrix} = \begin{pmatrix} q_1 \\ K(q_2 - q_1) \end{pmatrix}, \quad (5.30)$$

where the i-th component of z is the *elastic force* transmitted through joint i to the driven link.

For simplicity, only the model (5.27) is considered next. Solving the second equation in (5.27) for the motor acceleration and using (5.30) gives

$$\ddot{q}_2 = H_3^{-1}(u - z - H_2^T \ddot{q}_1). \quad (5.31)$$

Substituting (5.31) into the first equation in (5.27) yields

$$\Delta(q_1)\ddot{q}_1 + C_1(q_1, \dot{q}_1)\dot{q}_1 + g_1(q_1) - (I + H_2 H_3^{-1})z + H_2 H_3^{-1} u = 0, \quad (5.32)$$

where $\Delta(q_1) = H_1(q_1) - H_2 H_3^{-1} H_2^T$ is a block appearing on the diagonal of the inverse of inertia matrix, and thus is positive definite for all q_1. Combining (5.32) and (5.31) yields

$$\begin{aligned}
\ddot{z} &= K(\ddot{q}_2 - \ddot{q}_1) \\
&= K(I + H_3^{-1} H_2^T)\Delta^{-1}(q_1)(C_1(q_1, \dot{q}_1)\dot{q}_1 + g_1(q_1)) \\
&\quad - K\bigl((I + H_3^{-1} H_2^T)\Delta^{-1}(q_1)(I + H_2 H_3^{-1}) + H_3^{-1}\bigr)z \\
&\quad + K\bigl((I + H_3^{-1} H_2^T)\Delta^{-1}(q_1)H_2 H_3^{-1} + H_3^{-1}\bigr)u. \quad (5.33)
\end{aligned}$$

Notice that the matrix premultiplying z in (5.33) is always invertible, being the sum of a positive definite and a positive semi-definite matrix.

Since it is assumed that the diagonal matrix K has all large and similar elements, we can extract a large common scalar factor $1/\epsilon^2 \gg 1$ from K

$$K = \frac{1}{\epsilon^2}\hat{K} = \frac{1}{\epsilon^2}\text{diag}\{\hat{k}_1, \ldots, \hat{k}_n\}. \quad (5.34)$$

Then, eq. (5.33) can be compactly rewritten as

$$\epsilon^2 \ddot{z} = \hat{K} a_1(q_1, \dot{q}_1) + \hat{K} A_2(q_1)z + \hat{K} A_3(q_1)u. \quad (5.35)$$

Eqs. (5.32) and (5.35) take on the usual form of a *singularly perturbed* dynamic system once the fast time variable $\tau = t/\epsilon$ is introduced in (5.35), i.e.,

$$\epsilon^2 \ddot{z} = \epsilon^2 \frac{d^2 z}{dt^2} = \frac{d^2 z}{d\tau^2}. \quad (5.36)$$

Eq. (5.32) characterizes the *slow* dynamics of the rigid robot manipulator, while (5.35) describes the *fast* dynamics associated with the elastic joints.

We note that (5.32) and (5.35) are considerably simplified when the reduced model with $H_2 = 0$ is used. In that case, they become:

$$H_1(q_1)\ddot{q}_1 + C_1(q_1,\dot{q}_1)\dot{q}_1 + g_1(q_1) = z$$
$$\epsilon^2 \ddot{z} = \hat{K} H_1^{-1}(q_1)\big(C_1(q_1,\dot{q}_1)\dot{q}_1 + g_1(q_1)\big)$$
$$- \hat{K}\big(H_1^{-1}(q_1) + H_3^{-1}\big)z + \hat{K} H_3^{-1} u. \tag{5.37}$$

5.2 Regulation

We analyze first the problem of controlling the position of the end effector of a robot manipulator with joint elasticity in simple point-to-point tasks. As shown in the modelling section, this corresponds to regulation of the link variables q_1 to a desired *constant* value q_{1d}, achieved using control inputs u applied to the motor side of the elastic joints. A major aspect of the presence of joint elasticity is that the feedback part of the control law may depend in general on *four* variables for each joint; namely, the motor and link position, and the motor and link velocity. However, in the most common robot manipulator configurations only *two* sensors are available for joint measurements. We will study a single elastic joint with no gravity (leading to a linear model) to point out what are the control possibilities and the drawbacks in this situation. This provides some indications on how to handle the general multilink case in presence of gravity. In particular, it will be shown that a PD controller on the motor variables and a constant gravity compensation are sufficient to ensure global asymptotic stabilization of any manipulator configuration.

5.2.1 Single link

Consider a single link rotating on a horizontal plane and actuated with a motor through an elastic joint coupling. For the sake of simplicity, all friction or damping effects are neglected. Let ϑ_m and ϑ_ℓ be the motor and link angular positions, respectively. Then, the dynamic equations are

$$I_\ell \ddot{\vartheta}_\ell + k(\vartheta_\ell - \vartheta_m) = 0$$
$$I_m \ddot{\vartheta}_m + k(\vartheta_m - \vartheta_\ell) = u, \tag{5.38}$$

where I_m and I_ℓ are the motor and the link inertia about the rotation axis, and k is the joint stiffness. Assuming $y = \vartheta_\ell$ as system output, the open-loop transfer function is

$$\frac{y(s)}{u(s)} = \frac{k}{I_m I_\ell s^2 + (I_m + I_\ell)k} \frac{1}{s^2}, \qquad (5.39)$$

which has all poles on the imaginary axis. Note, however, that no zeros appear in (5.39).

In the following, one position variable and one velocity variable will be used for designing a linear stabilizing feedback. Since the desired position is given in terms of the link variable ($q_{1d} = \vartheta_{\ell d}$), the most natural choice is a feedback from the *link variables*

$$u = v_1 - (k_{P\ell}\vartheta_\ell + k_{D\ell}\dot{\vartheta}_\ell), \qquad (5.40)$$

where $k_{P\ell}, k_{D\ell} > 0$ and v_1 is the external input used for defining the set point. In this case, the closed-loop transfer function is

$$\frac{y(s)}{v_1(s)} = \frac{k}{I_m I_\ell s^4 + (I_m + I_\ell)ks^2 + kk_{D\ell}s + kk_{P\ell}}. \qquad (5.41)$$

No matter how the gain values are chosen, the system is still unstable due to the vanishing coefficient of s^3 in the denominator of (5.41). Indeed, if some viscous friction or spring damping were present, there would exist a small interval for the two positive gains $k_{P\ell}$ and $k_{D\ell}$ which guarantees closed-loop stability; however, in that case, the obtained performance would be very poor.

Another possibility is offered by a full feedback from the *motor variables*

$$u = v_2 - (k_{Pm}\vartheta_m + k_{Dm}\dot{\vartheta}_m), \qquad (5.42)$$

leading to the transfer function

$$\frac{y(s)}{v_2(s)} = \frac{k}{I_m I_\ell s^4 + I_\ell k_{Dm} s^3 + \left(I_\ell(k + k_{Pm}) + I_m k\right)s^2 + kk_{Dm}s + kk_{Pm}}. \qquad (5.43)$$

It is easy to see that strictly positive values for both k_{Pm} and k_{Dm} are necessary and sufficient for closed-loop stability. Notice that, in the absence of gravity, the equilibrium position ϑ_{md} for the motor variable coincides with the desired link position $\vartheta_{\ell d}$, and thus the reference value is $v_2 = k_{Pm}\vartheta_{\ell d}$. However, this is no longer true when gravity is present, deflecting the joint at steady state; in that case, the value of v_2 has to be computed using also the model parameters.

A third feedback strategy is to use the *motor velocity* and the *link position*

$$u = v_3 - (k_{P\ell}\vartheta_\ell + k_{Dm}\dot{\vartheta}_m). \qquad (5.44)$$

This combination is rather convenient since it corresponds to what is actually measured in a robotic drive, when a tachometer is mounted on the DC motor and an optical encoder senses position on the load shaft, without any knowledge about the relevance of joint elasticity. Use of (5.44) leads to

$$\frac{y(s)}{v_3(s)} = \frac{k}{I_m I_\ell s^4 + I_\ell k_{Dm} s^3 + (I_\ell + I_m) k s^2 + k k_{Dm} s + k k_{P\ell}} \quad (5.45)$$

which differs in practice from (5.43) only for the coefficient of the quadratic term in the denominator. Using Routh's criterion, asymptotic stability occurs if and only if the feedback gains are chosen as

$$0 < k_{P\ell} < k \qquad 0 < k_{Dm}. \quad (5.46)$$

Hence, the proportional feedback on the link variable should not "override" the spring stiffness. Even for a set-point task with gravity, there is no need to transform the desired reference with this scheme, since the link position error is directly available and the steady-state motor velocity is zero anyway; hence, it is $v_3 = k_{P\ell} \vartheta_{\ell d}$.

Following the same lines, it is immediate to see that the combination of motor position and link velocity feedback is always unstable. Note also that other combinations would be possible, depending on the available sensing devices. For instance, mounting a strain gauge on the transmission shaft provides a direct measure of the elastic force $z = k(\vartheta_m - \vartheta_\ell)$ for control use.

To summarize, the use of alternate output measures may be a critical issue in the presence of joint elasticity. Conversely, a full state feedback may certainly guarantee asymptotic stability; however, this would be obtained at the cost of additional sensors and would require a proper tuning of the four gains. The previous developments were presented for set-point regulation, but similar considerations apply also to tracking control. These simple facts should be carefully kept in mind when moving from this canonical linear example to the more complex nonlinear dynamics of articulated manipulators.

5.2.2 PD control using only motor variables

Following the results of the previous section, we focus here our attention on general multilink robot manipulators with elastic joints modelled by (5.9), i.e., with $H_2(q_1) \neq 0$. It has been shown that feeding back the motor position and velocity guarantees asymptotic stability for a single link with elastic joint and no gravity. Moving to robot manipulators under the action of gravity imposes some caution in the selection of the control gains. However, it can be shown that a simple PD controller with constant gravity compensation globally stabilizes any desired link reference position q_{1d}.

Theorem 5.1 *Consider the control law*

$$u = K_P(q_{2d} - q_2) - K_D \dot{q}_2 + g_1(q_{1d}) \qquad (5.47)$$

where K_P and K_D are $(n \times n)$ symmetric positive definite matrices and the motor reference position q_{2d} is chosen as

$$q_{2d} = q_{1d} + K^{-1} g_1(q_{1d}). \qquad (5.48)$$

If

$$\lambda_{\min}(K_q) = \lambda_{\min} \begin{pmatrix} K & -K \\ -K & K + K_P \end{pmatrix} > \alpha, \qquad (5.49)$$

with α as defined by Property 5.4, then

$$q_1 = q_{1d} \qquad q_2 = q_{2d} \qquad \dot{q} = 0$$

is a globally asymptotically stable equilibrium point for the closed-loop system (5.9) and (5.47).

⋄ ⋄ ⋄

Proof. The equilibrium positions of (5.9) and (5.47) are the solutions to

$$K(q_1 - q_2) + g_1(q_1) = 0 \qquad (5.50)$$
$$K(q_1 - q_2) - K_P(q_2 - q_{2d}) + g_1(q_{1d}) = 0. \qquad (5.51)$$

By recalling (5.25), we can add the null term $K(q_{2d} - q_{1d}) - g_1(q_{1d})$ to (5.50) and (5.51) leading to

$$K(q_1 - q_{1d}) - K(q_2 - q_{2d}) + g_1(q_1) - g_1(q_{1d}) = 0 \qquad (5.52)$$
$$K(q_1 - q_{1d}) - (K + K_P)(q_2 - q_{2d}) = 0, \qquad (5.53)$$

which can be rewritten in matrix form as

$$K_q(q - q_d) = g(q_{1d}) - g(q_1), \qquad (5.54)$$

where $q_d = (q_{1d}, q_{2d})$. The inequality (5.25) enables us to write, $\forall q \neq q_d$,

$$\|K_q(q - q_d)\| \geq \lambda_{\min}(K_q)\|q - q_d\| > \alpha\|q - q_d\| \geq \|g(q_{1d}) - g(q_1)\|. \qquad (5.55)$$

Hence, (5.54) has the unique solution $q = q_d$.

5.2. REGULATION

Define the position-dependent function

$$P_1(q) = \frac{1}{2}(q - q_d)^T K_q (q - q_d) + U_g(q_1) - q^T g(q_{1d}). \tag{5.56}$$

The stationary points of $P_1(q)$ are given by the solutions to

$$\left(\frac{\partial P_1(q)}{\partial q}\right)^T = 0 \tag{5.57}$$

which coincides with (5.54). Therefore, $P_1(q)$ has the unique stationary point $q = q_d$. Moreover,

$$\frac{\partial^2 P_1(q)}{\partial q^2} = K_q + \frac{\partial g(q_1)}{\partial q}. \tag{5.58}$$

By virtue of Property 5.4 and of (5.49), (5.58) is positive definite and thus $q = q_d$ is an absolute minimum for $P_1(q)$.

Consider now the Lyapunov function candidate

$$V(q, \dot{q}) = \frac{1}{2}\dot{q}^T H(q_1)\dot{q} + P_1(q) - P_1(q_d) \tag{5.59}$$

that is positive definite with respect to $q = q_d$, $\dot{q} = 0$. The time derivative of (5.59) along a closed-loop trajectory is given by

$$\dot{V}(q, \dot{q}) = \frac{1}{2}\dot{q}^T \dot{H}(q_1)\dot{q} - \dot{q}^T \left(C(q, \dot{q})\dot{q} + Kq + g(q_1)\right)$$
$$- \dot{q}_2^T K_P (q_2 - q_{2d}) - \dot{q}_2^T K_D \dot{q}_2 + \dot{q}_2^T g_1(q_{1d})$$
$$+ \dot{q}^T K_q (q - q_d) + \left(\frac{\partial U_g(q_1)}{\partial q}\right)^T \dot{q} - \dot{q}^T g(q_{1d}). \tag{5.60}$$

By recalling Property 5.1, (5.60) reduces to

$$\dot{V}(q, \dot{q}) = -\dot{q}_2^T K_D \dot{q}_2 + (\dot{q}_1 - \dot{q}_2)^T \left(K(q_{2d} - q_{1d}) - g_1(q_{1d})\right) \tag{5.61}$$

which, in turn, by virtue of (5.48) becomes

$$\dot{V}(q, \dot{q}) = -\dot{q}_2^T K_D \dot{q}_2. \tag{5.62}$$

Therefore, $\dot{V}$ is negative semi-definite and vanishes if and only if $\dot{q}_2 = 0$. Imposing this condition in (5.9) for all times and recalling the structure of terms from Property 5.2, we get

$$H_1(q_1)\ddot{q}_1 + C_{B1}(q_1, \dot{q}_1)\dot{q}_1 + Kq_1 + g_1(q_1) = Kq_2 = \text{const} \tag{5.63}$$

and

$$H_2^T(q_1)\ddot{q}_1 + C_{B3}(q_1,\dot{q}_1)\dot{q}_1 - Kq_1 = -Kq_2 - K_P(q_2 - q_{2d}) + g_1(q_{1d})$$
$$= \text{const.} \quad (5.64)$$

By taking (5.20) and (5.19) into account, the first scalar equation in (5.64) becomes

$$q_{11} = \text{const.} \quad (5.65)$$

Substitution of (5.65) into the second scalar equation in (5.64) yields $q_{12} = \text{const}$. Proceeding in the same way, we finally obtain

$$q_1 = \text{const.} \quad (5.66)$$

This, substituted into (5.63) and (5.64), leads to

$$K(q_1 - q_2) + g_1(q_1) = 0$$
$$K(q_2 - q_1) + K_P(q_2 - q_{2d}) - g_1(q_{1d}) = 0. \quad (5.67)$$

Since, as previously shown, (5.67) has the unique solution $q = q_d$, then $q = q_d$, $\dot{q} = 0$ is the largest invariant subset in the set $\dot{V} = 0$. The thesis is proved by applying La Salle's theorem.

◇

Remarks

- The assumption $\lambda_{\min}(K_q) > \alpha$ in the above theorem is not restrictive; in fact, joint stiffness K dominates gravity so that, by increasing the smallest eigenvalue of K_P, inequality (5.49) can always be satisfied.

- The PD control law (5.47) is robust with respect to some model uncertainty. In particular, asymptotic stability is guaranteed even though the inertial parameters of the manipulator are not known. Conversely, uncertainty on the gravitational and elastic parameters may affect the performance of the controller since these terms appear explicitly in the control law (see also (5.48)). However, it can be shown that the PD controller is still stable subject to uncertainty on these parameters, but the equilibrium point of the closed-loop system is, in general, different from the desired one. If $\hat{g}_1(q_1)$ and $\hat{K}$ are the available estimates of the gravity vector and of the stiffness matrix, then the control law

$$u = K_P(\hat{q}_{2d} - q_2) - K_D \dot{q}_2 + \hat{g}_1(q_{1d}) \quad (5.68)$$

with
$$\widehat{\bar{q}}_{2d} = q_{1d} + \widehat{K}^{-1}\widehat{g}_1(q_{1d}) \qquad (5.69)$$

asymptotically stabilizes the equilibrium point $q = \bar{q}_d$, $\dot{q} = 0$, where $\bar{q}_d$ is the solution to the steady-state equation

$$K_e(\bar{q}_d - q_d) = \begin{pmatrix} g_1(q_{1d}) - g_1(\bar{q}_{1d}) \\ K_P(\widehat{\bar{q}}_{2d} - q_{2d}) + \widehat{g}_1(q_{1d}) - g_1(q_{1d}) \end{pmatrix}. \qquad (5.70)$$

This solution is unique provided that $\lambda_{\min}(K_q) > \alpha$, as before. It is apparent from (5.70) that the better is the model estimate, the closer $\bar{q}_d$ will be to the desired q_d.

- Since uncertainty on gravity and elastic terms affect directly the reference value for the motor variables, inclusion of an integral term in the control law (5.68) is not useful for recovering regulation at the desired set point q_d. In order for the integral term to be effective, the proportional as well as the integral parts of the PID controller should be driven by the link error $q_{1d} - q_1$. However, as shown in the simple one-joint linear case, velocity should be still fed back at the motor level in order to prevent unstable behaviour, leading to

$$u = K_P(q_{1d} - q_1) - K_D \dot{q}_2 + K_I \int_0^t (q_{1d} - q_1) d\tau + \widehat{g}_1(q_{1d}). \qquad (5.71)$$

The asymptotic stability of such a controller has not yet been proved. In particular, the choice of the integral matrix gain K_I is a critical one.

5.3 Tracking control

As for rigid robot manipulators, also for manipulators with elastic joints the problem of tracking link (end-effector) trajectories is harder than achieving constant regulation. Nonlinear state feedback control may be useful in order to transform the closed-loop system into an equivalent linear and decoupled one for which the tracking task is easily accomplished. However, the application of this inverse dynamics control strategy is not straightforward in the presence of joint elasticity. Furthermore, it can be shown that the general dynamic model (5.9) may satisfy neither the necessary conditions for feedback linearization nor those for input–output decoupling.

On the other hand, the reduced model (5.28) (with $H_2 = 0$) is more tractable from this point of view, and it always allows exact linearization via *static state feedback*. Therefore, we will use the reduced model to illustrate this nonlinear control approach to the trajectory tracking problem. The same reduced model, in its format (5.37), will also be used to present a *two-time scale control* approach. In particular, the design of this approximate nonlinear control is fully carried out for a one-link arm with joint elasticity.

Two further control strategies will be presented with reference to the complete model. The use of a larger class of control laws, based on *dynamic state feedback*, allows us to recover exact linearization and decoupling results in general. As a preliminary step, it will be shown that the robot manipulator system (5.9), with the link position taken as output, displays *no zero dynamics*. In a nonlinear setting, this is equivalent to state that no internal motion is possible when the input is chosen so as to constrain the output to be constantly zero.

The second control approach is a simpler one, making use only of a feedforward command plus linear feedback from the full state. In this case, convergence to the desired trajectory is only locally guaranteed, i.e., the initial error should be small enough. This technique is referred to as *nonlinear regulation*.

5.3.1 Static state feedback

Consider the reduced model (5.28) and define as system output the link position vector

$$y = q_1, \quad (5.72)$$

i.e., the variables to be controlled for tracking purposes.

We will show next that the robot manipulator system with output (5.72) can be input–output decoupled with the use of a nonlinear static state feedback. Moreover, the same decoupling control law will automatically linearize the closed-loop system equations. Also, the coordinate transformation needed to display this linearity is provided as a byproduct of the same approach. The decoupling algorithm requires us to differentiate each output component y_i until the input torque u appears explicitly. The control law is then computed from the last set of obtained differential equations, under proper conditions. This procedure does not require transforming the manipulator dynamic model into the usual state space form, although it is completely equivalent.

By taking the first time derivative of the output

$$\dot{y} = \dot{q}_1 \quad (5.73)$$

and the second one

$$\ddot{y} = \ddot{q}_1 = -H_1^{-1}(C_1\dot{q}_1 + K(q_1 - q_2) + g_1), \quad (5.74)$$

5.3. TRACKING CONTROL

it is immediate to see that the link acceleration is not instantaneously dependent of the applied motor torque u. In (5.74), model dependence has been dropped for compactness. Proceeding further, we have

$$y^{(3)} = \frac{d^3 q_1}{dt^3} = -(\dot{H_1^{-1}})(C_1 \dot{q}_1 + K(q_1 - q_2) + g_1)$$
$$- H_1^{-1}(\dot{C}_1 \dot{q}_1 + C_1 \ddot{q}_1 + K(\dot{q}_1 - \dot{q}_2) + \dot{g}_1). \quad (5.75)$$

The right-hand side depends twice on the link acceleration, once directly and once through $\dot{C}_1$. By using (5.74), the link jerk can be rewritten as

$$y^{(3)} = a_3(q_1, \dot{q}_1, q_2) + H_1^{-1} K \dot{q}_2, \quad (5.76)$$

where

$$a_3 = H_1^{-1} \left(3 C_1 - \sum_{i=1}^{n} \frac{\partial c_{1i}}{\partial \dot{q}_1} \dot{q}_{1i} \right) H_1^{-1} \left(C_1 \dot{q}_1 + K(q_1 - q_2) + g_1 \right)$$
$$- H_1^{-1} \left(\sum_{i=1}^{n} \frac{\partial c_{1i}}{\partial q_1} \dot{q}_{1i} \dot{q}_1 + K \dot{q}_1 + \dot{g}_1 \right), \quad (5.77)$$

with c_{1i} denoting the i-th column of matrix C_1. Next, the fourth derivative of the output gives

$$y^{(4)} = \frac{d^4 q_1}{dt^4} = \dot{a}_3 + (\dot{H_1^{-1}}) K \dot{q}_2 + H_1^{-1} K \ddot{q}_2. \quad (5.78)$$

Substituting $\ddot{q}_2$ from the model, differentiating (5.77) with respect to time, and using again (5.74), yields finally

$$y^{(4)} = a_4(q_1, q_2, \dot{q}_1, \dot{q}_2) + H_1^{-1} K H_3^{-1} u \quad (5.79)$$

with

$$a_4 = \dot{a}_3 - H_1^{-1} \left(\dot{H}_1 H_1^{-1} K \dot{q}_2 - K H_3^{-1} K (q_1 - q_2) \right). \quad (5.80)$$

Since the matrix premultiplying u is always *nonsingular*, we can set $y^{(4)} = u_0$ (the external control input) in (5.79) and solve for the feedback control u as

$$u = H_3 K^{-1} H_1(q_1) \left(u_0 - a_4(q_1, q_2, \dot{q}_1, \dot{q}_2) \right). \quad (5.81)$$

The matrix $H_1^{-1} K H_3^{-1}$ premultiplying u in (5.79) is the so-called *decoupling matrix* of the system. Moreover, the *relative degree* r_i of output y_i is equal to 4, uniformly for all outputs. Thus, the sum of all relative degrees equals the state space dimension, i.e., $\sum_{i=1}^{n} r_i = 4n$, which is a sufficient condition for obtaining full linearization, both for the input–output and the state equations. This is obtained by using the same static state feedback decoupling control (5.81). The coordinate transformation which, after the application of (5.81), displays linearity is defined by (5.72) through (5.75). This global diffeomorphism has the inverse transformation given by

$$\begin{aligned} q_1 &= y \\ \dot{q}_1 &= \dot{y} \\ q_2 &= y + K^{-1}\big(H_1(y)\ddot{y} + C_1(y,\dot{y})\dot{y} + g_1(y)\big) \\ \dot{q}_2 &= \dot{y} + K^{-1}\big(H_1(y)y^{(3)} + \dot{H}_1(y)\ddot{y} + C_1(y,\dot{y})\ddot{y} \\ &\quad + \dot{C}_1(y,\dot{y})\dot{y} + \dot{g}_1(y)\big). \end{aligned} \qquad (5.82)$$

Notice that the linearizing coordinates are the link position, velocity, acceleration and jerk. However, in order to perform feedback linearization it is not needed to measure link acceleration and jerk since the control law (5.81) is completely defined in terms of the original states (including motor position and velocity).

By defining

$$z_1 = y \quad z_2 = \dot{y} \quad z_3 = \ddot{y} \quad z_4 = y^{(3)}, \qquad (5.83)$$

the transformed system is described by

$$\begin{aligned} \dot{z}_{1i} &= z_{2i} \\ \dot{z}_{2i} &= z_{3i} \\ \dot{z}_{3i} &= z_{4i} \qquad i = 1,\ldots,n, \\ \dot{z}_{4i} &= u_{0i} \\ y_i &= z_{1i} \end{aligned} \qquad (5.84)$$

that corresponds to n independent chains of 4 integrators. To complete a tracking controller for a desired trajectory $y_{di}(t)$ of joint i, we should design the new input u_{0i} as

$$u_{0i} = y_{di}^{(4)} + \sum_{j=0}^{3} \alpha_{ji}(y_{di}^{(j)} - z_{j+1,i}) \qquad (5.85)$$

where the scalar constants α_{ji}, $j = 0,\ldots,3$ are coefficients of a Hurwitz polynomial. Note that the control law (5.85) implicitly assumes that the reference trajectory is differentiable up to order four. If the model parameters are known and full state feedback is available, the control (5.81) and (5.85) guarantees trajectory tracking with exponentially decaying error. If the initial state $q_1(0)$, $q_2(0)$, $\dot{q}_1(0)$, $\dot{q}_2(0)$ is matched with the reference trajectory and its derivatives at time $t = 0$ —in this respect, equations (5.82) are to be used— exact reproduction of the reference trajectory is achieved.

5.3.2 Two-time scale control

A simpler strategy for trajectory tracking exploits the two-time scale nature of the flexible part and the rigid part of the dynamic equations. The use of this approach allows us to develop a *composite controller*, just by adding terms accounting for joint elasticity to any original control law designed for the rigid manipulator.

In the following, we consider only the reduced model in its singularly perturbed form (5.37). For simplicity, the control approach will be illustrated on a one-link elastic joint manipulator under gravity. All steps followed for this single-input case can be easily adapted to the general multi-input case, by replacing scalar terms with matrix expressions.

The dynamic equations of a robot manipulator having one revolute elastic joint and one link moving in the vertical plane are

$$I_\ell \ddot{q}_1 + mg\ell \sin q_1 + k(q_1 - q_2) = 0$$
$$I_m \ddot{q}_2 - k(q_1 - q_2) = u, \qquad (5.86)$$

where I_ℓ and I_m are the link and motor inertia, respectively, k is the joint stiffness, m is the link mass and ℓ is the distance of the link center of mass from the joint axis. By setting

$$z = k(q_2 - q_1) \qquad \epsilon^2 = \frac{1}{k}, \qquad (5.87)$$

the singularly perturbed model is written as

$$I_\ell \ddot{q}_1 + mg\ell \sin q_1 = z \qquad (5.88)$$

$$\epsilon^2 \ddot{z} = -\left(\frac{1}{I_\ell} + \frac{1}{I_m}\right) z + \frac{1}{I_\ell} mg\ell \sin q_1 + \frac{1}{I_m} u, \qquad (5.89)$$

with the link position q_1 as the slow variable and the joint elastic force z as the fast variable. Since the joint stiffness k is usually quite large, in the limit we can set $\epsilon = 0$ and obtain the approximate dynamic representation

$$I_\ell \ddot{q}_1 + mg\ell \sin q_1 = z \qquad (5.90)$$

$$0 = -\left(\frac{1}{I_\ell} + \frac{1}{I_m}\right) z + \frac{1}{I_\ell} mg\ell \sin q_1 + \frac{1}{I_m} u_s \qquad (5.91)$$

where $u_s = u|_{\epsilon=0}$. The first step in a singular perturbation approach requires solving (5.91) for z and substitute it into (5.90), so as to obtain a dynamic equation in terms of the slow variable only. Note that this can always be done when $u_s = 0$. When a nonzero control input is present in (5.91), its structure should still allow expressing (5.91) with respect to z.

To this purpose, it is sufficient to choose the dependence of the overall control input as

$$u = u_s(q_1, \dot{q}_1, t) + \epsilon u_f(z, \dot{z}, q_1, \dot{q}_1, t), \qquad (5.92)$$

where u_f does *not* contain terms of order $1/\epsilon$ or higher. Thus, a two-time scale control law is obtained which is composed of the *slow* part u_s, designed using only slow variables, and of the *fast* part ϵu_f (vanishing for $\epsilon = 0$) which counteracts the effects of joint elasticity.

Plugging (5.92) into (5.91), and solving for z gives

$$z = \frac{1}{I_\ell + I_m}(I_m mg\ell \sin q_1 + I_\ell u_s). \qquad (5.93)$$

This algebraic relation defines a control dependent manifold in the four-dimensional state space of the system. Substituting z in (5.90) yields the so-called *slow reduced system*

$$I_\ell \ddot{q}_1 + mg\ell \sin q_1 = \frac{1}{I_\ell + I_m}(I_m mg\ell \sin q_1 + I_\ell u_s) \qquad (5.94)$$

or else

$$(I_\ell + I_m)\ddot{q}_1 + mg\ell \sin q_1 = u_s, \qquad (5.95)$$

which is the equivalent rigid manipulator model. The synthesis of the slow control part is based only on this representation of the system. Given a desired trajectory $q_{1d}(t)$ for the link (the system output), a convenient choice could be an inverse dynamics control law

$$u_s = (I_\ell + I_m)u_{s0} + mg\ell \sin q_1, \qquad (5.96)$$

with the linear tracking part

$$u_{s0} = \ddot{q}_{1d} + k_D(\dot{q}_{1d} - \dot{q}_1) + k_P(q_{1d} - q_1). \qquad (5.97)$$

This is an exact feedback linearizing control law, performed only on the rigid equivalent model. Note that any other control strategy could be used for defining $u_s = u_s(q_1, \dot{q}_1, t)$ (time dependence is introduced through the reference trajectory), without affecting the subsequent steps.

Substituting the control structure (5.92) in the fast dynamics (5.89) yields

$$\epsilon^2 \ddot{z} = -\left(\frac{1}{I_\ell} + \frac{1}{I_m}\right)z + \frac{1}{I_\ell}mg\ell \sin q_1 + \frac{1}{I_m}u_s(q_1, \dot{q}_1, t)$$
$$+ \frac{1}{I_m}\epsilon u_f(q_1, \dot{q}_1, z, \dot{z}, t). \qquad (5.98)$$

5.3. TRACKING CONTROL

Due to the time scale separation, we can assume that slow variables are at steady state with respect to variations of the fast variable z and rewrite (5.98) as

$$\epsilon^2 \ddot{z} = -\left(\frac{1}{I_\ell} + \frac{1}{I_m}\right) z + \frac{1}{I_m}\epsilon u_f(q_1, \dot{q}_1, z, \dot{z}, t) + w_s(\hat{q}_1, \hat{\dot{q}}_1, \hat{t}), \qquad (5.99)$$

where

$$w_s(\hat{q}_1, \hat{\dot{q}}_1, \hat{t}) = \frac{1}{I_\ell} mg\ell \sin \hat{q}_1 + \frac{1}{I_m} u_s(\hat{q}_1, \hat{\dot{q}}_1, \hat{t}) \qquad (5.100)$$

and a hat characterizes steady-state values. Note that $\hat{t}$ stands for the slow nature of the reference trajectory for the link variable.

By comparing (5.100) with the expression of the manifold (5.93), we have

$$w_s(\hat{q}_1, \hat{\dot{q}}_1, \hat{t}) = \left(\frac{1}{I_\ell} + \frac{1}{I_m}\right) \hat{z}, \qquad (5.101)$$

with $\hat{z}$ as a parameter in the fast time scale. Defining $\zeta = z - \hat{z}$, the *fast error* dynamics becomes

$$\epsilon^2 \ddot{\zeta} = \left(\frac{1}{I_\ell} + \frac{1}{I_m}\right) \zeta + \frac{1}{I_m}\epsilon u_f. \qquad (5.102)$$

The fast control u_f should stabilize this linear error dynamics, which means that the fast variable z asymptotically converges to its *boundary layer* behaviour $\hat{z}$. A possible choice is

$$u_f = -k_f \dot{\zeta} = -k_f \dot{z} \qquad k_f > 0. \qquad (5.103)$$

This yields

$$\epsilon^2 \ddot{\zeta} + \frac{k_f}{I_m}\epsilon\dot{\zeta} + \left(\frac{1}{I_\ell} + \frac{1}{I_m}\right)\zeta = 0 \qquad (5.104)$$

or, by setting $\tau = t/\epsilon$ as the fast time scale,

$$\frac{d^2\zeta}{d\tau^2} + a\frac{d\zeta}{d\tau} + b\zeta = 0 \qquad a, b > 0, \qquad (5.105)$$

which is exponentially stable.

The final composite *two-time scale* control law is

$$u = u_s(q_1, \dot{q}_1, t) - \epsilon k_f \dot{z}. \qquad (5.106)$$

For example, using the inverse dynamics control law (5.96) and (5.97) as the slow controller, (5.106) becomes in the original link and motor variables

$$u = (I_m + I_\ell)(\ddot{q}_{1d} + k_D(\dot{q}_{1d} - \dot{q}_1) + k_P(q_{1d} - q_1))$$
$$+ mg\ell \sin q_1 - k_f \sqrt{k}(\dot{q}_2 - \dot{q}_1), \qquad (5.107)$$

since $\epsilon = 1/\sqrt{k}$. The fast control part is just a damping action on the relative motion of the motor and the link. In order to keep the time scale separation between the rigid and elastic dynamics, the gain k_f should be chosen so that $k_f \ll 1/\epsilon = \sqrt{k}$.

In the above analysis, the slow control part has been designed so as to suitably work for the case $\epsilon = 0$. Its action around the manifold (5.93) is only an approximate one. At the expense of a greater complexity, this approach can be improved by adding correcting terms in ϵ which expand the validity of the slow control also beyond $\epsilon = 0$, i.e.,

$$u_s = u_0 + \epsilon u_1 + \epsilon^2 u_2 + \ldots \quad (5.108)$$

where u_0 is the previously designed slow control term. For large values of k the correcting terms are small with respect to u_0. Associated with u_s in (5.108), a modified control dependent manifold can be defined which characterizes the slow behaviour, similarly to (5.93). It can be shown that a second-order expansion in (5.108) is enough to guarantee that this manifold becomes an *invariant* one; if the initial state is on this manifold, the control u_s will keep the system evolution within this manifold. In particular, this means that the robot manipulator will exactly track the desired link trajectory if the initial state is properly set. The fast control is then needed to counteract mismatched initial conditions and/or disturbances.

5.3.3 Dynamic state feedback

We turn now our attention to the general case of robot manipulators with elastic joints, described by the complete dynamic model (5.9). It will be shown next that input–output decoupling in this case is generically impossible using only static state feedback. We remind that the necessary and sufficient condition for this is that the system decoupling matrix is nonsingular. It is then convenient to rewrite the model in the following partitioned form, where dependence is dropped for compactness:

$$\begin{pmatrix} H_1 & H_2 \\ H_2^T & H_3 \end{pmatrix} \begin{pmatrix} \ddot{q}_1 \\ \ddot{q}_2 \end{pmatrix} + \begin{pmatrix} C_1 \dot{q} \\ C_2 \dot{q} \end{pmatrix} + \begin{pmatrix} g_1 \\ 0 \end{pmatrix} + \begin{pmatrix} K(q_1 - q_2) \\ -K(q_1 - q_2) \end{pmatrix} = \begin{pmatrix} 0 \\ u \end{pmatrix}. \quad (5.109)$$

Solving the second set of equations for $\ddot{q}_2$ and substituting it into the first set yields

$$(H_1 - H_2 H_3^{-1} H_2^T)\ddot{q}_1 + (C_1 - H_2 H_3^{-1} C_2)\dot{q}$$
$$+ (I + H_2 H_3^{-1})K(q_1 - q_2) + g_1 + H_2 H_3^{-1} u = 0. \quad (5.110)$$

If the link position is chosen as output

$$y = q_1, \quad (5.111)$$

5.3. TRACKING CONTROL

application of the decoupling algorithm requires, as before, differentiation of each component of (5.111) as many times as until the input explicitly appears. Using (5.110), it is immediate to show that after *two steps*, we obtain

$$\ddot{y} = \ddot{q}_1 = a_2(q, \dot{q}) - (H_1 - H_2 H_3^{-1} H_2^T)^{-1} H_2 H_3^{-1} u. \tag{5.112}$$

Provided that the matrix

$$A(q_1) = -(H_1 - H_2 H_3^{-1} H_2^T)^{-1} H_2 H_3^{-1} \tag{5.113}$$

has *at least one nonzero element for each row*, this will be exactly the decoupling matrix of the system. The first and last matrices on the right-hand side of (5.113) are $(n \times n)$ nonsingular matrices, being respectively the first diagonal block of the inverse of inertia matrix and the inverse of the second diagonal block of the inertia matrix. Thus, nonsingularity of the decoupling matrix depends only on H_2. However, this matrix is always singular since its structure is given by (5.20). As a consequence, input–output decoupling via static state feedback is impossible on the above assumption. Indeed, if one row of (5.113) is identically zero, the associated output component should be differentiated further in order to obtain an explicit dependence from the input u. Notice that for the reduced dynamic model, $H_2 \equiv 0$ implies that no input appears in the second time derivative of the output (see also (5.74)), and so the decoupling matrix will be completely different from (5.113).

Unfortunately, no general conclusion can be inferred on the rank of the decoupling matrix for the full dynamic model, because its structure strongly depends on the kinematic arrangement of the manipulator with elastic joints. For instance, the single elastic joint case and the 2-revolute-joint polar arm have a nonsingular decoupling matrix (both in fact have $H_2 \equiv 0$). The same considerations apply also to the case of prismatic elastic joints: the cylindric manipulator (prismatic-revolute-prismatic joints), with all joints being elastic, has a nonsingular decoupling matrix. On the other hand, common structures such as the two-revolute-joint planar arm, the 3-revolute-joint anthropomorphic manipulator, as well as manipulators with *mixed* rigid and elastic joints have a structurally singular decoupling matrix.

Similar arguments can be used for the analysis of the feedback linearization property (i.e., the existence of a static state feedback that transforms the closed-loop system into a linear one, not taking into account the output functions), which is also found to depend on the specific kinematic arrangement of the robot manipulator.

For both the input–output decoupling and the exact state linearization problems, a more general class of control laws can be considered. As a matter of fact, we may try to design a *dynamic state feedback* law of the form

$$u = \alpha(q, \dot{q}, \xi) + \beta(q, \dot{q}, \xi)u_0$$
$$\dot{\xi} = \gamma(q, \dot{q}, \xi) + \delta(q, \dot{q}, \xi)u_0 \qquad (5.114)$$

where the ($\nu \times 1$) vector ξ is the state of the dynamic compensator, α, β, γ, δ are suitable nonlinear vector functions, and the ($n \times 1$) vector u_0 (n is the dimension of the joint space) is the new external input used for trajectory tracking purposes. In the multi-input case, the conditions for obtaining noninteraction and/or exact linearization in the closed-loop system using (5.114) are indeed weaker than those based on static state feedback. Note that the latter is a special case of (5.114) for $\nu = 0$.

In particular, a sufficient condition for the existence of a linearizing and input–output decoupling dynamic controller is that the given system has *no zero dynamics*, so that no internal motion is compatible with the output being kept at a fixed (zero) value. Such an interpretation of zero dynamics allows us to generalize the concept of transfer function zeros of a linear system. In the following, we will show that the complete dynamic model (5.109) with output chosen as

$$y = q_1 - q_{10}, \qquad (5.115)$$

for any constant q_{10}, is a nonlinear system with no zero dynamics. As just said, this will be the nonlinear analogue of the fact that the transfer function (5.39) from motor torque to link position has no zeros. Imposing $y \equiv 0$ implies

$$q_1 = q_{10} \qquad \dot{q}_1 = 0 \qquad \ddot{q}_1 = 0 \qquad (5.116)$$

which, substituted into the first set of equations in (5.109), gives

$$H_2(q_{10})\ddot{q}_2 + K(q_{10} - q_2) + g_1(q_{10}) = 0, \qquad (5.117)$$

where the expressions (5.14) and (5.15) of the velocity terms have been used. Due to the strict upper triangular structure of matrix H_2, the set of n equations (5.117) can be analyzed starting from the last one which is

$$k_n(q_{10n} - q_{2n}) + g_{1n}(q_{10}) = 0, \qquad (5.118)$$

or

$$q_{2n} = q_{10n} + \frac{1}{k_n}g_{1n}(q_{10}) = \text{const.} \qquad (5.119)$$

Proceeding backward, all components of q_2 are found to be constant and equal to

$$q_2 = q_{10} + K^{-1}g_1(q_{10}) = q_{20} \qquad \dot{q}_2 = 0. \qquad (5.120)$$

5.3. TRACKING CONTROL

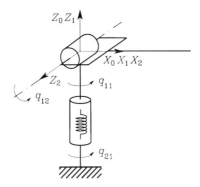

Figure 5.2: A 2-revolute-joint polar robot arm with the first joint being elastic.

Therefore, no internal motion is possible when the output is constantly zero. Notice also that the input needed for keeping this equilibrium condition is computed from the second set of equations in (5.109) as

$$u = K(q_{20} - q_{10}) = g_1(q_{10}). \quad (5.121)$$

As a result, all robot manipulators with elastic joints can be fully linearized and input–output decoupled provided that dynamic state feedback control is allowed.

Two-revolute-joint polar arm

In order to illustrate the synthesis of such a controller, we will use, as an example, one of the simplest structures where dynamic feedback is needed. Consider a 2-revolute-joint polar arm as in Fig. 5.2, displaying relevant elasticity only in the first joint, whose axis is vertical; the second joint, whose axis is horizontal, is assumed to be perfectly rigid. For simplicity, links with uniform mass distribution are considered.

Let q_{11} and q_{12} be the two link variables, defined through the Denavit–Hartenberg notation, and q_{21} be the first joint motor variable (beyond the reduction gearbox).

The total kinetic energy is the sum of the four contributions associated with the two motors and the two links (no additional coordinate is needed to describe the second motor position, due to the rigidity assumption for the second joint):

$$T_{m1} = \frac{1}{2} I_{m1zz} \dot{q}_{21}^2$$

$$T_{\ell 1} = \frac{1}{2} I_{\ell 1 zz} \dot{q}_{11}^2$$

$$T_{m2} = \frac{1}{2} I_{m2zz} \dot{q}_{12}^2 + \frac{1}{2} I_{m2xx} \dot{q}_{11}^2$$

$$T_{\ell 2} = \frac{1}{2} I_{\ell 2zz} \dot{q}_{12}^2 + \frac{1}{2} (I_{\ell 2xx} \sin^2 q_{12} + I_{\ell 2yy} \cos^2 q_{12}) \dot{q}_{11}^2$$

$$+ \frac{1}{2} m_2 r_{2x}^2 (\dot{q}_{12}^2 + \cos^2 q_{12} \dot{q}_{11}^2), \tag{5.122}$$

where m_2 is the mass of the second link, r_2 is the position vector of the center of mass of the second link expressed in its frame, and I_{mi} and $I_{\ell i}$ are respectively the constant rigid body inertia matrices of motor i and link i (diagonal in the associated frames). The inertia matrix is readily computed from the above expressions, resulting in a diagonal form; Coriolis and centrifugal terms are then obtained by proper differentiation. The total potential energy is the sum of the elastic energy of the first joint and of the gravitational energy of the second link:

$$U_{e1} = \frac{1}{2} k_1 (q_{21} - q_{11})^2$$

$$U_{g2} = g m_2 r_{2x} \sin q_{12}, \tag{5.123}$$

where k_1 is the elastic constant of the first joint. By introducing the following parameters:

$$\pi_1 = I_{\ell 1 yy} + I_{m2xx} + I_{\ell 2 zz}$$

$$\pi_2 = I_{\ell 2 yy} - I_{\ell 2 xx} + m_2 r_{2x}^2$$

$$\pi_3 = I_{m2zz} + I_{\ell 2 zz} + m_2 r_{2x}^2$$

$$\pi_4 = I_{m1zz}$$

$$\pi_5 = g m_2 r_{2x}, \tag{5.124}$$

the dynamic equations can be finally written as

$$(\pi_1 + \pi_2 \cos^2 q_{12}) \ddot{q}_{11} - 2\pi_2 \sin q_{12} \cos q_{12} \dot{q}_{11} \dot{q}_{12} + k_1 (q_{11} - q_{21}) = 0$$

$$\pi_3 \ddot{q}_{12} + \pi_2 \sin q_{12} \cos q_{12} \dot{q}_{11}^2 + \pi_5 \cos q_{12} = u_2$$

$$\pi_4 \ddot{q}_{21} + k_1 (q_{21} - q_{11}) = u_1. \tag{5.125}$$

Note that *three* second-order differential equations result, due to the mixed nature of rigid and elastic joints. Also, it should be mentioned that for the same reason the general model structure investigated in Section 5.1 cannot be directly applied to (5.125).

Choosing as output the link positions

$$y_1 = q_{11}$$

$$y_2 = q_{12}, \tag{5.126}$$

5.3. TRACKING CONTROL

the application of the input–output decoupling algorithm requires, in this case, to differentiate three times y_1 and two times y_2 in order to have the input explicitly appearing:

$$y_1^{(3)} = \frac{d^3 q_{11}}{dt^3} = \frac{d}{dt}\left(\frac{k_1(q_{21} - q_{11}) + 2\pi_2 \sin q_{12} \cos q_{12} \dot{q}_{11} \dot{q}_{12}}{\pi_1 + \pi_2 \cos^2 q_{12}}\right)$$

$$= a_3(q_{11}, q_{12}, q_{21}, \dot{q}_{11}, \dot{q}_{12}, \dot{q}_{21}) + \frac{2\pi_2 \sin q_{12} \cos q_{12} \dot{q}_{11}}{\pi_3(\pi_1 + \pi_2 \cos^2 q_{12})} u_2$$

$$\ddot{y}_2 = \ddot{q}_{12} = -\frac{\pi_2 \sin q_{12} \cos q_{12} \dot{q}_{11}^2 + \pi_5 \cos q_{12}}{\pi_3} + \frac{1}{\pi_3} u_2, \qquad (5.127)$$

where the accelerations are obtained from the dynamic model (5.125). As a result, the decoupling matrix has the form

$$A(q_{12}, \dot{q}_{11}) = \begin{pmatrix} 0 & \dfrac{1}{\pi_3} \dfrac{2\pi_2 \sin q_{12} \cos q_{12} \dot{q}_{11}}{\pi_1 + \pi_2 \cos^2 q_{12}} \\ 0 & \dfrac{1}{\pi_3} \end{pmatrix} \qquad (5.128)$$

which is *always* singular. This means that the second input appears "too soon" in both outputs, before the action of the first input torque is felt through the natural path of joint elasticity. Therefore, decoupling can never be achieved without the use of dynamic components which slow down the action of the second input. In fact, consider the addition of *two integrators* on the second input channel. Denote by ξ_1, ξ_2 the corresponding states, by u_2' the input to the second integrator, and by $u_1' = u_1$ the other input which does not change. The system equations are rewritten as:

$$(\pi_1 + \pi_2 \cos^2 q_{12})\ddot{q}_{11} - 2\pi_2 \sin q_{12} \cos q_{12}\, \dot{q}_{11}\dot{q}_{12} + k_1(q_{11} - q_{21}) = 0$$
$$\pi_3 \ddot{q}_{12} + \pi_2 \sin q_{12} \cos q_{12}\, \dot{q}_{11}^2 + \pi_5 \cos q_{12} - \xi_1 = 0$$
$$\dot{\xi}_1 = u_2'$$
$$\pi_4 \ddot{q}_{21} + k_1(q_{21} - q_{11}) = u_1'. \qquad (5.129)$$

The problem is now turned to control the link positions by acting on the torque of the first motor and on the second derivative of the torque of the second motor. By applying the decoupling algorithm to the *extended system* (5.129), it is immediate to check that neither the second nor the third derivatives of both outputs depend on the new inputs u_1' and u_2', which appear instead in $y^{(4)}$. To see this dependence, it is convenient to take the second derivative of the first set of two equations in (5.129)

$$(\pi_1 + \pi_2 \cos^2 q_{12})q_{11}^{(4)} + 2\pi_2 \sin q_{12} \cos q_{12}\, \dot{q}_{12} q_{11}^{(3)}$$

$$+2\pi_2(\cos^2 q_{12} - \sin^2 q_{12})\ddot{q}_{11}\dot{q}_{12}$$
$$-2\pi_2\frac{d^2}{dt^2}(\sin q_{12}\cos q_{12}\,\dot{q}_{11}\dot{q}_{12}) + k_1(\ddot{q}_{11} - \ddot{q}_{21}) = 0$$
$$\pi_3 q_{12}^{(4)} + \frac{d^2}{dt^2}(\pi_2 \sin q_{12}\cos q_{12}\,\dot{q}_{11}^2 + \pi_5 \cos q_{12}) - \ddot{\xi}_1 = 0 \quad (5.130)$$

and substitute therein $\ddot{\xi}_1$ and $\ddot{q}_{21}$, as obtained from the second set of two equations:

$$\ddot{\xi}_1 = u_2'$$
$$\ddot{q}_{21} = \frac{1}{\pi_4}u_1' + \frac{1}{\pi_4}k_1(q_{11} - q_{21}). \quad (5.131)$$

This yields

$$\begin{pmatrix} q_{11}^{(4)} \\ q_{12}^{(4)} \end{pmatrix} = \begin{pmatrix} a_{41}(q_{11}, q_{12}, q_{21}, \dot{q}_{11}, \dot{q}_{12}, \dot{q}_{21}, \xi_1, \xi_2) \\ a_{42}(q_{11}, q_{12}, q_{21}, \dot{q}_{11}, \dot{q}_{12}, \dot{q}_{21}, \xi_1, \xi_2) \end{pmatrix}$$
$$+ \begin{pmatrix} \dfrac{k_1}{\pi_4(\pi_1 + \pi_2 \cos^2 q_{12})} & 0 \\ 0 & \dfrac{1}{\pi_3} \end{pmatrix} \begin{pmatrix} u_1' \\ u_2' \end{pmatrix}, \quad (5.132)$$

from which it is apparent that the decoupling matrix for the extended system is *always nonsingular*. Therefore, a *static* state feedback decoupling control law for the *extended* system is obtained as

$$\begin{pmatrix} u_1' \\ u_2' \end{pmatrix} = \begin{pmatrix} \dfrac{\pi_4(\pi_1 + \pi_2 \cos^2 q_{12})}{k_1} & 0 \\ 0 & \pi_3 \end{pmatrix} \begin{pmatrix} u_{01} - a_{41} \\ u_{02} - a_{42} \end{pmatrix}. \quad (5.133)$$

Moreover, the relative degrees for the two outputs are $r_1 = r_2 = 4$, so that their sum is equal to the dimension of the extended state space (the six states of the robot arm plus the two states of the compensator). Thus, the same control law (5.133) will also fully linearize the closed-loop system. In terms of the original system, the combination of the control law (5.133) and of the dynamic extension performed with the addition of the two integrators is equivalent to the following *dynamic* state feedback controller

$$\dot{\xi}_1 = \xi_2$$
$$\dot{\xi}_2 = \pi_3(u_{02} - a_{42})$$
$$u_1 = \frac{\pi_4(\pi_1 + \pi_2 \cos^2 q_{12})}{k_1}(u_{01} - a_{41})$$
$$u_2 = \xi_1. \quad (5.134)$$

To summarize, the polar robot arm with the first joint being elastic has been transformed under the action of control (5.134) into two decoupled chains of four integrators. The tracking control problem can then be solved using standard linear techniques for the synthesis of u_{01} and u_{02}. Notice that it is sufficient to have a four times differentiable reference link trajectory in order to obtain its exact reproduction.

5.3.4 Nonlinear regulation

All previous control approaches for trajectory tracking are based on the use of nonlinear state feedback, which is in general rather complex. This is true both in the static case (e.g., when using the reduced model) and in the dynamic case (i.e., when the decoupling matrix of the system is singular). Roughly speaking, the purpose of these controllers is to set up a way to predict the evolution of the robot manipulator state by enforcing a linear behaviour through model-based feedback. On the other hand, given a desired link trajectory $q_1 = q_{1d}(t)$, it is always possible to compute the nominal trajectory of the state variables which is associated with the given output behaviour. Similarly, the nominal torque producing this robot manipulator motion can also be computed in closed form. This allows us to design a simpler tracking controller based on feedforward plus linear feedback of the computed state error. Such a control scheme is called *nonlinear regulation* since the error linear feedback stabilizes the system around the desired trajectory while computation of the feedforward term and of the reference state trajectory is based on the full nonlinear robot manipulator dynamics.

The approach can be developed directly for the complete dynamic model (5.9), using conveniently its partitioned form (5.109), and applies to the static linearizable as well as to the dynamic linearizable case. In the following, we will refer to the model (5.27), i.e., to the case $H_2 = \text{const}$, for ease of exposition.

Once the output is assigned, we have immediately a specified behaviour for the $2n$ state variables

$$q_1 = q_{1d}(t) \qquad \dot{q}_1 = \dot{q}_{1d}(t). \tag{5.135}$$

Our objective is to compute also the nominal trajectory for the remaining $2n$ state variables. To this purpose, notice that we cannot use the simple coordinate transformation (5.82) when the model is not linearizable by static state feedback. Using (5.135) in the first set of equations (5.27) yields

$$H_1(q_{1d})\ddot{q}_{1d} + H_2\ddot{q}_2 + C_1(q_{1d}, \dot{q}_{1d})\dot{q}_{1d} + K(q_{1d} - q_2) + g_1(q_{1d}) = 0, \tag{5.136}$$

which can be compactly rewritten as

$$H_2 \ddot{q}_2 + K(q_{1d} - q_2) + w_d(t) = 0. \tag{5.137}$$

The term $w_d(t)$ depends only on the reference trajectory and its derivatives, collecting all known quantities. By exploiting the structure (5.20) of matrix H_2, we can solve for q_{2n} the n-th equation in (5.137) as

$$q_{2dn} = q_{1dn} + \frac{1}{k_n} w_{dn}. \tag{5.138}$$

Differentiating twice q_{2dn}

$$\ddot{q}_{2dn} = \ddot{q}_{1dn} + \frac{1}{k_n} \ddot{w}_{dn}, \tag{5.139}$$

and substituting it into the $(n-1)$-th equation in (5.137) provides the evolution for $q_{2,n-1}$ as

$$q_{2d,n-1} = q_{1d,n-1} + \frac{1}{k_{n-1}} (w_{d,n-1} + H_{2,n-1,n} \ddot{q}_{2d,n}). \tag{5.140}$$

Proceeding backward recursively, it is then possible to define the nominal evolution of all motor variables

$$q_2 = q_{2d}(t) \qquad \dot{q}_2 = \dot{q}_{2d}(t). \tag{5.141}$$

At this point, the nominal torque for the given trajectory is computed in closed form using the second set of equations in (5.27), i.e.,

$$u_d(t) = H_2^T \ddot{q}_{1d} + H_3 \ddot{q}_{2d} + K(q_{2d} - q_{1d}). \tag{5.142}$$

We notice that the above algebraic computation is allowed by the absence of zero dynamics in the system. Otherwise, the derivation of a state reference trajectory from an output trajectory would require the integration of some differential equations.

To complete the design of a nonlinear regulator we need to find a stabilizing matrix gain F for a linear approximation of the robot manipulator system. This approximation may be derived around a fixed equilibrium point or around the nominal reference trajectory, leading respectively to a linear time-invariant or to a linear time-varying system. In any case, the existence of a (possibly time-varying) stabilizing feedback matrix is guaranteed by the controllability of the linear approximation. The resulting controller becomes

$$u = u_d + F \begin{pmatrix} q_{1d} - q_1 \\ q_{2d} - q_2 \\ \dot{q}_{1d} - \dot{q}_1 \\ \dot{q}_{2d} - \dot{q}_2 \end{pmatrix}. \tag{5.143}$$

Remarks

- The validity of this approach is only *local* in nature and the region of convergence depends both on the given trajectory and on the robustness of the designed linear feedback. On the other hand, the final control structure (5.143) is quite simple.

- A special pattern may be selected for the feedback matrix F, so as to avoid the measurement of the full robot manipulator state. Following the previous set-point regulation result for elastic joint manipulators, we can attempt using

$$F = \begin{pmatrix} 0 & K_P & 0 & K_D \end{pmatrix}, \qquad (5.144)$$

where only motor variables are fed back. However, there is no proof of global validity for this choice in the tracking case.

- Similar arguments can be used to derive a *dynamic* nonlinear regulator, based only on the measure of the link positions. This controller includes a model-based state observer and therefore requires the assumption that the linear approximation is observable. This is certainly true for robot manipulators with elastic joints, when the output is the link position.

5.4 Further reading

An early study on the inclusion of joint elasticity in the modelling of robot manipulators is due to [32]. The general dynamic model of manipulators with elastic joints can be generated automatically using symbolic manipulation programs [4]. Subsequent investigations include, e.g., [45], while the detailed analysis of the model structure presented in Section 5.1 comes from [47]. In [31], the special case of motors mounted on the driven links is treated. Modelling and control analysis of robot manipulators having some joint rigid and some other elastic is presented in [8]. The relevant mechanical considerations involved in the design of robot manipulators as well as in the evaluation of their compliant elements are collected in [39].

A large interest for the control problem of manipulators with elastic joints was excited by the experimental findings of [46, 15] on the GE P-50 robot manipulator. Since then, a number of conventional linear controllers have been proposed, see, e.g., [24], [17], and [26]. However, schemes with proved convergence characteristics have appeared only recently: the linear PD controller with constant compensation of Section 5.2 is a contribution of [48]. An iterative scheme that learns the desired gravity compensation at the set point has been developed in [11].

The reduced model was first introduced in [41], where its exact linearizability via static state feedback is shown. Results on feedback linearization and decoupling for special classes of manipulators had already been found in [13] and [14]; indeed, all these robot manipulators display the reduced model format. The robustness of feedback linearization (or inverse dynamics) control was studied in [16]. A comparative study on the errors induced by inverse dynamics control used on robotic systems which are not linearizable by static state feedback has been carried out in [33]. Practical implementation of inverse dynamics control in discrete time can be found in [20].

The observation that joints with limited elasticity lead to a singularly perturbed dynamic model dates back to the work in [29]. Nonlinear controllers based on the two-time scale separation property were then proposed by [23] and [44]. The corrective control is an outcome of these singular perturbation methods.

When considering the complete dynamic model, feedback linearizability and input–output decoupling were investigated by [28], reporting many negative results (most significantly, on the 3-revolute-joint articulated manipulator). On the other hand, it was found that robot manipulators with elastic joints possess nice structural properties such as nonlinear controllability [3]. Other interesting differential geometric results and a classification of the control characteristics of robot manipulators with different kinematic arrangements can be found in [6].

The use of dynamic state feedback was first proposed in [9] for the 3-revolute-joint robot manipulator, while the general approach to dynamic linearization and decoupling is described in [5]. The proof that absence of zero dynamics in an invertible nonlinear system is a sufficient condition for full linearization via dynamic feedback can be found in [18]. Checking of these sufficient conditions for the general models of robot manipulators with elastic joints is given in [10]. The analysis of manipulator zero dynamics in the presence of damping at the joints can be found in [7]. On the other hand, the reduced model of robot manipulators with mixed elastic/rigid joints may require either static or dynamic feedback for linearization and decoupling [8].

The nonlinear regulation approach for robot manipulators with joint elasticity has been introduced in [7]. The same strategy of state trajectory and feedforward computation can be found in [25] and [36].

Although not treated in this chapter, different state observers have been presented, starting from an approximate one in [38] up to the exact ones in [34] and [47], where a tracking controller based on the estimated state is also tested. In all the above cases, link position and velocity measurements are assumed. When only link positions are measured, instead, regulation schemes have been proposed in [1, 21] while the tracking problem has been considered in [37].

Adaptive control results for robot manipulators with elastic joints include approximate schemes based either on high-gain [42] or on singular perturbations [22], as well as the global solution obtained in [27] and extended in [2]; all analyzed using the reduced dynamic model. Another approach valid for the scalar case can be found in [35]. Moreover, robust control schemes have been proposed in [40] and later in [49], while iterative learning has been used in [12].

Finally, an interesting problem concerns force control of manipulators with elastic joints in constrained tasks, which is discussed in [43] and [30], following a singular perturbation technique, and in [19], using the inverse dynamics approach.

References

[1] A. Ailon and R. Ortega, "An observer-based set-point controller for robot manipulators with flexible joints," *Systems & Control Lett.*, vol. 21, pp. 329–335, 1993.

[2] B. Brogliato, R. Ortega, and R. Lozano, "Global tracking controllers for flexible-joint manipulators: a comparative study," *Automatica*, vol. 31, pp. 941–956, 1995.

[3] G. Cesareo and R. Marino, "On the controllability properties of elastic robots," in *Analysis and Optimization of Systems*, A. Bensoussan and J.L. Lions (Eds.), Lecture Notes in Control and Information Sciences, Springer-Verlag, Berlin, D, vol. 63, pp. 352–363, 1984.

[4] G. Cesareo, F. Nicolò, and S. Nicosia, "DYMIR: A code for generating dynamic model of robots," *Proc. 1984 IEEE Int. Conf. on Robotics and Automation*, Atlanta, GA, pp. 115–120, 1984.

[5] A. De Luca, "Dynamic control of robots with joint elasticity," *Proc. 1988 IEEE Int. Conf. on Robotics and Automation*, Philadelphia, PA, pp. 152–158, 1988.

[6] A. De Luca, "Control properties of robot arms with joint elasticity," in *Analysis and Control of Nonlinear Systems*, C.I. Byrnes, C.F. Martin, R.E. Saeks (Eds.), North-Holland, Amsterdam, NL, pp. 61–70, 1988.

[7] A. De Luca, "Nonlinear regulation of robot motion," *Proc. 1st European Control Conf.*, Grenoble, F, pp. 1045–1050, 1991.

[8] A. De Luca, "Decoupling and feedback linearization of robots with mixed rigid/elastic joints," *Proc. 1996 IEEE Int. Conf. on Robotics and Automation*, Minneapolis, MN, pp. 816–821, 1996.

[9] A. De Luca, A. Isidori, and F. Nicolò, "Control of robot arm with elastic joints via nonlinear dynamic feedback," *Proc. 24th IEEE Conf. on Decision and Control*, Ft. Lauderdale, FL, pp. 1671–1679, 1985.

[10] A. De Luca and L. Lanari, "Robots with elastic joints are linearizable via dynamic feedback," *Proc. 34th IEEE Conf. on Decision and Control*, New Orleans, LA, pp. 3895–3897, 1995.

[11] A. De Luca and S. Panzieri, "Learning gravity compensation in robots: Rigid arms, elastic joints, flexible links," *Int. J. of Adaptive Control and Signal Processing*, vol. 7, pp. 417–433, 1993.

[12] A. De Luca and G. Ulivi, "Iterative learning control of robots with elastic joints," *Proc. 1992 IEEE Int. Conf. on Robotics and Automation*, Nice, F, pp. 1920–1926, 1992.

[13] C. De Simone and F. Nicolò, "On the control of elastic robots by feedback decoupling," *IASTED Int. J. of Robotics and Automation*, vol. 1, no. 2, pp. 64–69, 1986.

[14] M.G. Forrest-Barlach and S.M. Babcock, "Inverse dynamics position control of a compliant manipulator," *IEEE J. of Robotics and Automation*, vol. 3, pp. 75–83, 1987.

[15] M.C. Good, L.M. Sweet, and K.L. Strobel, "Dynamic models for control system design of integrated robot and drive systems," *ASME J. of Dynamic Systems, Measurement, and Control*, vol. 107, pp. 53–59, 1985.

[16] W.M. Grimm, "Robustness analysis of nonlinear decoupling for elastic-joint robots," *IEEE Trans. on Robotics and Automation*, vol. 6, pp. 373–377, 1990.

[17] M.G. Hollars and R.H. Cannon, "Experiments on the end-point control of a two-link robot with elastic drives," *Proc. AIAA Guidance, Navigation and Control Conf.*, Williamsburg, VA, pp. 19–27, 1986.

[18] A. Isidori, C.H. Moog, and A. De Luca, "A sufficient condition for full linearization via dynamic state feedback," *Proc. 25th IEEE Conf. on Decision and Control*, Athina, GR, pp. 203–208, 1986.

[19] K.P. Jankowski and H.A. ElMaraghy, "Dynamic control of flexible joint robots with constrained end-effector motion," *Prepr. 3rd IFAC Symp. on Robot Control*, Vienna, A, pp. 345–350, 1991.

[20] K.P. Jankowski and H. Van Brussel, "An approach to discrete inverse dynamics control of flexible-joint robots," *IEEE Trans. on Robotics and Automation*, vol. 8, pp. 651–658, 1992.

[21] R. Kelly, R. Ortega, A. Ailon, and A. Loria, "Global regulation of flexible joint robots using approximate differentiation," *IEEE Trans. on Automatic Control*, vol. 39, pp. 1222–1224, 1994.

[22] K. Khorasani, "Adaptive control of flexible-joint robots," *IEEE Trans. on Robotics and Automation*, vol. 8, pp. 250–267, 1992.

[23] K. Khorasani and P.V. Kokotovic, "Feedback linearization of a flexible manipulator near its rigid body manifold," *Systems & Control Lett.*, vol. 6, pp. 187–192, 1985.

[24] H.B. Kuntze and A.H.K. Jacubasch, "Control algorithms for stiffening an elastic industrial robot," *IEEE J. of Robotics and Automation*, vol. 1, pp. 71–78, 1985.

[25] L. Lanari and J.T. Wen, "Feedforward calculation in tracking control of flexible robots," *Proc. 30th IEEE Conf. on Decision and Control*, Brighton, UK, pp. 1403–1408, 1991.

[26] S.H. Lin, S. Tosunoglu, and D. Tesar, "Control of a six-degree-of-freedom flexible industrial manipulator," *IEEE Control Systems Mag.*, vol. 11, no. 2, pp. 24–30, 1991.

[27] R. Lozano and B. Brogliato, "Adaptive control of robot manipulators with flexible joints," *IEEE Trans. on Automatic Control*, vol. 37, pp. 174–181, 1992.

[28] R. Marino and S. Nicosia, "On the feedback control of industrial robots with elastic joints: A singular perturbation approach," rep. R-84.01, Dipartimento di Ingegneria Elettronica, Università degli Studi di Roma "Tor Vergata", 1984.

[29] R. Marino and S. Nicosia, "Singular perturbation techniques in the adaptive control of elastic robots," *Prepr. 1st IFAC Symp. on Robot Control*, Barcelona, E, pp. 11–16, 1985.

[30] J.K. Mills, "Control of robotic manipulators with flexible joints during constrained motion task execution," *Proc. 28th IEEE Conf. on Decision and Control*, Tampa, FL, pp. 1676–1681, 1989.

[31] S.H. Murphy, J.T. Wen, and G.N. Saridis, "Simulation and analysis of flexibly jointed manipulators," *Proc. 29th IEEE Conf. on Decision and Control*, Honolulu, HI, pp. 545–550, 1990.

[32] S. Nicosia, F. Nicolò, and D. Lentini, "Dynamical control of industrial robots with elastic and dissipative joints," *Proc. 8th IFAC World Congr.*, Kyoto, J, pp. 1933–1939, 1981.

[33] S. Nicosia and P. Tomei, "On the feedback linearization of robots with elastic joints," *Proc. 27th IEEE Conf. on Decision and Control*, Austin, TX, pp. 180–185, 1988.

[34] S. Nicosia and P. Tomei, "A method for the state estimation of elastic joint robots by global position measurements," *Int. J. of Adaptive Control and Signal Processing*, vol. 4, pp. 475–486, 1990.

[35] S. Nicosia and P. Tomei, "A method to design adaptive controllers for flexible joint robots," *Proc. 1992 IEEE Int. Conf. on Robotics and Automation*, Nice, F, pp. 701–706, 1992.

[36] S. Nicosia and P. Tomei, "Design of global tracking controllers for flexible joint robots," *J. of Robotic Systems*, vol. 10, pp. 835–846, 1993.

[37] S. Nicosia and P. Tomei, "A tracking controller for flexible joint robots using only link position feedback," *IEEE Trans. on Automatic Control*, vol. 40, pp. 885–890, 1995.

[38] S. Nicosia, P. Tomei, and A. Tornambè, "A nonlinear observer for elastic robots," *IEEE J. of Robotics and Automation*, vol. 4, pp. 45–52, 1988.

[39] E. Rivin, *Mechanical Design of Robots*, McGraw-Hill, New York, NY, 1988.

[40] H. Sira-Ramirez and M.W. Spong, "Variable structure control of flexible joint manipulators," *IASTED Int. J. of Robotics and Automation*, vol. 3, no. 2, pp. 57–64, 1988.

[41] M.W. Spong, "Modeling and control of elastic joint robots," *ASME J. of Dynamic Systems, Measurement, and Control*, vol. 109, pp. 310–319, 1987.

[42] M.W. Spong, "Adaptive control of flexible joint manipulators," *Systems & Control Lett.*, vol. 13, pp. 15–21, 1989.

[43] M.W. Spong, "On the force control problem for flexible joint manipulators," *IEEE Trans. on Automatic Control*, vol. 34, pp. 107–111, 1989.

[44] M.W. Spong, K. Khorasani, and P.V. Kokotovic, "An integral manifold approach to the feedback control of flexible joint robots," *IEEE J. of Robotics and Automation*, vol. 3, pp. 291–300, 1987.

[45] H. Springer, P. Lugner, and K. Desoyer, "Equations of motion for manipulators including dynamic effects of active elements," *Proc. 1st IFAC Symp. on Robot Control*, Barcelona, E, pp. 425–430, 1985.

[46] L.M. Sweet and M.C. Good, "Redefinition of the robot motion control problem," *IEEE Control Systems Mag.*, vol. 5, no. 3, pp. 18–24, 1985.

[47] P. Tomei, "An observer for flexible joint robots," *IEEE Trans. on Automatic Control*, vol. 35, pp. 739–743, 1990.

[48] P. Tomei, "A simple PD controller for robots with elastic joints," *IEEE Trans. on Automatic Control*, vol. 36, pp. 1208–1213, 1991.

[49] P. Tomei, "Tracking control of flexible joint robots with uncertain parameters and disturbances," *IEEE Trans. on Automatic Control*, vol. 39, pp. 1067–1072, 1994.

Chapter 6

Flexible links

This chapter is devoted to modelling and control of robot manipulators with *flexible links*. This class of robots includes lightweight manipulators and/or large articulated structures that are encountered in a variety of conventional and nonconventional settings. From the point of view of applications, we can think about very long arms needed for accessing hostile environments (nuclear sites, underground waste deposits, deep sea, space, etc.) or automated crane devices for building construction. The ultimate challenge is the design of mechanical arms made of light materials that are suitable for typical industrial manipulation tasks, such as pick-and-place, assembly, or surface finishing. Lightweight structures are expected to improve performance of robots with typically low payload-to-arm weight ratio. As opposed to slow and bulky motion of conventional industrial manipulators, such robotic designs are expected to achieve fast and dexterous motion.

In order to fully exploit the potential offered by flexible robot manipulators, we must explicitly consider the effects of structural link flexibility and properly deal with active and/or passive control of vibrational behaviour. In this context, it is highly desirable to have an explicit, complete, and accurate *dynamic model* at disposal. The model should be explicit to provide a clear understanding of dynamic interaction and couplings, to be useful for control design, and to guide reduction or simplification of terms. The model should be complete in that, even if it is simple, it inherits the most relevant properties of the system. The model should be accurate as required for simulation purposes, design of advanced model-based nonlinear controllers, and off-line optimal trajectory planning. These general guidelines are even more important in the modelling of flexible robotic structures, where schemata or approximations of the modes of link deformation are unavoidably introduced. Symbolic manipulation packages may prove useful to derive dynamic models in a systematic error-free way.

Once a dynamic model for a flexible manipulator is available, its validation goes through the experimental identification of the relevant parameters. Besides those parameters that are inherited from the rigid case (mass, inertia, etc.), we should also identify the set of structural resonant frequencies and of associated deformation profiles. In the following, we assume that the analytical model matches the experimental data up to the desired order of model accuracy.

On the other hand, the control problem for flexible robot manipulators belongs to the class of mechanical systems where the number of controlled variables is strictly less than the number of mechanical degrees of freedom; this complicates the control design, ruling out a number of solutions that work in the rigid case. Further, the linear effects of flexibility are not separated from the typical nonlinear effects of multibody rigid dynamics. Although an effective control system could take advantage of some partitioning of rigid and flexible dynamics, the analysis of its behaviour should face in general the overall nonlinearities; in this respect, the linear dynamics resulting in the case of a single-link arm is a remarkable exception.

In order to tackle problems of increasing difficulty, a convenient classification of control targets can be introduced. As a minimum requirement, we should be able to control *rapid positioning* of the flexible arm. This is not a trivial problem since it requires both the synthesis of optimal feedforward commands, i.e., limiting the excitement of flexibility, and the active suppression of residual vibrations. For a high-performance flexible manipulator, a more demanding task is that of *tracking* a smooth *trajectory* of motion. This can be assigned at the joint level, as if the manipulator were rigid; provided that link deformation is kept limited, satisfactory results may be obtained also at the end-effector level. Last, the most difficult problem is that of reproducing trajectories defined directly for the end effector of the flexible manipulator.

The material is organized in two parts, covering modelling issues and control problems and algorithms.

First, the dynamic modelling of a single flexible link is presented. This simple case is representative of the complexity induced by the distributed nature of flexibility, and thus it has been extensively investigated in the literature. On the usual assumptions of the Euler-Bernoulli beam, an *infinite-dimensional model* is derived which is exact for deflections in the range of linearity. The relevant properties of the zero/pole structure of this linear model are discussed and *finite-dimensional approximations* are introduced. Next, the basic steps for obtaining dynamic models for the general multilink case are illustrated, leading to fully nonlinear equations of motion. The energy-based Lagrange formulation is adopted as the most convenient approach for describing the coupling of rigid and flexible body dynamics.

On the basis of the above models, a series of control strategies are investigated. For the problem of *point-to-point motion*, *linear controllers* provide satisfactory performance. A joint co-located PD control is shown to achieve asymptotic stabilization of any given manipulator configuration. We also discuss the use of linear control laws with feedback from the whole state, i.e., including deflection variables, for improved damping of vibration around the terminal position. The efficacy of *nonlinear control* methods is then emphasized for the accurate *tracking* of *joint trajectories* in multi-link flexible robot manipulators. In particular, we present the design of *inversion control* for input–output decoupling, and of *two-time scale control* based on a singularly perturbed model reformulation. Finally, the *end-effector trajectory tracking* problem is considered. Differently from the rigid manipulator and from the elastic joint case, the use of pure inversion strategies may lead here to closed-loop instabilities. The nature of this problem is analyzed in connection with the nonminimum phase characteristics of the system zero dynamics —a concept which has also a nonlinear counterpart. Two solutions are presented; namely, an inversion procedure defined in the *frequency domain*, suitable for the single link linear case, and a combined feedforward/feedback strategy based on *nonlinear regulation* theory.

6.1 Modelling of a single-link arm

The derivation of a linear dynamic model of a single-link flexible arm is presented below. Basic assumptions for the validity of the model are stated first, leading to the so-called Euler-Bernoulli beam equations of motion, in which terms of second or higher order in the deformation variables are neglected. The *modal analysis* is then accomplished as an eigenvalue problem for the resulting infinite-dimensional system; besides the exact *unconstrained* mode method, also the usual *constrained* mode approximation is considered. The possibility of obtaining a distributed parameter input–output transfer function is discussed. Finite-dimensional models are finally derived from a frequency-based truncation procedure. Some comments on other feasible approximations of link deformation using different sets of assumed modes conclude this section.

6.1.1 Euler-Bernoulli beam equations

Consider a robot arm with a single *flexible link* as in Fig. 6.1, moving on a horizontal plane. The arm is clamped at the base to a rigid hub that is driven in rotation by an ideal torque actuator, and has no tip payload. In

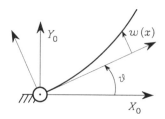

Figure 6.1: Schematic representation of single-link flexible arm.

order to derive a mathematical model, the following usual assumptions are made:

Assumption 6.1 The arm is a slender beam with uniform geometric characteristics and homogeneous mass distribution.

□

Assumption 6.2 The arm is flexible in the lateral direction, being stiff with respect to axial forces, torsion, and bending forces due to gravity; further, only elastic deformations are present.

□

Assumption 6.3 Nonlinear deformations as well as internal friction or other external disturbances are negligible effects.

□

Assumption 6.1 is the one that mainly characterizes *Euler-Bernoulli beam theory*, implying that the deflection of a section along the arm is due only to bending and not to shear. Moreover, the contribution of the rotary inertia of an arm section to the total energy is negligible. Notice that the analysis and the resulting partial differential equation will not represent those very high oscillation frequencies, whose wavelengths become comparable with the cross section of the arm. Assumption 6.2 is conveniently enforced by a suitable mechanical construction of the real flexible arm. Note that in Fig. 6.1 the small extension of the link along its neutral axis is neglected in the bending description; inclusion of this effect tends to stiffen the arm behaviour. Concerning Assumption 6.3, inclusion of nonlinear deformation terms is possible, while an accurate model of internal friction is usually difficult to obtain.

The system physical parameters of interest are: the linear density ρ, the flexural rigidity EI, the arm length ℓ, and the hub inertia I_h.

6.1. MODELLING OF A SINGLE-LINK ARM

In order to derive the equations of motion for this system which is a combination of a lumped parameter part (the hub rotation) and of a distributed parameter part (the link deformation), an energy-based method is the most convenient, e.g., the Lagrange formulation or the Hamilton principle. Therefore, the kinetic energy T and the potential energy U of the system have to be computed.

For describing the arm kinematics, let t denote the time variable and x the space coordinate along the neutral axis of the beam; then, $\vartheta(t)$ is the angle of hub rotation and $w(x,t)$ is the beam deflection from the neutral axis. The absolute position vector of a point along the beam is then described by

$$p = \begin{pmatrix} p_x \\ p_y \end{pmatrix} = \begin{pmatrix} x\cos\vartheta(t) - w(x,t)\sin\vartheta(t) \\ x\sin\vartheta(t) + w(x,t)\cos\vartheta(t) \end{pmatrix}. \tag{6.1}$$

In the following, primes will denote differentiation with respect to x and dots differentiation with respect to t. Thus, $w'(x,t)$ is the angle of deflection of an arm section at distance x from the base. Moreover, since the beam is *clamped* at the base, we have the geometric boundary conditions

$$w(0,t) = w'(0,t) = 0. \tag{6.2}$$

The kinetic energy $T = T_h + T_\ell$ has contributions from the hub

$$T_h = \frac{1}{2} I_h \dot\vartheta(t)^2, \tag{6.3}$$

and from the link

$$T_\ell = \frac{1}{2}\rho \int_0^\ell (\dot p_x^2 + \dot p_y^2) dx \tag{6.4}$$

$$= \frac{1}{2}\rho \int_0^\ell (x^2 \dot\vartheta(t)^2 + y^2(x,t)\dot\vartheta(t)^2 + \dot y^2(x,t) + 2x\dot\vartheta(t)\dot w(x,t)) dx.$$

The potential energy is

$$U = \frac{1}{2} EI \int_0^\ell (w''(x,t))^2 dx. \tag{6.5}$$

According to Hamilton principle, the system equations are obtained from the variational condition

$$\int_{t_1}^{t_2} (\delta T(t) - \delta U(t) + \delta W(t)) dt = 0, \tag{6.6}$$

where $\delta W(t) = u(t)\delta\vartheta(t)$ is the virtual work performed by the actuator driving torque $u(t)$. Grouping terms in (6.6) with respect to the independent variations $\delta\vartheta(t)$, $\delta w(x,t)$, $\delta w(\ell,t)$, and $\delta w'(\ell,t)$, and using calculus of variations arguments gives:

$$I_t\ddot{\vartheta}(t) + \rho \int_0^\ell x\ddot{w}(x,t)dx = u(t) \tag{6.7}$$

$$EIw''''(x,t) + \rho\ddot{w}(x,t) + \rho x\ddot{\vartheta}(t) = 0 \tag{6.8}$$

$$w(0,t) = w'(0,t) = 0 \tag{6.9}$$

$$w''(\ell,t) = w'''(\ell,t) = 0, \tag{6.10}$$

where $I_t = I_h + \rho\ell^3/3$. Eqs. (6.7) and (6.8) have been derived neglecting all second-order terms (products of state variables). The first equation can be attributed to the hub dynamics, while the second equation is associated with the flexible link. Note that, besides the geometric boundary conditions (6.9), the dynamic boundary conditions (6.10) arise at the tip representing balance of moment and of shearing force; in the absence of a payload, these are usually called *free* boundary conditions.

Integrating with respect to x eq. (6.8) multiplied by x and substituting in (6.7) yields

$$I_h\ddot{\vartheta}(t) - EIw''(0,t) = u(t) \tag{6.11}$$

that can be used in place of (6.7).

Eqs. (6.8)–(6.11) constitute the basis for the modal analysis of deformation in the Euler-Bernoulli beam. The distributed nature of the system is evidenced by the presence of partial differential equations.

6.1.2 Constrained and unconstrained modal analysis

As the system described by (6.8)–(6.11) is linear, we can proceed to compute eigenvalues and eigenvectors for the homogeneous system obtained by setting the forcing input $u(t) \equiv 0$. For the flexible arm, these will represent respectively the *resonant frequencies* and the associated *mode shapes*; this procedure is denoted in the literature as the *unconstrained mode* method. However, beam modal analysis is often performed by assuming also $\vartheta(t) \equiv 0$ (or $\ddot{\vartheta} = 0$), as if the rigid hub had infinite inertia and thus would be always at rest; this corresponds to the so-called *constrained mode* method.

We present first the constrained method because of its intrinsic simplicity. In this case, the link equation (6.8) becomes

$$w''''(x,t) + \frac{\rho}{EI}\ddot{w}(x,t) = 0 \tag{6.12}$$

6.1. MODELLING OF A SINGLE-LINK ARM

with the same boundary conditions (6.9) and (6.10). This problem can be solved by separation of variables, setting

$$w(x,t) = \psi(x)\eta(t) \tag{6.13}$$

and substituting in (6.12). This gives in time and space

$$\ddot{\eta}(t) + \zeta^2 \eta(t) = 0 \tag{6.14}$$

$$\psi''''(x) - \frac{\rho \zeta^2}{EI}\psi(x) = 0, \tag{6.15}$$

where ζ^2 is the eigenvalue and $\psi(x)$ is the *eigenfunction* of this *self-adjoint* boundary value problem. The general time solution to (6.14) is

$$\eta(t) = \eta(0)\cos(\zeta t) + \dot{\eta}(0)\sin(\zeta t), \tag{6.16}$$

representing an undamped harmonic oscillation at angular frequency ζ. From (6.9) and (6.10) the boundary conditions for $\psi(x)$ become

$$\psi(0) = \psi'(0) = 0 \tag{6.17}$$
$$\psi''(\ell) = \psi'''(\ell) = 0. \tag{6.18}$$

The general solution to (6.15) has the form

$$\psi(x) = A\sin(\beta x) + B\cos(\beta x) + C\sinh(\beta x) + D\cosh(\beta x) \tag{6.19}$$

for $x \in [0,\ell]$, where $\beta^4 = \rho\zeta^2/EI$, and the constants A, B, C, D are determined from (6.17) and (6.18) up to a scaling factor. From (6.17), it is $C = -A$ and $D = -B$. Then, in order to obtain a nontrivial solution for $\psi(x)$, the characteristic equation

$$1 + \cos(\beta\ell)\cosh(\beta\ell) = 0 \tag{6.20}$$

must hold, which follows from imposing (6.18). Accordingly, the solution takes on the form

$$\psi(x) = \psi_0\Big((\cos(\beta\ell) + \cosh(\beta\ell))(\sinh(\beta x) - \sin(\beta x))$$
$$+ (\sin(\beta\ell) + \sinh(\beta\ell))(\cos(\beta x) - \cosh(\beta x))\Big), \tag{6.21}$$

where ψ_0 is a constant that is determined through a suitable normalization condition on the eigenfunction. Eq. (6.20) has a countable infinity of positive solutions $\{\beta_i, i = 1, 2, \ldots\}$; an angular frequency $\zeta_i = \beta_i^2\sqrt{EI/\rho}$, a mode shape $\psi_i(x)$, and a time evolution $\eta_i(t)$ are obtained for each solution β_i.

The exact modal analysis is accomplished without forcing to zero $\ddot{\vartheta}(t)$. Hence, in the unconstrained mode method, we assume for $\vartheta(t)$ a solution of the form

$$\vartheta(t) = \alpha(t) + kv(t), \tag{6.22}$$

where $\alpha(t)$ describes the motion of the link center of mass, and for $w(x,t)$ a solution of the form

$$w(x,t) = \phi(x)v(t), \tag{6.23}$$

where k is chosen so as to satisfy

$$I_t k + \rho \int_0^\ell x\phi(x)dx = 0. \tag{6.24}$$

Note that eq. (6.24) guarantees that no perturbation of the motion of the center of mass occurs, i.e.,

$$\ddot{\alpha}(t) = 0. \tag{6.25}$$

Substituting (6.22) and (6.23) in (6.8) and taking into account (6.24) yields

$$\ddot{v}(t) + \omega^2 v(t) = 0 \tag{6.26}$$

$$\phi''''(x) - \frac{\rho\omega^2}{EI}(\phi(x) + kx) = 0, \tag{6.27}$$

where ω^2 is the eigenvalue and $\bar{\phi}(x) = \phi(x) + kx$ is the eigenfunction of the boundary value problem. The time solution to (6.26) is

$$v(t) = v(0)\cos(\omega t) + \dot{v}(0)\sin(\omega t). \tag{6.28}$$

The boundary conditions for $\phi(x)$ are

$$\phi(0) = \phi'(0) = 0 \tag{6.29}$$
$$\phi''(\ell) = \phi'''(\ell) = 0. \tag{6.30}$$

The general solution to (6.27) has the form

$$\phi(x) = A\sin(\gamma x) + B\cos(\gamma x) + C\sinh(\gamma x) + D\cosh(\gamma x) \tag{6.31}$$

for $x \in [0, \ell]$, where $\gamma^4 = \rho\omega^2/EI$, and the constants A, B, C, D are determined from (6.29) and (6.30) up to a scaling factor. It can be shown that the solution $\phi(x)$ has the form

$$\phi(x) = \phi_0 \Big(\big(\cos(\gamma\ell)\sinh(\gamma\ell) - \sin(\gamma\ell)\cosh(\gamma\ell)\big)\big(\cos(\gamma x) - \cosh(\gamma x)\big)$$
$$+ \big(1 + \sin(\gamma\ell)\sinh(\gamma\ell) + \cos(\gamma\ell)\cosh(\gamma\ell)\big)\big(\sin(\gamma x) - \sinh(\gamma x)\big)$$
$$+ 2\big(1 + \cos(\gamma\ell)\cosh(\gamma\ell)\big)\big(\sinh(\gamma x) - \gamma x\big)\Big), \tag{6.32}$$

6.1. MODELLING OF A SINGLE-LINK ARM

where ϕ_0 is determined through a normalization condition, and γ has to satisfy the characteristic equation

$$I_h\gamma^3\bigl(1 + \cos(\gamma\ell)\cosh(\gamma\ell)\bigr) + \rho\bigl(\sin(\gamma\ell)\cosh(\gamma\ell) - \cos(\gamma\ell)\sinh(\gamma\ell)\bigr) = 0. \tag{6.33}$$

This equation derives from imposing a nontrivial solution to the linear homogeneous system obtained from (6.30) and (6.24). Notice that (6.33) has a double solution for $\gamma = 0$, accounting for the unconstrained (rigid body) motion at the link base. Also, when the hub inertia $I_h \to \infty$, this characteristic equation reduces to (6.20) with $\gamma \equiv \beta$ and the constrained model is fully recovered. This effect is enforced when closing a proportional control loop at the joint level with increasingly large gain.

As before, eq. (6.33) has a countable infinity of positive solutions $\{\gamma_i, i = 1, 2, \ldots\}$; an angular frequency $\omega_i = \gamma_i^2\sqrt{EI/\rho}$, a mode shape $\phi_i(x)$, a constant k_i, and a time evolution $v_i(t)$ (and then $\vartheta_i(t)$ from (6.22)) are obtained for each γ_i. In particular, from (6.11) and (6.24) it follows that

$$k_i = -\frac{EI}{I_h\omega_i^2}\phi_i''(0) = \bar{\phi}_i'(0). \tag{6.34}$$

From a system point of view, it is clear that the angular frequencies ω_i obtained with the above free evolution analysis will be the poles of the transfer function between the Laplace-transform $U(s)$ of the input and the Laplace-transform $Y(s)$ of any output taken for the flexible arm, e.g., hub rotation or tip position. It is worth noticing that distributed transfer functions can be derived, with no need to discretize a priori the system equations. In particular, taking the Laplace-transform of (6.11)

$$s^2 I_h\Theta(s) - EIW''(0,s) = U(s), \tag{6.35}$$

it can be shown that

$$W''(0,s) = -s^2\Theta(s)\frac{W_2(s)}{W_1(s)} \tag{6.36}$$

with

$$W_1(s) = \cos^2(\sigma\ell) + \cosh^2(\sigma\ell) \tag{6.37}$$

$$W_2(s) = \frac{\rho}{EI\sigma^3}\bigl(\sinh(\sigma\ell)\cosh(\sigma\ell) - \sin(\sigma\ell)\cos(\sigma\ell)\bigr), \tag{6.38}$$

and $2\sigma^2 = \sqrt{\rho/EI}s$. As a result, the transfer function from torque input to hub rotation output is

$$\frac{\Theta(s)}{U(s)} = \frac{W_1(s)}{s^2(I_h W_1(s) + EIW_2(s))}, \tag{6.39}$$

with the double pole at $s = 0$ accounting for the rigid body rotation. We point out the existence of a strict relation between the zeros of (6.39) and the characteristic solutions of the constrained modal analysis. In fact, from (6.37) the equation $W_1(s) = 0$ has no solution for $\sigma \in R$. On the other hand, by letting $\sigma = (\xi \pm j\xi)/2$ with $\xi \in R$, we find that $W_1(s) = 0$ when

$$1 + \cos(\xi\ell)\cosh(\xi\ell) = 0 \tag{6.40}$$

which coincides with (6.20). Then its positive roots are β_i, and the zeros of the transfer function are

$$s_i = 2\sqrt{\frac{EI}{\rho}}\sigma_i^2 = \pm j\sqrt{\frac{EI}{\rho}}\beta_i^2 \qquad i = 1, 2, \ldots \tag{6.41}$$

lying on the imaginary axis. The presence of some structural damping would move these zeros to the left complex half-plane. The location of zeros will be very important for the control problem, since high-gain output feedback as well as inversion control cause the open-loop zeros to become poles for the closed-loop system. In the above case of output taken at the joint level, the system is (marginally) *minimum-phase* and closed-loop stability is preserved.

6.1.3 Finite-dimensional models

From the previous development we can easily obtain *finite-dimensional* approximated models by including only a finite number of eigenvalues/eigenvectors. Considering the first n_e roots of (6.33), link deformation can be expressed in terms of n_e mode shapes as

$$w(x,t) = \sum_{i=1}^{n_e} \phi_i(x) v_i(t), \tag{6.42}$$

and accordingly

$$\vartheta(t) = \alpha(t) + \sum_{i=1}^{n_e} \phi_i'(0) v_i(t). \tag{6.43}$$

This allows transforming the nonhomogeneous equations (6.7)–(6.10) into a set of $n_e + 1$ second-order ordinary differential equations of the form:

$$I_t \ddot{\alpha}(t) = u(t) \tag{6.44}$$

$$\ddot{v}_i(t) + \omega_i^2 v_i(t) = \phi_i'(0) u(t) \qquad i = 1, \ldots, n_e \tag{6.45}$$

6.1. MODELLING OF A SINGLE-LINK ARM

that can be rearranged in matrix form as

$$\begin{pmatrix} I_t & 0 \\ 0 & I \end{pmatrix} \begin{pmatrix} \ddot{\alpha} \\ \ddot{v} \end{pmatrix} + \begin{pmatrix} 0 & 0 \\ 0 & K \end{pmatrix} \begin{pmatrix} \alpha \\ v \end{pmatrix} = \begin{pmatrix} 1 \\ \phi'(0) \end{pmatrix} u, \qquad (6.46)$$

where the zero blocks and the identity matrix I have proper dimensions. The decoupled nature of these equations is a consequence of the inherent orthogonality of the eigenfunctions $\phi_i(x)$. In particular, a diagonal $(n_e \times n_e)$ stiffness matrix $K = \text{diag}\{\omega_i^2\}$ is obtained. Link structural damping can be introduced in (6.46) either on the basis of the values in the stiffness matrix or observing experimentally the time decay of the system excited at each frequency of deformation; this leads to a term $D\dot{v}$ appearing in the lower equations of (6.46), with a diagonal $(n_e \times n_e)$ damping matrix D.

At this point, any similarity transformation can be performed on the vector $(\alpha \ v^T)^T$ producing equivalent equations. For instance, in order to refer to a directly measurable quantity (the hub angle), eq. (6.46) can be represented as

$$\begin{pmatrix} I_t & -I_t \phi'(0)^T \\ -I_t \phi'(0) & I + I_t \phi'(0) \phi'(0)^T \end{pmatrix} \begin{pmatrix} \ddot{\vartheta} \\ \ddot{v} \end{pmatrix} + \begin{pmatrix} 0 & 0 \\ 0 & D \end{pmatrix} \begin{pmatrix} \dot{\vartheta} \\ \dot{v} \end{pmatrix}$$
$$+ \begin{pmatrix} 0 & 0 \\ 0 & K \end{pmatrix} \begin{pmatrix} \vartheta \\ v \end{pmatrix} = \begin{pmatrix} 1 \\ 0 \end{pmatrix} u \qquad (6.47)$$

with a full inertia matrix, diagonal damping and stiffness terms, and the input torque appearing only in the first equation.

Standard $2(n_e + 1)$-dimensional state space representations can be immediately obtained from the above second-order dynamic models; for instance, by considering the hub rotation as system output and the state $x(t) = (\alpha \ v^T \ \dot{\alpha} \ \dot{v}^T)^T$, the equations associated with (6.46) are of the form

$$\dot{x}(t) = Ax(t) + bu(t) \qquad (6.48)$$
$$y(t) = c^T x(t) \qquad (6.49)$$

with the triple (A, b, c^T) given by

$$A = \begin{pmatrix} 0 & 0 & 1 & 0 \\ 0 & 0 & 0 & I \\ -1/I_t & 0 & 0 & 0 \\ 0 & -K & 0 & -D \end{pmatrix} \qquad b = \begin{pmatrix} 0 \\ 0 \\ 1/I_t \\ \phi'(0) \end{pmatrix} \qquad (6.50)$$

$$c^T = (1 \ 0 \ 0 \ 0).$$

From (6.48), a finite-dimensional transfer function (i.e., with $n_e + 1$ poles and a finite number of zeros) replacing the distributed one is obtained as $Y(s)/U(s) = c^T (sI - A)^{-1} b$.

Other finite-dimensional models can be obtained by considering approximations other than simple truncation of the number of eigenfunctions. These are aimed at avoiding the complicated mode shape analysis involved by the exact (unconstrained) approach. A first example is represented by the constrained approximation in the above treated modal analysis; by assuming the finite expansion

$$w(x,t) = \sum_{i=1}^{n_e} \psi_i(x)\eta_i(t) \tag{6.51}$$

in place of (6.13) and proceeding as before, we get

$$\begin{pmatrix} I_t & \rho\mu^T \\ \rho\mu & \rho I \end{pmatrix} \begin{pmatrix} \ddot{\vartheta} \\ \ddot{\eta} \end{pmatrix} + \begin{pmatrix} 0 & 0 \\ 0 & \rho\hat{D} \end{pmatrix} \begin{pmatrix} \dot{\vartheta} \\ \dot{\eta} \end{pmatrix} + \begin{pmatrix} 0 & 0 \\ 0 & \rho\hat{K} \end{pmatrix} \begin{pmatrix} \vartheta \\ \eta \end{pmatrix} = \begin{pmatrix} 1 \\ 0 \end{pmatrix} u, \tag{6.52}$$

where $\hat{K} = \text{diag}\{\zeta_i^2\}$, damping $\hat{D}$ was added similarly to (6.47), and

$$\mu = \begin{pmatrix} \dfrac{\psi_1''(0)}{\beta_1^4} & \cdots & \dfrac{\psi_{n_e}''(0)}{\beta_{n_e}^4} \end{pmatrix}^T. \tag{6.53}$$

Again, suitable orthogonality relations among the eigenfunctions $\psi_i(x)$ were used in (6.52). This approximate model, usually referred to as *clamped-free*, indeed displays a slightly different pattern of eigenvalues with respect to the exact one. However, the simplicity of the underlying computations needed to provide the coefficients in (6.52) makes the constrained approach quite appealing.

More in general, we can pursue simpler modelling techniques by assuming a finite number of mode shapes for the deformation, i.e.,

$$w(x,t) = \sum_{i=1}^{n_e} \varphi_i(x)\delta_i(t), \tag{6.54}$$

where the spatial *assumed modes* $\varphi_i(x)$ satisfy only a reduced set of boundary conditions, but no dynamic equations of motion like (6.8). When the assumed modes satisfy purely geometric boundary conditions, they are denoted as *admissible functions*; when the chosen deformation modes comply also with natural boundary conditions, they are called *comparison functions*. Incidentally, we remark that finite-element descriptions with concentrated elasticity as well as Ritz-Kantorovich expansions belong to the latter class of methods. The use of this type of assumed modes becomes necessary for treating more complex multilink flexible structures, far beyond the simple one-link arm considered in this section. On the other hand, nice features like dynamic orthogonality are lost and coupled equations typically result for the flexible motion.

6.2 Modelling of multilink manipulators

In order to obtain the dynamic model of a *multilink* robot manipulator, it is necessary to introduce a convenient kinematic description of the manipulator which takes into account the deformation of the links. For determining position and orientation of relevant link frames, a recursive procedure can be established similarly to the rigid case. Kinematic relationships are then used for computing kinetic and potential energy of the system within a Lagrange formulation. The modelling results of the previous section will be embedded in the description of each flexible link. In order to limit the complexity of the derivation, we will assume that rigid motion and link deformation occur in the same plane.

6.2.1 Direct kinematics

Consider a *planar n*-link flexible arm with revolute joints subject only to bending deformations without torsional effects; Fig. 6.2 shows a two-link example. The following coordinate frames are established: the inertial frame $(\hat{X}_0, \hat{Y}_0)$, the rigid body moving frame associated with link i (X_i, Y_i), and the flexible body moving frame associated with link i $(\hat{X}_i, \hat{Y}_i)$. The rigid motion is described by the joint angles ϑ_i, while $w_i(x_i)$ denotes the transversal deflection of link i at x_i with $0 \leq x_i \leq \ell_i$, being ℓ_i the link length.

Let ${}^i p_i(x_i) = (\; x_i \quad w_i(x_i)\;)^T$ be the position of a point along the deflected link i with respect to frame (X_i, Y_i) and p_i be the absolute position of the same point in frame $(\hat{X}_0, \hat{Y}_0)$. Also, ${}^i r_{i+1} = {}^i p_i(\ell_i)$ indicates the position of the origin of frame (X_{i+1}, Y_{i+1}) with respect to frame (X_i, Y_i), and r_{i+1} its absolute position in frame $(\hat{X}_0, \hat{Y}_0)$.

The joint (rigid) rotation matrix R_i and the rotation matrix E_i of the (flexible) link at the end point are, respectively,

$$R_i = \begin{pmatrix} \cos\vartheta_i & -\sin\vartheta_i \\ \sin\vartheta_i & \cos\vartheta_i \end{pmatrix} \qquad E_i = \begin{pmatrix} 1 & -w'_{ie} \\ w'_{ie} & 1 \end{pmatrix}, \qquad (6.55)$$

where $w'_{ie} = (\partial w_i / \partial x_i)|_{x_i = \ell_i}$, and the linear approximation $\arctan w'_{ie} \simeq w'_{ie}$, valid for small deflections, has been made. This also implies that all second-order terms involving products of deformations will be neglected. Therefore, the above absolute position vectors can be expressed as

$$p_i = r_i + W_i {}^i p_i, \qquad r_{i+1} = r_i + W_i {}^i r_{i+1} \qquad (6.56)$$

where W_i is the global transformation matrix from $(\hat{X}_0, \hat{Y}_0)$ to (X_i, Y_i), which obeys to the recursive equation

$$W_i = W_{i-1} E_{i-1} R_i = \hat{W}_{i-1} R_i \qquad \hat{W}_0 = I. \qquad (6.57)$$

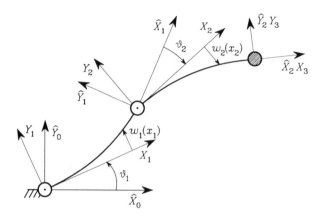

Figure 6.2: Schematic representation of a two-link flexible arm.

On the basis of the above relations, the kinematics of any point along the arm is fully characterized.

For later use in the arm kinetic energy, also the differential kinematics is needed. In particular, the (scalar) absolute angular velocity of frame (X_i, Y_i) is

$$\dot{\alpha}_i = \sum_{j=1}^{i} \dot{\vartheta}_j + \sum_{k=1}^{i-1} \dot{y}'_{ke}, \qquad (6.58)$$

where the upper dot denotes time derivative. Moreover, the absolute linear velocity of an arm point is

$$\dot{p}_i = \dot{r}_i + \dot{W}_i{}^i p_i + W_i{}^i \dot{p}_i, \qquad (6.59)$$

and ${}^i r_{i+1} = {}^i p_i(\ell_i)$. Since we neglect link axis extension ($\dot{x}_i = 0$), then ${}^i p_i(x_i) = (\,0\ \ \dot{w}_i(x_i)\,)^T$. The computation of (6.59) takes advantage of the recursions

$$\dot{W}_i = \dot{\hat{W}}_{i-1} R_i + \hat{W}_{i-1} \dot{R}_i \qquad \dot{\hat{W}}_i = \dot{W}_i E_i + W_i \dot{E}_i. \qquad (6.60)$$

Also, note that

$$\dot{R}_i = S R_i \dot{\vartheta}_i \qquad \dot{E}_i = S \dot{w}'_{ie} \qquad S = \begin{pmatrix} 0 & -1 \\ 1 & 0 \end{pmatrix}. \qquad (6.61)$$

6.2.2 Lagrangian dynamics

The dynamic equations of motion of a planar n-link flexible arm can be derived following the standard *Lagrange formulation*, i.e., by computing the kinetic energy T and the potential energy U of the system and then forming the Lagrangian $L = T - U$.

The total *kinetic energy* is given by the sum of the following contributions:

$$T = \sum_{i=1}^{n} T_{hi} + \sum_{i=1}^{n} T_{\ell i} + T_p. \tag{6.62}$$

The kinetic energy of the rigid body located at hub i of mass m_{hi} and moment of inertia I_{hi} is

$$T_{hi} = \frac{1}{2} m_{hi} \dot{r}_i^T \dot{r}_i + \frac{1}{2} I_{hi} \dot{\alpha}_i^2 \tag{6.63}$$

with $\dot{\alpha}_i$ as in (6.58), the kinetic energy pertaining to the slender link i of linear density ρ_i is

$$T_{\ell i} = \frac{1}{2} \int_0^{\ell_i} \rho_i(x_i) \dot{p}_i(x_i)^T \dot{p}_i(x_i) dx_i, \tag{6.64}$$

and the kinetic energy associated with a payload of mass m_p and moment of inertia I_p located at the end of link n is

$$T_p = \frac{1}{2} m_p \dot{r}_{n+1}^T \dot{r}_{n+1} + \frac{1}{2} I_p (\dot{\alpha}_n + \dot{w}'_{ne})^2. \tag{6.65}$$

Remarkably, the evaluation of the expressions in (6.63) and (6.64) exploits the following identities:

$$R_i^T R_i = E_i^T E_i = S^T S = I \tag{6.66}$$

$$R_i^T \dot{R}_i = S \dot{\vartheta}_i \qquad E_i^T \dot{E}_i = \dot{w}'_{ie}(w'_{ie} I + S). \tag{6.67}$$

The *potential energy* is given by the sum of the following contributions:

$$U = \sum_{i=1}^{n} U_{ei} + \sum_{i=1}^{n} U_{ghi} + \sum_{i=1}^{n} U_{g\ell i} + U_{gp}. \tag{6.68}$$

The elastic energy stored in link i is

$$U_{ei} = \frac{1}{2} \int_0^{\ell_i} (EI)_i(x_i) \left(\frac{d^2 w_i(x_i)}{dx_i^2} \right)^2 dx_i, \tag{6.69}$$

being $(EI)_i$ its flexural rigidity. The gravitational energy of hub i is

$$U_{ghi} = -m_{hi}g_0^T r_i, \tag{6.70}$$

that of link i is

$$U_{g\ell i} = -g_0^T \int_0^{\ell_i} \rho_i(x_i)p_i(x_i)dx_i, \tag{6.71}$$

and that of the payload is

$$U_{gp} = -m_p g_0^T r_{n+1}, \tag{6.72}$$

being g_0 the gravity acceleration vector.

Notice that no discretization of structural link flexibility has been made so far. The Lagrangian L can be shown to generate an infinite-dimensional *nonlinear* model, which is of limited use for simulation and/or control purposes. Hence, in order to obtain a finite-dimensional dynamic model, the assumed modes link approximation (6.54) can be used. On the basis of this discretization, the Lagrangian L becomes a function of a set of $n+\sum_{i=1}^{n} n_{ei}$ generalized coordinates $\{\vartheta_i(t), \delta_{ij}(t)\}$, and the dynamic model is obtained by satisfying the Lagrange's equations

$$\frac{d}{dt}\frac{\partial L}{\partial \dot\vartheta_i} - \frac{\partial L}{\partial \vartheta_i} = u_i \qquad i = 1, \ldots, n \tag{6.73}$$

$$\frac{d}{dt}\frac{\partial L}{\partial \dot\delta_{ij}} - \frac{\partial L}{\partial \delta_{ij}} = 0 \qquad j = 1, \ldots, n_{ei},\ i = 1, \ldots, n, \tag{6.74}$$

where u_i is the joint torque at hub i. Note that no input torque appears on the right-hand side of the flexible equations, as can be inferred from the single-link equations (6.47) and (6.52). Thus, both unconstrained and constrained modes (or any other approximation) can be used in the model.

As a result, the equations of motion for a planar n-link flexible arm can be written in the closed form

$$\begin{pmatrix} H_{\vartheta\vartheta}(\vartheta,\delta) & H_{\vartheta\delta}(\vartheta,\delta) \\ H_{\vartheta\delta}^T(\vartheta,\delta) & H_{\delta\delta}(\vartheta,\delta) \end{pmatrix} \begin{pmatrix} \ddot\vartheta \\ \ddot\delta \end{pmatrix} + \begin{pmatrix} c_\vartheta(\vartheta,\delta,\dot\vartheta,\dot\delta) \\ c_\delta(\vartheta,\delta,\dot\vartheta,\dot\delta) \end{pmatrix}$$
$$+ \begin{pmatrix} g_\vartheta(\vartheta,\delta) \\ g_\delta(\vartheta,\delta) \end{pmatrix} + \begin{pmatrix} 0 \\ D\dot\delta + K\delta \end{pmatrix} = \begin{pmatrix} u \\ 0 \end{pmatrix}; \tag{6.75}$$

the positive definite symmetric inertia matrix H has been partitioned in blocks according to the rigid and flexible components, c is the vector of Coriolis and centrifugal forces, g is the vector of gravitational forces, while K and D are the system stiffness and damping diagonal matrices of proper

6.2. MODELLING OF MULTILINK MANIPULATORS

dimensions. We note that the linear expression $K\delta$ follows from rewriting the elastic energy U_e as —see also (6.69)—

$$U_e = \frac{1}{2}\delta^T K \delta. \qquad (6.76)$$

By defining the configuration vector $q = (\vartheta^T \;\; \delta^T)^T$, eqs. (6.75) can be compacted into

$$H(q)\ddot{q} + c(q,\dot{q}) + g(q) + \begin{pmatrix} 0 \\ D\dot{\delta} + K\delta \end{pmatrix} = \begin{pmatrix} u \\ 0 \end{pmatrix}. \qquad (6.77)$$

As for the components of c, these can be evaluated as in the rigid case through the Christoffel symbols, i.e.,

$$c_i = \sum_j \sum_k \left(\frac{\partial H_{ij}}{\partial q_k} - \frac{1}{2}\frac{\partial H_{jk}}{\partial q_i} \right) \dot{q}_j \dot{q}_k. \qquad (6.78)$$

6.2.3 Dynamic model properties

In the following, we list useful properties of the dynamic model (6.75).

Property 6.1 The spatial dependence present in the link kinetic and potential energy terms (6.64) and (6.71) can be resolved by the introduction of a number of constant parameters, characterizing the mechanical properties of the (uniform density) links:

$$\begin{aligned}
m_i &= \int_0^{\ell_i} \rho_i dx_i = \rho_i \ell_i \\
r_i &= \frac{1}{m_i}\int_0^{\ell_i} \rho_i x_i dx_i = \frac{1}{2}\ell_i \\
I_{oi} &= \int_0^{\ell_i} \rho_i x_i^2 dx_i = \frac{1}{3}m_i\ell_i^2 \\
\mu_{ij} &= \int_0^{\ell_i} \rho_i \varphi_{ij}(x_i) dx_i \qquad (6.79)\\
\nu_{ij} &= \int_0^{\ell_i} \rho_i \varphi_{ij}(x_i) x_i dx_i \\
\varrho_{ijk} &= \int_0^{\ell_i} \rho_i \varphi_{ij}(x_i)\varphi_{ik}(x_i) dx_i \\
k_{ijk} &= \int_0^{\ell_i} (EI)_i \varphi''_{ij}(x_i)\varphi''_{ik}(x_i) dx_i.
\end{aligned}$$

Therefore, m_i is the mass of link i, r_i is the distance of center of mass of link i from joint i axis, I_{oi} is the inertia of link i about joint i axis, μ_{ij} and ν_{ij} are deformation moments of order zero and one of mode j of link i, and ϱ_{ijk} is the cross moment of modes j and k of link i. Also, k_{ijk} is the cross elasticity coefficient of modes j and k of link i (zero for $j \neq k$, when orthogonal modes are used). Although eq. (6.75) is in general highly nonlinear, it is not difficult to show that the left-hand side can be given a *linear* factorization of the type $Y(\vartheta, \delta, \dot\vartheta, \dot\delta, \ddot\vartheta, \ddot\delta)a$, where the vector a contains all the above pre-computable parameters, while the regressor matrix Y has a known structure.

□

Property 6.2 A factorization of vector c in (6.77) exists

$$c(q,\dot q) = C(q,\dot q)\dot q = \begin{pmatrix} C_{\vartheta\vartheta} & C_{\vartheta\delta} \\ C_{\delta\vartheta} & C_{\delta\delta} \end{pmatrix} \begin{pmatrix} \dot\vartheta \\ \dot\delta \end{pmatrix} \qquad (6.80)$$

such that the matrix $N = \dot H - 2C$ is skew-symmetric. A possible choice is the one induced by (6.78). Moreover, a similar result holds for the block elements in (6.75); for instance, the matrix $N_{\delta\delta} = \dot H_{\delta\delta} - 2C_{\delta\delta}$ enjoys the same property.

□

Property 6.3 The choice of specific assumed modes implies convenient simplifications in the block $H_{\delta\delta}$ of the inertia matrix. In particular, orthonormality of the modes of each link induces a decoupled structure for the diagonal inertia subblocks of $H_{\delta\delta}$ which in turn may reduce to a constant diagonal one.

□

Property 6.4 A rather common approximation is to evaluate the total kinetic energy of the system in the undeformed configuration $\delta = 0$. This implies that the inertia matrix, and thus also c_ϑ and c_δ, are independent of δ. It can be shown that the velocity terms c_δ will then lose the quadratic dependence on $\dot\delta$. Moreover, if $H_{\delta\delta}$ is constant, c_ϑ too loses the quadratic dependence on $\dot\delta$, while each component of c_δ becomes a quadratic function of $\dot\vartheta$ only. Finally, if also $H_{\vartheta\delta}$, representing the coupling between the rigid body and the flexible body dynamics, is approximated by a constant matrix then $c_\delta \equiv 0$ and $c_\vartheta = C_{\vartheta\vartheta}(\vartheta, \dot\vartheta)\dot\vartheta$, i.e., a quadratic function of $\dot\vartheta$ only.

□

Property 6.5 Due to the assumption of small deformation of each link, the dependence of the gravity term in the flexible equation is simply $g_\delta = g_\delta(\vartheta)$.

□

Property 6.6 It is expected that the flexible body dynamics is much faster than the rigid body dynamics, so that the dynamic model can be cast in a singularly perturbed form, similarly to the flexible joint case. Furthermore, if a clear frequency separation exists among the arm vibration modes, then the system can be effectively described on a multiple time-scale basis.

□

6.3 Regulation

We start the presentation of control laws for flexible arms by considering the classical *regulation* problem, i.e., the case of a *constant* desired equilibrium configuration $q_d = (\vartheta_d^T \ \delta_d^T)^T$. Indeed, the arm deformation is zero for any value of the joint variables only in the absence of gravity. It is assumed that the full state of the flexible arm is available for feedback. However, when some structural damping is present, a linear feedback using only the joint variables will be shown to asymptotically stabilize the system. Otherwise, active vibration damping can be realized by designing a full state stabilizing control on the basis of the linear approximation of the arm dynamics around the terminal point. Although they are conceived for point-to-point tasks, the following schemes can also be applied in the terminal phase of a trajectory around the steady-state configuration.

6.3.1 Joint PD control

For the design of a *joint PD control* law, some considerations are in order concerning the terms in (6.75) deriving from the potential energy U.

In view of Assumption 6.2, we have that

$$U_e \leq U_{e\max} < \infty. \tag{6.81}$$

In view of (6.76), a direct consequence is that

$$\|\delta\| \leq \sqrt{\frac{2U_{e\max}}{\lambda_{\min}(K)}}. \tag{6.82}$$

Concerning the gravity contribution, the vector g satisfies the inequality

$$\left\|\frac{\partial g}{\partial q}\right\| \leq \alpha_0 + \alpha_1 \|\delta\| \leq \alpha_0 + \alpha_1 \sqrt{\frac{2U_{e\max}}{\lambda_{\min}(K)}} =: \alpha, \tag{6.83}$$

where $\alpha_0, \alpha_1, \alpha > 0$. This can be easily proved by observing that the gravity term contains only trigonometric functions of ϑ and linear/trigonometric

functions of δ, and using (6.82). As a direct consequence of (6.83), we have for any q_1, q_2:
$$\|g(q_1) - g(q_2)\| \leq \alpha \|q_1 - q_2\|. \tag{6.84}$$

Theorem 6.1 *Consider the control law*
$$u = K_P(\vartheta_d - \vartheta) - K_D \dot{\vartheta} + g_\vartheta(\vartheta_d, \delta_d), \tag{6.85}$$

where K_P and K_D are $(n \times n)$ symmetric positive definite matrices and δ_d is defined as
$$\delta_d = -K^{-1} g_\delta(\vartheta_d). \tag{6.86}$$

If
$$\lambda_{\min}(K_q) = \lambda_{\min}\begin{pmatrix} K_P & 0 \\ 0 & K \end{pmatrix} > \alpha, \tag{6.87}$$

with α as in (6.83), then
$$q = q_d \qquad \dot{q} = 0$$

is a globally asymptotically stable equilibrium point for the closed-loop system (6.75) and (6.85)

◇ ◇ ◇

Proof. The equilibrium points of the closed-loop system (6.75) and (6.85) satisfy the equations
$$g_\vartheta(\vartheta, \delta) = K_P(\vartheta_d - \vartheta) + g_\vartheta(\vartheta_d, \delta_d) \tag{6.88}$$
$$g_\delta(\vartheta) = -K\delta. \tag{6.89}$$

It is easy to recognize that (6.89) has a unique solution δ for any value of ϑ. Adding $K\delta_d + g_\delta(\vartheta_d) = 0$ to the right-hand side of (6.89) yields
$$K_q(q_d - q) = \begin{pmatrix} K_P & 0 \\ 0 & K \end{pmatrix} \begin{pmatrix} \vartheta_d - \vartheta \\ \delta_d - \delta \end{pmatrix}$$
$$= \begin{pmatrix} g_\vartheta(\vartheta, \delta) - g_\vartheta(\vartheta_d, \delta_d) \\ g_\delta(\vartheta) - g_\delta(\vartheta_d) \end{pmatrix} = g(q) - g(q_d). \tag{6.90}$$

In view of (6.87), we have that, for $q \neq q_d$,
$$\|K_q(q_d - q)\| \geq K_{qm} \|q_d - q\| > \alpha \|q_d - q\| \geq \|g(q) - g(q_d)\|, \tag{6.91}$$

where the last inequality follows from (6.84). This implies that $\{q = q_d, \dot{q} = 0\}$ is the *unique* equilibrium point of the closed-loop system (6.75) and (6.85).

6.3. REGULATION

Consider the energy-based Lyapunov function candidate

$$V = \frac{1}{2}\dot{q}^T H \dot{q} + \frac{1}{2}(q_d - q)^T K_q (q_d - q)$$
$$+ U_g(q) - U_g(q_d) + (q_d - q)^T g(q_d) \geq 0, \tag{6.92}$$

where $U_g = \sum_{i=1}^{n}(U_{ghi} + U_{g\ell i}) + U_{gp}$ is the total gravitational energy. The function (6.92) vanishes only at the desired equilibrium point, due to (6.88)–(6.91). The time derivative of (6.92) along the trajectories of the closed-loop system (6.75) and (6.85) is

$$\dot{V} = \dot{q}^T \left(H\ddot{q} + \frac{1}{2}\dot{H}\dot{q}\right) - \dot{q}^T K_q (q_d - q) + \dot{q}^T (g(q) - g(q_d))$$
$$= \dot{q}^T \left(\begin{pmatrix} K_P(\vartheta_d - \vartheta) - K_D \dot{\vartheta} + g_\vartheta(q_d) \\ -(D\dot{\delta} + K\delta) \end{pmatrix} - g(q)\right)$$
$$- \dot{q}^T \begin{pmatrix} K_P(\vartheta_d - \vartheta) \\ K(\delta_d - \delta) \end{pmatrix} + \dot{q}^T \left(g(q) - \begin{pmatrix} g_\vartheta(q_d) \\ g_\delta(\vartheta_d) \end{pmatrix}\right), \tag{6.93}$$

where identity (6.80) and the skew-symmetry of the matrix N have been used. Simplifying terms yields

$$\dot{V} = -\dot{\vartheta}^T K_D \dot{\vartheta} - \dot{\delta}^T D \dot{\delta} \leq 0, \tag{6.94}$$

where (6.86) has been utilized. When $\dot{V} = 0$, it is $\dot{q} = 0$ and the closed-loop system (6.75) and (6.85) becomes

$$H\ddot{q} = \begin{pmatrix} K_P(\vartheta_d - \vartheta) + g_\vartheta(q_d) - g_\vartheta(q) \\ -(K\delta + g_\delta(\vartheta)) \end{pmatrix}. \tag{6.95}$$

In view of the previous equilibrium analysis and of (6.87), it is $\ddot{q} = 0$ if and only if $q = q_d$, or $\vartheta = \vartheta_d$ and $\delta = \delta_d$. Invoking La Salle's invariant set theorem, global asymptotic stability of the desired point follows.

◇

Remarks

- The above control law does not require any feedback from the deflection variables, and is composed of a *linear* term plus a *constant feedforward* action which includes only part of the gravity force appearing in the model (6.75). The satisfaction of the structural assumption $\lambda_{\min}(K) > \alpha$ is not restrictive in general, and depends on the relative importance of stiffness vs. gravity.

- Condition (6.87) will automatically be satisfied, provided that the assumption on the structural link flexibility

$$\lambda_{\min}(K) > \alpha \qquad (6.96)$$

holds, and that the proportional control gain is chosen so that the condition

$$\lambda_{\min}(K_p) > \alpha \qquad (6.97)$$

is verified.

- The knowledge of the link stiffness K and of the complete gravity term g is needed in the definition of the steady-state deformation δ_d. Uncertainty in the associated model parameters produces a different asymptotically stable equilibrium point, which can be made arbitrarily close to the desired one by increasing K_P.

- It can be shown that stability is guaranteed even in the absence of link internal damping ($D = 0$). Physically, some small damping will always exist but the regulation transients could be very slow. If desired, a passive increase of damping can be achieved by structural modification, e.g., viscoelastic layer damping treatment.

- If the tip location is of interest, $p = k(\vartheta, \delta)$, then ϑ_d can be computed by inverting for ϑ the direct kinematics equation

$$k(\vartheta, -K^{-1}g_\delta(\vartheta)) = p_d, \qquad (6.98)$$

so as to achieve end-effector regulation at steady-state.

6.3.2 Vibration damping control

During the execution of a point-to-point task, the arm typically undergoes dynamic deformations which are excited by the imposed accelerations. At the terminal point, the residual arm deformation is expected to vanish in accordance with the damping characteristics of the mechanical structure. These can be enhanced by the choice of suitable materials and surface treatment of the links. When passive damping is too low, oscillations will persist in the flexible arm long after the completion of the useful motion. To overcome this drawback, we need to have measurements related to arm deflection and use them properly in a feedback stabilizing control. *Active vibration linear control* can be designed as follows.

The first step is to linearize the nonlinear system dynamics (6.75) around the given (final) arm configuration q_d. Since at the equilibrium (6.86) holds,

6.3. REGULATION

setting $\Delta q = q_d - q$, $\Delta u = u_d - u$ with $u_d = g_\vartheta(q_d)$, and neglecting second and higher-order terms yields

$$\begin{pmatrix} H_{\vartheta\vartheta}(q_d) & H_{\vartheta\delta}(q_d) \\ H_{\vartheta\delta}^T(q_d) & H_{\delta\delta}(q_d) \end{pmatrix} \begin{pmatrix} \Delta\ddot\vartheta \\ \Delta\ddot\delta \end{pmatrix} + \begin{pmatrix} 0 & 0 \\ 0 & D \end{pmatrix} \begin{pmatrix} \Delta\dot\vartheta \\ \Delta\dot\delta \end{pmatrix}$$
$$+ \begin{pmatrix} \dfrac{\partial g_\vartheta}{\partial \vartheta}(q_d) & \dfrac{\partial g_\vartheta}{\partial \delta}(q_d) \\ \dfrac{\partial g_\delta}{\partial \vartheta}(q_d) & K \end{pmatrix} \begin{pmatrix} \Delta\vartheta \\ \Delta\delta \end{pmatrix} = \begin{pmatrix} \Delta u \\ 0 \end{pmatrix}. \tag{6.99}$$

A suitable state space representation of (6.99) is obtained by choosing

$$\Delta x = \begin{pmatrix} \Delta q \\ H(q_d)\Delta\dot q \end{pmatrix}, \tag{6.100}$$

i.e., variations of positions and of generalized momenta. By defining

$$M = \begin{pmatrix} M_{\vartheta\vartheta} & M_{\vartheta\delta} \\ M_{\vartheta\delta}^T & M_{\delta\delta} \end{pmatrix} = H^{-1}(q_d), \tag{6.101}$$

the linear state equations become

$$\Delta\dot x = \begin{pmatrix} 0 & 0 & M_{\vartheta\vartheta} & M_{\vartheta\delta} \\ 0 & 0 & M_{\vartheta\delta}^T & M_{\delta\delta} \\ -\dfrac{\partial g_\vartheta}{\partial \vartheta} & -\dfrac{\partial g_\vartheta}{\partial \delta} & 0 & 0 \\ -\dfrac{\partial g_\delta}{\partial \vartheta} & -K & -DM_{\vartheta\delta}^T & -DM_{\delta\delta} \end{pmatrix} \Delta x + \begin{pmatrix} 0 \\ 0 \\ I \\ 0 \end{pmatrix} \Delta u$$

$$= A\Delta x + B\Delta u. \tag{6.102}$$

The linear stabilizing full state feedback can be designed as

$$\Delta u = F\Delta x = K_P \Delta\vartheta + K_D \Delta\dot\vartheta + K_{P\delta}\Delta\delta + K_{D\delta}\Delta\dot\delta \tag{6.103}$$

according to well-established methods, e.g., a pole placement technique. In order to assign any desired set of stable poles, it is necessary and sufficient that the pair (A, B) be controllable; it is not difficult to check that this condition always holds.

A convenient choice is to have the feedback matrix gains in (6.103) in block-diagonal form, corresponding to a decentralized strategy in which each stabilizing input component Δu_i uses only local information from link i; this can be achieved typically under more restrictive controllability conditions. Moreover, a pure damping action can be performed on the deformation modes by setting $K_{P\delta} = 0$; in this way, uncertainty in the knowledge of the static deflection δ_d due to gravity —see (6.86)— turns only into a mismatch in the constant feedforward component u_d, while the system is still asymptotically stabilized around a different (perturbed) configuration.

We remark that the whole procedure can be applied also by linearizing the dynamics along a nominal trajectory, which gives a linear time-varying system whose stabilization by constant feedback is much more critical, though.

6.4 Joint tracking control

When trajectory tracking is of concern, performance of the above class of linear controllers is usually quite poor. It is then necessary to resort to *nonlinear control* strategies that take into account the largely varying dynamic couplings of the flexible robot arm during motion. The goal is to reproduce a time-varying smooth reference trajectory $q_d(t) = (\vartheta_d^T(t) \quad \delta_d^T(t))^T$ characterizing the desired behaviour. However, due to the fact that fewer control inputs than configuration variables are available, it is in general not possible to achieve simultaneous tracking of both joint and deflection variables, unless the latter are defined in a consistent manner with the former. Upon this premise, an input–output *inversion* controller will be presented that guarantees exact reproduction of a desired joint trajectory $\vartheta_d(t)$, while inducing a bounded evolution for $\delta(t)$. An alternative design will be introduced that exploits the typical *two-time scale* nature of the rigid plus flexible equations of motion, providing approximate reproduction of $\vartheta_d(t)$. Both these model-based nonlinear control laws ensure closed-loop stability without any additional action required.

6.4.1 Inversion control

Trajectory tracking in nonlinear systems is typically achieved by input–output *inversion control* techniques. Once a meaningful output has been defined for the system, a nonlinear state feedback is designed so that the resulting closed-loop system is transformed into a linear and decoupled one, with the possible appearance of an unobservable internal dynamics. On the assumption of stability of the resulting closed-loop system, *exact* reproduction of smooth desired output trajectories is feasible.

For the purpose of control derivation, it is convenient to extract the

6.4. JOINT TRACKING CONTROL   243

flexible accelerations from (6.75) as

$$\ddot{\delta} = -H_{\delta\delta}^{-1}\left(c_\delta + g_\delta + D\dot{\delta} + K\delta + H_{\vartheta\delta}^T \ddot{\vartheta}\right) \quad (6.104)$$

which, substituted into the upper part of (6.75), gives

$$(H_{\vartheta\vartheta} - H_{\vartheta\delta}H_{\delta\delta}^{-1}H_{\vartheta\delta}^T)\ddot{\vartheta} + c_\vartheta + g_\vartheta - H_{\vartheta\delta}H_{\delta\delta}^{-1}(c_\delta + g_\delta + D\dot{\delta} + K\delta) = u. \quad (6.105)$$

Notice that eq. (6.105) describes the modification that undergoes the rigid body dynamics due to the effects of link flexibility. The matrix $H_{\vartheta\vartheta} - H_{\vartheta\delta}H_{\delta\delta}^{-1}H_{\vartheta\delta}^T$ has full rank as can be seen from the following identity

$$\begin{pmatrix} H_{\vartheta\vartheta} & H_{\vartheta\delta} \\ H_{\vartheta\delta}^T & H_{\delta\delta} \end{pmatrix} \begin{pmatrix} I & 0 \\ -H_{\delta\delta}^{-1}H_{\vartheta\delta}^T & I \end{pmatrix} = \begin{pmatrix} H_{\vartheta\vartheta} - H_{\vartheta\delta}H_{\delta\delta}^{-1}H_{\vartheta\delta}^T & H_{\vartheta\delta} \\ 0 & H_{\delta\delta} \end{pmatrix} \quad (6.106)$$

and the positive definiteness of the inertia matrix.

We define the system output as the vector of joint variables ϑ. Following the inversion algorithm, this output needs to be differentiated as many times as until the input explicitly appearing. Inspection of eq. (6.105) suggests that the joint accelerations $\ddot{\vartheta}$ are at the same differential level as the torque inputs u. Therefore, the *relative degree* is uniform for all outputs and equal to *two*, and the input u can be fully recovered from (6.105).

Let u_0 denote a joint acceleration vector. Setting $\ddot{\vartheta} = u_0$ in (6.105) and solving for u yields the feedback law

$$u = (H_{\vartheta\vartheta} - H_{\vartheta\delta}H_{\delta\delta}^{-1}H_{\vartheta\delta}^T)u_0 + c_\vartheta + g_\vartheta - H_{\vartheta\delta}H_{\delta\delta}^{-1}(c_\delta + g_\delta + D\dot{\delta}) + K\delta, \quad (6.107)$$

where $(H_{\vartheta\vartheta} - H_{\vartheta\delta}H_{\delta\delta}^{-1}H_{\vartheta\delta}^T)^{-1}$ is the so-called *decoupling matrix* of the system and is nonsingular. From (6.107) it can be recognized that only the inversion of the block relative to the flexible variables is required for control law implementation. Hence, the complexity of this nonlinear feedback strategy increases only with the number of flexible variables; in the limit, no inertia matrix inversion is required for the rigid case and the inversion control law reduces to the well-known *inverse dynamics control*. Further, if $H_{\delta\delta}$ is constant, its inversion can be conveniently performed once for all off-line.

The control (6.107) transforms the closed-loop system (6.105) into the input–output linearized form

$$\ddot{\vartheta} = u_0 \quad (6.108)$$
$$\ddot{\delta} = -H_{\delta\delta}^{-1}(H_{\vartheta\delta}^T u_0 + c_\delta + g_\delta + D\dot{\delta} + K\delta). \quad (6.109)$$

In order to achieve tracking of a desired trajectory $\vartheta_d(t)$, the control design is completed by choosing the joint acceleration as

$$u_0 = \ddot{\vartheta}_d + K_D(\dot{\vartheta}_d - \dot{\vartheta}) + K_P(\vartheta_d - \vartheta), \quad (6.110)$$

where $K_P > 0$, $K_D > 0$ are feedback matrix gains that allow pole placement in the open left-hand complex half plane for the linear input–output part (6.108). From (6.110) it is clear that the desired trajectory must be differentiable at least once for having exact reproduction.

As previously mentioned, the applicability of the inversion controller (6.107) is based on the stability of the induced unobservable dynamics (6.109). The analysis can be carried out by studying the so-called *zero dynamics* associated with the system (6.108) and (6.109). This dynamics is obtained by constraining the output ϑ of the system to be a constant, without loss of generality zero. Hence, from (6.109) we obtain

$$\ddot{\delta} = -H_{\delta\delta}^{-1}(c_\delta + g_\delta + D\dot{\delta} + K\delta), \qquad (6.111)$$

where functional dependence is dropped but it is intended that terms are evaluated for $\dot{\vartheta} = 0$.

A sufficient condition that guarantees at least local stability of the overall closed-loop system is that the zero dynamics (6.111) is *asymptotically stable*.

Lemma 6.1 *The state*

$$\delta = \delta_d = -K^{-1} g_\delta(\vartheta_d) \qquad \dot{\delta} = 0$$

is a globally asymptotically stable equilibrium point for system (6.111).

⋄ ⋄ ⋄

Proof. The proof goes through a similar Lyapunov direct method argument as in Section 6.3.1, using the energy-based candidate function — see (6.92)—

$$\begin{aligned} V &= \frac{1}{2}\dot{\delta}^T B_{\delta\delta}\dot{\delta} + \frac{1}{2}(\delta_d - \delta)^T K(\delta_d - \delta) \\ &\quad + U_g(\vartheta_d, \delta) - U_g(\vartheta_d, \delta_d) + (\delta_d - \delta)^T g_\delta(\vartheta_d), \end{aligned} \qquad (6.112)$$

and exploiting the skew-symmetry of $N_{\delta\delta} = \dot{H}_{\delta\delta} - 2C_{\delta\delta}$.

⋄

This lemma ensures also the overall closed-loop stability of system (6.108)–(6.110) when the regulation case is considered.

The above result is useful also in the trajectory tracking case ($\dot{\vartheta}_d \neq 0$). For instance, consider the simpler case of an inertia matrix independent of δ. When $\vartheta = \vartheta_d(t)$ is imposed by the inversion control, from (6.109) the flexible variables satisfy the following *linear time-varying* equation

$$\ddot{\delta} = f_\delta(t) - A_2(t)\dot{\delta} - A_1(t)\delta, \qquad (6.113)$$

6.4. JOINT TRACKING CONTROL

where

$$f_\delta(t) = -H_{\delta\delta}^{-1}(\vartheta_d)\left(H_{\vartheta\delta}^T(\vartheta_d)\ddot{\vartheta}_d + c_\delta(\vartheta_d,\dot{\vartheta}_d) + g_\delta(\vartheta_d)\right) \qquad (6.114)$$

is a known function of time, and

$$A_1(t) = -H_{\delta\delta}^{-1}(\vartheta_d)K, \qquad (6.115)$$
$$A_2(t) = -H_{\delta\delta}^{-1}(\vartheta_d)D. \qquad (6.116)$$

Then, as long as all time-varying functions are bounded, stability is ensured by Lemma 6.1 even during trajectory tracking. However, in general, we have to check a further condition; namely, the input-to-state stability of the closed-loop system.

In the above derivation of the inversion control, it was assumed that full state feedback is available, i.e., ϑ, $\dot{\vartheta}$, δ, $\dot{\delta}$ should be measurable. In general, joint positions and velocities are measured via ordinary encoders and tachometers mounted on the actuators, while for link deflection different apparatus can be used ranging from strain gauges to accelerometers or optical devices. Nevertheless, in spite of the availability of direct measurements of link flexibility, it may be convenient to avoid their use within the computation of the nonlinear part of the controller. The joint-based approach lends itself to a cheap implementation in terms of joint variable measures only, obtained by keeping the robustifying linear feedback (6.110) and performing the nonlinear compensation as a *feedforward* action. This generalizes in a natural way the linear PD control (6.85) to the tracking case.

Specifically, given a differentiable joint trajectory $\vartheta_d(t)$, forward integration of the flexible dynamics

$$\ddot{\delta} = -H_{\delta\delta}^{-1}(\vartheta_d,\delta)\left(c_\delta(\vartheta_d,\delta,\dot{\vartheta}_d,\dot{\delta})+g_\delta(\vartheta_d)+D\dot{\delta}+K\delta+H_{\vartheta\delta}^T(\vartheta_d,\delta)\ddot{\vartheta}_d\right) \quad (6.117)$$

from initial conditions $\delta(0)=\delta_0$, $\dot{\delta}(0)=\dot{\delta}_0$ provides the nominal evolutions $\delta_d(t)$, $\dot{\delta}_d(t)$ of the flexible variables. Hence, evaluation of the nonlinearities in (6.107) along the computed state trajectory gives a control law in the form

$$u = u_d(t) + K_P(t)(\vartheta_d - \vartheta) + K_D(t)(\dot{\vartheta}_d - \dot{\vartheta}), \qquad (6.118)$$

where (6.110) has been used, and

$$\begin{aligned}u_d(t) = {}& H_{\vartheta\vartheta}(\vartheta_d,\delta_d)\ddot{\vartheta}_d + c_\vartheta(\vartheta_d,\delta_d,\dot{\vartheta}_d,\dot{\delta}_d) + g_\vartheta(\vartheta_d,\delta_d)\\
& - H_{\vartheta\delta}(\vartheta_d,\delta_d)H_{\delta\delta}^{-1}(\vartheta_d,\delta_d)\Big(H_{\vartheta\delta}^T(\vartheta_d,\delta_d)\ddot{\vartheta}_d + c_\delta(\vartheta_d,\delta_d,\dot{\vartheta}_d,\dot{\delta}_d)\\
& + g_\delta(\vartheta_d) + K\delta_d + D\dot{\delta}_d\Big) \end{aligned} \qquad (6.119)$$

$$K_P(t) = \left(H_{\vartheta\vartheta}(\vartheta_d,\delta_d) - H_{\vartheta\delta}(\vartheta_d,\delta_d)H_{\delta\delta}^{-1}(\vartheta_d,\delta_d)H_{\vartheta\delta}^T(\vartheta_d,\delta_d)\right)K_P \quad (6.120)$$

$$K_D(t) = \left(H_{\vartheta\vartheta}(\vartheta_d,\delta_d) - H_{\vartheta\delta}(\vartheta_d,\delta_d)H_{\delta\delta}^{-1}(\vartheta_d,\delta_d)H_{\vartheta\delta}^T(\vartheta_d,\delta_d)\right)K_D. \quad (6.121)$$

The initial conditions for numerical integration of (6.117) are typically $\delta_0 = \dot{\delta}_0 = 0$ associated with an undeformed rest configuration for the arm; however, any set of initial values could be used since this dynamics is asymptotically stable, as previously demonstrated.

An even simpler implementation of (6.118) is

$$u = u_d(t) + K_P(\vartheta_d - \vartheta) + K_D(\dot{\vartheta}_d - \dot{\vartheta}), \quad (6.122)$$

with constant feedback matrices.

6.4.2 Two-time scale control

An alternative nonlinear approach to the problem of joint trajectory tracking can be devised using the different time scale between the flexible body dynamics and the rigid body dynamics. The equations of motion of the robot arm can be separated into the rigid and flexible parts by considering the inverse M of the inertia matrix H, which can be partitioned as —see also (6.101)—

$$H^{-1}(\vartheta,\delta) = M(\vartheta,\delta) = \begin{pmatrix} M_{\vartheta\vartheta}(\vartheta,\delta) & M_{\vartheta\delta}(\vartheta,\delta) \\ M_{\vartheta\delta}^T(\vartheta,\delta) & M_{\delta\delta}(\vartheta,\delta) \end{pmatrix}. \quad (6.123)$$

Then, the model (6.75) becomes

$$\ddot{\vartheta} = M_{\vartheta\vartheta}(\vartheta,\delta)(u - c_\vartheta(\vartheta,\delta,\dot{\vartheta},\dot{\delta}) - g_\vartheta(\vartheta,\delta))$$
$$\quad - M_{\vartheta\delta}(\vartheta,\delta)(c_\delta(\vartheta,\delta,\dot{\vartheta},\dot{\delta}) + g_\delta(\vartheta) + D\dot{\delta} + K\delta) \quad (6.124)$$

$$\ddot{\delta} = M_{\vartheta\delta}^T(\vartheta,\delta)(u - c_\vartheta(\vartheta,\delta,\dot{\vartheta},\dot{\delta}) - g_\vartheta(\vartheta,\delta))$$
$$\quad - M_{\delta\delta}(\vartheta,\delta)(c_\delta(\vartheta,\delta,\dot{\vartheta},\dot{\delta}) + g_\delta(\vartheta) + D\dot{\delta} + K\delta). \quad (6.125)$$

Time scale separation is intuitively determined by the values of oscillation frequencies of flexible links and in turn by the magnitudes of the elements of K (see Section 6.1). In detail, consider the smallest stiffness coefficient K_m so that the diagonal matrix K can be factored as $K = K_m \bar{K}$. The *elastic force* variables can be defined as

$$z = K_m \bar{K}\delta = \frac{1}{\epsilon^2}\bar{K}\delta, \quad (6.126)$$

6.4. JOINT TRACKING CONTROL

and accordingly the model (6.124) and (6.125) can be rewritten as

$$\ddot{\vartheta} = M_{\vartheta\vartheta}(\vartheta, \epsilon^2 z)\left(u - c_\vartheta(\vartheta, \epsilon^2 z, \dot{\vartheta}, \epsilon^2 \dot{z}) - g_\vartheta(\vartheta, \epsilon^2 z)\right)$$
$$- M_{\vartheta\delta}(\vartheta, \epsilon^2 z)\left(c_\delta(\vartheta, \epsilon^2 z, \dot{\vartheta}, \epsilon^2 \dot{z}) + g_\delta(\vartheta) + \epsilon^2 D\bar{K}^{-1}\dot{z} + z\right) \quad (6.127)$$

$$\epsilon^2 \ddot{z} = \bar{K} M_{\vartheta\delta}^T(\vartheta, \epsilon^2 z)\left(u - c_\vartheta(\vartheta, \epsilon^2 z, \dot{\vartheta}, \epsilon^2 \dot{z}) - g_\vartheta(\vartheta, \epsilon^2 z)\right)$$
$$- \bar{K} M_{\delta\delta}(\vartheta, \epsilon^2 z)\left(c_\delta(\vartheta, \epsilon^2 z, \dot{\vartheta}, \epsilon^2 \dot{z}) + g_\delta(\vartheta) + \epsilon^2 D\bar{K}^{-1}\dot{z} + z\right). \quad (6.128)$$

Eqs. (6.127) and (6.128) represent a *singularly perturbed* form of the flexible arm model, where ϵ is the so-called *perturbation parameter*. Although the model can be always recast as above, this form is significant only when ϵ is small, meaning that an effective time scale separation occurs; in particular, the flexural rigidity EI and the length ℓ of each link concur to determine the magnitude of the perturbation parameter and influence the validity of the approach.

When $\epsilon \to 0$, the model of an *equivalent* rigid arm is recovered. In fact, setting $\epsilon = 0$ and solving for z in (6.128) gives

$$z_s = M_{\delta\delta}^{-1}(\vartheta_s, 0) M_{\vartheta\delta}^T(\vartheta_s, 0)\left(u_s - c_\vartheta(\vartheta_s, 0, \dot{\vartheta}_s, 0) - g_\vartheta(\vartheta_s, 0)\right)$$
$$- c_\delta(\vartheta_s, 0, \dot{\vartheta}_s, 0) - g_\delta(\vartheta_s), \quad (6.129)$$

where subscript s indicates that the system is considered in the *slow* time scale. Plugging (6.129) into (6.127) with $\epsilon = 0$ yields

$$\ddot{\vartheta}_s = \left(M_{\vartheta\vartheta}(\vartheta_s, 0) - M_{\vartheta\delta}(\vartheta_s, 0) M_{\delta\delta}^{-1}(\vartheta_s, 0) M_{\vartheta\delta}^T(\vartheta_s, 0)\right) \cdot$$
$$\left(u_s - c_\vartheta(\vartheta_s, 0, \dot{\vartheta}_s, 0) - g_\vartheta(\vartheta_s, 0)\right). \quad (6.130)$$

It is not difficult to check that

$$M_{\vartheta\vartheta}(\vartheta_s, 0) - M_{\vartheta\delta}(\vartheta_s, 0) M_{\delta\delta}^{-1}(\vartheta_s, 0) M_{\vartheta\delta}^T(\vartheta_s, 0) = H_{\vartheta\vartheta}^{-1}(\vartheta_s, 0) \quad (6.131)$$

where $H_{\vartheta\vartheta}(\vartheta_s, 0)$ is the inertia matrix of the equivalent rigid arm, so that eq. (6.130) becomes

$$H_{\vartheta\vartheta}(\vartheta_s, 0)\ddot{\vartheta}_s + c_\vartheta(\vartheta_s, 0, \dot{\vartheta}_s, 0) + g_\vartheta(\vartheta_s, 0) = u_s. \quad (6.132)$$

In order to study the dynamics of the system in the *fast* time scale, the so-called *boundary-layer* subsystem has to be identified. This can be obtained by setting $\tau = t/\epsilon$, treating the slow variables as constants in the fast time scale, and introducing the fast variables $z_f = z - z_s$; thus, the fast subsystem of (6.128) is

$$\frac{d^2 z_f}{d\tau^2} = -\bar{K}M_{\delta\delta}(\vartheta_s, 0)z_f + \bar{K}M_{\vartheta\delta}^T(\vartheta_s, 0)u_f, \qquad (6.133)$$

where the fast control $u_f = u - u_s$ has been introduced accordingly.

On the basis of the above two-time scale model, the design of a feedback controller for the system (6.127) and (6.128) can be performed according to a *composite control* strategy, i.e.,

$$u = u_s(\vartheta_s, \dot{\vartheta}_s) + u_f(z_f, dz_f/d\tau) \qquad (6.134)$$

with the constraint that $u_f(0,0) = 0$, so that u_f is inactive along the equilibrium manifold specified by (6.129).

The slow control for the rigid nonlinear subsystem (6.132) can be designed according to the well-known inverse dynamics concept used for rigid manipulators. On the other hand, the fast subsystem is a marginally stable linear slowly time-varying system that can be stabilized by a fast control, e.g., based on classical pole placement techniques; an output feedback design is typically required to face the lack of measurements for the derivatives of flexible variables.

By applying Tikhonov's theorem, it can be shown that under the composite control (6.134), the goal of tracking a desired joint trajectory together with stabilizing the deflections around the equilibrium manifold, naturally established by the rigid subsystem under the slow control, is achieved with an order of ϵ approximation. A robustness analysis relative to the magnitude of the perturbation parameter can be performed. When ϵ is not small enough, the use of integral manifolds to obtain a more accurate slow subsystem that accounts for the effects of flexibility up to a certain order of ϵ represents a viable strategy to make the composite control perform satisfactorily.

Finally, comparing (6.107) with the inversion control law based on (6.132) clearly points out the difference between the synthesis at the joint level using the full dynamics and the one restricted to the equivalent rigid arm model.

6.5 End-effector tracking control

The problem of accurate end-effector trajectory tracking is the most difficult one for flexible manipulators. Even when an effective control strategy has been designed at the joint level, still considerable vibration at the arm tip

6.5. END-EFFECTOR TRACKING CONTROL 249

may be observed during motion for a lightly damped structure. From an input–output point of view, we may try to use the n available joint torque inputs for controlling n output variables characterizing the tip position, i.e., for reproducing a desired trajectory $y_d(t)$. We remark that the choice of suitable reference trajectories that do not induce large deflections and excite arm vibrations in the frequency range of interest is more critical than in the joint tracking case.

In this respect, the direct extension of the inversion strategy to the end-effector output typically leads to closed-loop instabilities, nominally generating unbounded deformations and/or very large applied torques. In practice, joint actuators saturate and a jerky arm behaviour is observed, while control of tip motion is completely lost.

This phenomenon is present both in the linear (one-link) and nonlinear (multilink) cases. In fact, according to the modelling presented in Section 6.1, the transfer function of a one-link flexible arm contains stable zeros when the output position is co-located with the joint actuation, but will present pairs of real stable/unstable zeros when the output is relocated at the tip; this qualifies a *nonminimum phase* transfer function. Then, when the system is inverted, feedback *cancellation* of nonminimum phase zeros with unstable poles produces internal instability, not observable in the input–output map. Similarly, it can be shown that the *end-effector zero dynamics* of a multilink flexible arm, i.e., the unobservable dynamics associated with an input action which attempts to keep the end-effector output at a constant (zero) value, is *unstable*; by analogy to the linear case, this situation is referred to as nonminimum phase. Again, since inversion control is designed for cancelling this zero dynamics, closed-loop instability occurs.

Following the above discussion, we can infer that the trajectory tracking problem for nonminimum phase systems has to be handled either by resorting to a suitable feedforward strategy or by avoiding exact feedback cancellation. In both cases, it is convenient to define the desired arm behaviour not in terms of $(\vartheta_d(t), \delta_d(t))$ but equivalently in terms of $(y_d(t), \delta_d(t))$, where y is one-to-one related to the tip position. A bounded evolution $\delta_d(t)$ of the arm deformation variables will then be computed on the basis of the given output trajectory $y_d(t)$, as a natural motion of the system.

An open-loop solution, suitable for the *one-link linear case*, will be derived defining the inversion procedure in the *frequency domain* by assuming that the given trajectory $y_d(t)$ is part of a periodic signal. This solution leads to a *noncausal* command, since the obtained input $u_d(t)$ anticipates the actual start of the reference output trajectory. A more general technique will be presented that computes the bounded evolution $\delta_d(t)$ through the solution of a set of (nonlinear) partial differential equations depending on the desired output trajectory; a time-varying reference state is then constructed and used in a combined feedforward/feedback strategy, where the feedback part stabilizes the system around the state trajectory. Thus, *nonlinear regulation* will be achieved with simultaneous closed-loop stabilization and output trajectory tracking; the latter is in general obtained only asymptotically, depending on the initial conditions of the arm.

6.5.1 Frequency domain inversion

Consider a one-link flexible arm model in the generic form

$$\begin{pmatrix} h_{\vartheta\vartheta} & h_{\delta\vartheta}^T \\ h_{\delta\vartheta} & H_{\delta\delta} \end{pmatrix} \begin{pmatrix} \ddot{\vartheta}(t) \\ \ddot{\delta}(t) \end{pmatrix} + \begin{pmatrix} 0 & 0 \\ 0 & D \end{pmatrix} \begin{pmatrix} \dot{\vartheta}(t) \\ \dot{\delta}(t) \end{pmatrix} + \begin{pmatrix} 0 & 0 \\ 0 & K \end{pmatrix} \begin{pmatrix} \vartheta(t) \\ \delta(t) \end{pmatrix} = \begin{pmatrix} u(t) \\ 0 \end{pmatrix} \tag{6.135}$$

that can be any of the models derived in Section 6.1. The tip position output associated with (6.135) can be written as

$$y(t) = \begin{pmatrix} 1 & c_e^T \end{pmatrix} \begin{pmatrix} \vartheta(t) \\ \delta(t) \end{pmatrix}, \tag{6.136}$$

where $c_e = \varphi'(\ell)$ expresses the contributions of each mode shape to the tip angular position.

In order to reproduce a desired smooth time profile $y_d(t)$, a stable inversion of the system can be set up in the Fourier domain by regarding both the input and the output as periodic functions. In this way, provided that the involved signals are Fourier-transformable, all quantities will automatically be bounded over time. The intrinsic assumption of this method leads to the *open-loop* computation of an input command $u_d(t)$ that will be used as a feedforward on the real system.

We start by rewriting eq. (6.135) as

$$\begin{pmatrix} h_{\vartheta\vartheta} & h_{\delta\vartheta}^T - h_{\vartheta\vartheta}c_e^T \\ h_{\delta\vartheta} & H_{\delta\delta} - h_{\delta\vartheta}c_e^T \end{pmatrix} \begin{pmatrix} \ddot{y}(t) \\ \ddot{\delta}(t) \end{pmatrix} + \begin{pmatrix} 0 & 0 \\ 0 & D \end{pmatrix} \begin{pmatrix} \dot{y}(t) \\ \dot{\delta}(t) \end{pmatrix}$$
$$+ \begin{pmatrix} 0 & 0 \\ 0 & K \end{pmatrix} \begin{pmatrix} y(t) \\ \delta(t) \end{pmatrix} = \begin{pmatrix} u(t) \\ 0 \end{pmatrix}, \tag{6.137}$$

where (6.136) has been used. Let then

$$\ddot{Y}(\omega) = \int_{-\infty}^{\infty} \exp(-j\omega t) \ddot{y}(t) dt \tag{6.138}$$

be the bilateral Fourier transform of the output acceleration; hence, eq.

6.5. END-EFFECTOR TRACKING CONTROL

(6.137) can be rewritten in the frequency domain as

$$\begin{pmatrix} h_{\vartheta\vartheta} & h_{\delta\vartheta}^T - h_{\vartheta\vartheta}c_e^T \\ h_{\delta\vartheta} & H_{\delta\delta} - h_{\delta\vartheta}c_e^T + \frac{1}{j\omega}D - \frac{1}{\omega^2}K \end{pmatrix} \begin{pmatrix} \ddot{Y}(\omega) \\ \ddot{\Delta}(\omega) \end{pmatrix} = \begin{pmatrix} U(\omega) \\ 0 \end{pmatrix}, \quad (6.139)$$

where $\ddot{\Delta}(\omega)$ and $U(\omega)$ are the Fourier transforms of the flexible accelerations and of the input torque, respectively. Inverting eq. (6.139) gives

$$\begin{pmatrix} \ddot{Y}(\omega) \\ \ddot{\Delta}(\omega) \end{pmatrix} = \begin{pmatrix} g_{11}(\omega) & g_{12}^T(\omega) \\ g_{21}(\omega) & G_{22}(\omega) \end{pmatrix} \begin{pmatrix} U(\omega) \\ 0 \end{pmatrix}, \quad (6.140)$$

and thus

$$U(\omega) = \frac{1}{g_{11}(\omega)}\ddot{Y}(\omega) = r(\omega)\ddot{Y}(\omega). \quad (6.141)$$

When a desired symmetric zero-mean acceleration profile $\ddot{y}_d(t)$ is given such that $\ddot{y}_d(t) = 0$ for $t \leq -T/2$ and $t \geq T/2$, this signal can be embedded in a periodic one and the Fourier transform (6.138) can be truncated outside the interval $[-T/2, T/2]$. Plugging $\ddot{Y}_d(\omega)$ in (6.141) yields $U_d(\omega)$, which generates the required input torque $u_d(t)$ through finite inverse Fourier transformation.

The obtained time profile expands beyond the interval $[-T/2, T/2]$, since the inverse system is *noncausal*. In fact, the command $u_d(t)$ can be also expressed as

$$u_d(t) = \int_{-\infty}^{\infty} r(t-\tau)\ddot{y}_d(\tau)d\tau = \int_{-T/2}^{T/2} r(t-\tau)\ddot{y}_d(\tau)d\tau, \quad (6.142)$$

where $r(t)$ is the impulsive time response of the inverse system corresponding to $r(\omega)$ in (6.141); since here $r(t-\tau) \neq 0$ also for $t < \tau$, then $u_d(t) \neq 0$ also for $t < -T/2$. In practice, the energy content of the required input torque will rapidly decay to zero outside the desired interval duration T of the output motion, and a time truncation can be reasonably performed.

It should be noted that eq. (6.140) provides also the frequency-based description of the flexible variables associated with the given output motion, which can be converted as well into a time evolution $\delta_d(t)$. In particular, the initial values $\delta_d(-T/2)$ and $\dot{\delta}_d(-T/2)$ correspond to the specific initial arm deflection state which gives overall bounded deformation under inversion control.

Finally, we remark that a noncausal behaviour of the inverse system is obtained whenever the original system presents a finite time delay between the input action and the observed output effects.

The above technique is inherently based on linear concepts, and the extension to the nonlinear multilink case can be performed only by iterating its application to repeated linear approximations of the system along the nominal trajectory.

6.5.2 Nonlinear regulation

The problem of achieving output tracking, at least asymptotically, while enforcing internal state stability is a classical *regulation problem,* whose solution is widely known in the linear case under full state feedback. The extension of the same approach to the nonlinear setting requires some technical assumptions and involves a larger amount of off-line computation, but it is rather straightforward in the robotic case.

The key point is to compute a *bounded* reference state trajectory for the flexible arm so as to produce the desired output motion. In connection with any bounded and smooth output trajectory $y_d(t)$, typically generated by an autonomous dynamic system (an *exosystem*), such a trajectory $(\vartheta_d(t), \delta_d(t))$ surely exists. The control law can then be chosen as formed by two contributions; namely, a *feedforward* term which drives the system trajectories along their desired evolution, and a *state feedback* term necessary to stabilize the closed-loop dynamics around the nominal trajectory. This stabilizing action can be typically designed using only first-order information, i.e., as a linear feedback, in view of the local controllability of the manipulator equations. Once the flexible arm is in the proper deformation state, only the feedforward term is active.

For convenience, we consider a generalization of (6.136) as system output, i.e.,

$$y = \vartheta + C\delta, \qquad (6.143)$$

where the elements in the constant matrix C include the contributions of link mode shapes. On the assumption of small deformation, this output is one-to-one related to the Cartesian position and orientation of the tip through the standard direct kinematics of the arm.

Moreover, we assume that the output reference trajectory is generated by an exosystem that can be chosen, without loss of generality, as a linear system in observable canonical form with

$$Y_d = \{y_{di}, \dot{y}_{di}, \ddot{y}_{di}, \ldots, y_{di}^{r_i-1}; \; i = 1, \ldots, n\} \qquad (6.144)$$

taken as its state; here, r_i is the smoothness degree of the i-th output trajectory.

6.5. END-EFFECTOR TRACKING CONTROL

The design of a nonlinear regulator is simplified by rewriting the manipulator dynamics (6.75) in terms of the new coordinates (y, δ) as

$$H_{\vartheta\vartheta}(y - C\delta, \delta)\ddot{y} + \big(H_{\vartheta\delta}(y - C\delta, \delta) - H_{\vartheta\vartheta}(y - C\delta, \delta)C\big)\ddot{\delta}$$
$$+ c_{\vartheta}(y - C\delta, \delta, \dot{y} - C\dot{\delta}, \dot{\delta}) + g_{\vartheta}(y - C\delta, \delta) = u \quad (6.145)$$
$$H_{\vartheta\delta}^{T}(y - C\delta, \delta)\ddot{y} + \big(H_{\delta\delta}(y - C\delta, \delta) - H_{\vartheta\delta}^{T}(y - C\delta, \delta)C\big)\ddot{\delta}$$
$$+ c_{\delta}(y - C\delta, \delta, \dot{y} - C\dot{\delta}, \dot{\delta}) + g_{\delta}(y - C\delta) + D\dot{\delta} + K\delta = 0, \quad (6.146)$$

where (6.143) has been used.

In order to determine the reference state evolution associated with $y_d(t)$, it is sufficient to specify only $\delta_d(t)$ and its derivative $\dot{\delta}_d(t)$. Hence, the problem reduces to determining the vector functions $\delta_d = \pi(Y_d)$ and $\dot{\delta}_d = (\partial \pi/\partial Y_d)\dot{Y}_d$. In particular, the function $\pi(Y_d)$ should satisfy eq. (6.146), evaluated along the reference evolution of the output, i.e.,

$$H_{\vartheta\delta}^{T}(y_d, \pi(Y_d))\ddot{y}_d + \big(H_{\delta\delta}(y_d, \pi(Y_d)) - H_{\vartheta\delta}^{T}(y_d, \pi(Y_d))C\big)\ddot{\pi}(Y_d, \dot{Y}_d, \ddot{Y}_d)$$
$$+ c_{\delta}(y_d, \pi(Y_d), \dot{y}_d, \dot{\pi}(Y_d, \dot{Y}_d)) + g_{\delta}(y_d, \pi(Y_d))$$
$$+ D\dot{\pi}(Y_d, \dot{Y}_d) + K\pi(Y_d) = 0. \quad (6.147)$$

This equation is independent of the applied torque and should be considered as a dynamic constraint for $\pi(Y_d)$. Being (6.147) a *nonlinear time-varying* differential equation, it is usually impossible to determine a bounded solution in closed form. We notice that this equation reduces to a matrix format in the case of linear systems, as for the one-link flexible arm.

A feasible approach is to build an approximate solution $\hat{\pi}(Y_d)$ for the arm deformation by using basis elements which are bounded functions of their arguments, e.g., polynomials in Y_d. As long as each component $y_{di}(t)$ of the desired trajectory and its derivatives up to order $(r_i + 1)$ are bounded, the approximation $\hat{\pi}(Y_d)$ is necessarily a bounded function of time. The problem of determining the constant coefficients in the expansion can then be solved through a recursive procedure using the polynomial identity principle.

Once a solution $\hat{\pi}(Y_d) \approx \pi(Y_d)$ is obtained up to any desired accuracy, back-substitution of the reference deformation δ_d, of the desired output trajectory y_d, and of their time derivatives into (6.145) will give the nominal feedforward regulation term as

$$u_d = H_{\vartheta\vartheta}(y_d, \delta_d)\ddot{y}_d + \big(H_{\vartheta\delta}(y_d, \delta_d) - H_{\vartheta\vartheta}(y_d, \delta_d)C\big)\ddot{\delta}_d$$
$$+ c_{\vartheta}(y_d, \delta_d, \dot{y}_d, \dot{\delta}_d) + g_{\vartheta}(y_d, \delta_d)$$
$$= \gamma(Y_d, \dot{Y}_d, \ddot{Y}_d). \quad (6.148)$$

The nonlinear regulator law takes on the final form

$$u = u_d + \hat{F}\begin{pmatrix} y_d - y \\ \dot{y}_d - \dot{y} \\ \delta_d - \delta \\ \dot{\delta}_d - \dot{\delta} \end{pmatrix} = u_d + F\begin{pmatrix} \vartheta_d - \vartheta \\ \dot{\vartheta}_d - \dot{\vartheta} \\ \delta_d - \delta \\ \dot{\delta}_d - \dot{\delta} \end{pmatrix}, \qquad (6.149)$$

where the feedback matrix F stabilizes the linear approximation of the system dynamics. Indeed, if a global stabilizer is available, convergence to the reference state behaviour can be achieved from *any* initial state. In particular, a PD feedback on the joint variables ϑ alone was shown to stabilize any arm configuration; as a result, a simplified form of F can be used in (6.149), and the overall regulator can be implemented using only *partial state* feedback.

It is worth noticing that solving eq. (6.147) for $\delta_d = \pi(Y_d)$ provides as a byproduct the unique initial state at time $t = 0$ that is bounded over the whole desired output motion. This should be related to the approach of the previous subsection.

While the stability property of the regulator approach rules out inversion techniques for end-effector tracking, it should be stressed that only *asymptotic* output reproduction can be guaranteed if the initial state of the flexible arm does not coincide with the solution to (6.147) at time $t = 0$. Moreover, during error transients, the input–output behaviour of the closed-loop system is still described by a fully nonlinear dynamics, contrary to the inversion case.

6.6 Further reading

A recent tutorial on modelling, design, and control of flexible manipulator arms is available in [9].

Modal analysis based on Euler-Bernoulli beam equations can be found in classical mechanics textbooks, covering flexible-body dynamics, e.g., [36]. The issue of unconstrained vs. constrained mode expansion was discussed in [3] and further in [7]. The dependence of the fundamental frequency of a single link on its shape and structure has been studied in [56]. The combined Lagrangian-assumed modes method for modelling multilink flexible arms is due to [8]; discussion on infinite-dimensional models can be found in [25], and the effect of mode truncation was studied in [31]. Alternative approaches to the assumed modes method are the finite-element approach [52] and the Ritz-Kantorovich expansions [53]. The use of symbolic software packages for dynamic modelling of manipulators with both joint and link flexibility was proposed in [11]. Early work on modelling,

6.6. FURTHER READING

identification and control of a single-link flexible arm was carried out in, e.g., [10, 45, 27, 57, 58]. Analysis of the finite- and infinite-dimensional transfer functions was performed in [29, 55]. Two-link flexible arms were studied in, e.g., [37, 38].

Performance of co-located joint feedback controllers was investigated in [12]. The importance of dynamic models for control of flexible manipulators was pointed out in [22], and a symbolic closed-form model was derived in [21]. The problem of joint regulation of flexible arms under gravity was solved in [23] where also the case of no damping is covered; the control scheme was extended to tip regulation in [46]. An iterative learning scheme for positioning the end effector without knowledge of the gravity term is given in [19]. For enhancement of structural passive damping, the reader is referred to [1]. Active vibration linear control was proposed by [50, 51] and further refined by [14]. Inversion-based controllers that achieve input–output linearization were derived in [24]; the use of the feedforward strategy was tested in [15]. Two-time scale control was proposed by [47] and extended to the output feedback setting in [49]; a refinement of the singularly perturbed model via the use of integral manifolds can be found in [48]. A combined feedback linearization/singular perturbation approach was recently presented in [54]. An overview on the use of perturbation techniques in control of flexible manipulators is given in [26]. Robust schemes using H_∞ control for a one-link flexible arm were proposed in [42, 2]. Alternative techniques for stabilization and trajectory tracking were proposed in [33] using direct strain feedback, and in [30] using acceleration feedback. A state observer for a two-link flexible arm has been designed and implemented in [39].

The problem of suitable trajectory generation for flexible arms was investigated in [5, 28]. The frequency domain approach was presented in [4] for the single-link flexible arm, and was extended to the multilink case in [6]. Performance of non-colocated tip feedback controllers was investigated in [13]. The use of inversion control for stable tracking of an output chosen along the links was proposed in [18], and a control scheme for a single-link flexible arm was proposed in [20]. The problem of output regulation of a flexible robot arm was studied in [16] and a comparison of approaches based on regulation theory is given in [17]. Related work on the use of a partial state feedback regulator can be found in [40]. Stable tip trajectory control for multilink flexible arms is studied in the time domain in [32, 59].

Only planar deformation of flexible arms has been considered in this chapter. A challenging problem is the control of both bending and torsional vibrations for multilink flexible arms; satisfactory results for a single-link arm can be found in [44, 34]. A new line of research concerns the problem of position and force control of flexible link manipulators; preliminary work is reported in [43, 35].

References

[1] T.E. Alberts, L.J. Love, E. Bayo, and H. Moulin, "Experiments with end-point control of a flexible link using the inverse dynamics approach and passive damping," *Proc. 1990 American Control Conf.*, San Diego, CA, pp. 350–355, 1990.

[2] R.N. Banavar and P. Dominic, "An LQG/H_∞ controller for a flexible manipulator," *IEEE Trans. on Control Systems Technology*, vol. 3, pp. 409–416, 1995.

[3] E. Barbieri and Ü. Özgüner, "Unconstrained and constrained mode expansions for a flexible slewing link," *ASME J. of Dynamic Systems, Measurement, and Control*, vol. 110, pp. 416–421, 1988.

[4] E. Bayo, "A finite-element approach to control the end-point motion of a single-link flexible robot," *J. of Robotic Systems*, vol. 4, pp. 63–75, 1987.

[5] E. Bayo and B. Paden, "On trajectory generation for flexible robots," *J. of Robotic Systems*, vol. 4, pp. 229–235, 1987.

[6] E. Bayo, M.A. Serna, P. Papadopoulus, and J. Stubbe, "Inverse dynamics and kinematics of multi-link elastic robots: An iterative frequency domain approach," *Int. J. of Robotics Research*, vol. 8, no. 6, pp. 49–62, 1989.

[7] F. Bellezza, L. Lanari, and G. Ulivi, "Exact modeling of the slewing flexible link," *Proc. 1990 IEEE Int. Conf. on Robotics and Automation*, Cincinnati, OH, pp. 734–739, May 1990.

[8] W.J. Book, "Recursive Lagrangian dynamics of flexible manipulator arms," *Int. J. of Robotics Research*, vol. 3, no. 3, pp. 87–101, 1984.

[9] W.J. Book, "Controlled motion in an elastic world," *ASME J. of Dynamic Systems, Measurement, and Control*, vol. 115, pp. 252–261, 1993.

[10] R.H. Cannon and E. Schmitz, "Initial experiments on the end-point control of a flexible one-link robot," *Int. J. of Robotics Research*, vol. 3, no. 3, pp. 62–75, 1984.

[11] S. Cetinkunt and W.J. Book, "Symbolic modeling and dynamic simulation of robotic manipulators with compliant links and joints," *Robotics & Computer-Integrated Manufacturing*, vol. 5, pp. 301–310, 1989.

[12] S. Cetinkunt and W.J. Book, "Performance limitations of joint variable-feedback controllers due to manipulator structural flexibility," *IEEE Trans. on Robotics and Automation*, vol. 6, pp. 219–231, 1990.

[13] S. Cetinkunt and W.L. Yu, "Closed-loop behavior of a feedback controlled flexible arm: A comparative study," *Int. J. of Robotics Research*, vol. 10, pp. 263-275, 1991.

[14] A. Das and S.N. Singh, "Dual mode control of an elastic robotic arm: Nonlinear inversion and stabilization by pole assignment," *Int. J. of Systems Science*, vol. 21, pp. 1185–1204, 1990.

[15] A. De Luca, L. Lanari, P. Lucibello, S. Panzieri, and G. Ulivi, "Control experiments on a two-link robot with a flexible forearm," *Proc. 29th IEEE Conf. on Decision and Control*, Honolulu, HI, pp. 520–527, 1990.

[16] A. De Luca, L. Lanari, and G. Ulivi, "Output regulation of a flexible robot arm," in *Analysis and Optimization of Systems*, A. Bensoussan and J.L. Lions (Eds.), Lecture Notes in Control and Information Sciences, Springer-Verlag, Berlin, D, vol. 144, pp. 833–842, 1990.

[17] A. De Luca, L. Lanari, and G. Ulivi, "End-effector trajectory tracking in flexible arms: Comparison of approaches based on regulation theory," in *Advanced Robot Control — Proc. Int. Workshop on Nonlinear and Adaptive Control: Issues in Robotics*, C. Canudas de Wit (Ed.), Lecture Notes in Control and Information Sciences, Springer-Verlag, Berlin, D, vol. 162, pp. 190–206, 1991.

[18] A. De Luca, P. Lucibello, and G. Ulivi, "Inversion techniques for trajectory control of flexible robot arms," *J. of Robotic Systems*, vol. 6, pp. 325–344, 1989.

[19] A. De Luca and S. Panzieri, "End-effector regulation of robots with elastic elements by an iterative scheme," *Int. J. of Adaptive Control and Signal Processing*, vol. 10, pp. 379–393, 1996.

[20] A. De Luca and B. Siciliano, "Trajectory control of a non-linear one-link flexible arm," *Int. J. of Control*, vol. 50, pp. 1699–1716, 1989.

[21] A. De Luca and B. Siciliano, "Closed-form dynamic model of planar multilink lightweight robots," *IEEE Trans. on Systems, Man, and Cybernetics*, vol. 21, pp. 826–839, 1991.

[22] A. De Luca and B. Siciliano, "Relevance of dynamic models in analysis and synthesis of control laws for flexible manipulators," in *Robotics and Flexible Manufacturing Systems*, S.G. Tsafestas and J.C. Gentina (Eds.), Elsevier, Amsterdam, NL, pp. 161–168, 1992.

[23] A. De Luca and B. Siciliano, "Regulation of flexible arms under gravity," *IEEE Trans. on Robotics and Automation*, vol. 9, pp. 463–467, 1993.

[24] A. De Luca and B. Siciliano, "Inversion-based nonlinear control of robot arms with flexible links," *AIAA J. of Guidance, Control, and Dynamics*, vol. 16, pp. 1169–1176, 1993.

[25] X. Ding, T.J. Tarn, and A.K. Bejczy, "A novel approach to the modelling and control of flexible robot arms," *Proc. 27th IEEE Conf. on Decision and Control*, Austin, TX, pp. 52–57, 1988.

[26] A.R. Fraser and R.W. Daniel, *Perturbation Techniques for Flexible Manipulators*, Kluwer Academic Publishers, Boston, MA, 1991.

[27] G.G. Hastings and W.J. Book, "A linear dynamic model for flexible robotic manipulators," *IEEE Control Systems Mag.*, vol. 7, no. 1, pp. 61–64, 1987.

[28] J.M. Hyde and W.P. Seering, "Using input command pre-shaping to suppress multiple mode vibration," *Proc. 1991 IEEE Int. Conf. on Robotics and Automation*, Sacramento, CA, pp. 2604–2609, 1991.

[29] H. Kanoh, "Distributed parameter models of flexible robot arms," *Advanced Robotics*, vol. 5, pp. 87–99, 1991.

[30] F. Khorrami and S. Jain, "Nonlinear control with end-point acceleration feedback for a two-link flexible manipulator: Experimental results," *J. of Robotic Systems*, vol. 10, pp. 505–530, 1993.

[31] J. Lin and F.L. Lewis, "Enhanced measurement and estimation methodology for flexible arm control," *J. of Robotic Systems*, vol. 11, pp. 367–385, 1994.

[32] P. Lucibello and M.D. Di Benedetto, "Output tracking for a nonlinear flexible arm," *ASME J. of Dynamic Systems, Measurement, and Control*, vol. 115, pp. 78–85, 1993.

[33] Z.-H. Luo, "Direct strain feedback control of flexible robot arms: New theoretical and experimental results," *IEEE Trans. on Automatic Control*, vol. 38, pp. 1610–1622, 1993.

[34] F. Matsuno, T. Murachi, and Y. Sakawa, "Feedback control of decoupled bending and torsional vibrations of flexible beams," *J. of Robotic Systems*, vol. 11, pp. 341–353, 1994.

[35] F. Matsuno and K. Yamamoto, "Dynamic hybrid position/force control of a two degree-of-freedom flexible manipulator," *J. of Robotic Systems*, vol. 11, pp. 355–366, 1994.

[36] L. Meirovitch, *Analytical Methods in Vibrations*, Macmillan, New York, NY, 1967.

[37] C.M. Oakley and R.H. Cannon, "Initial experiments on the control of a two-link manipulator with a very flexible forearm," *Proc. 1988 American Control Conf.*, Atlanta, GA, pp. 996–1002, 1988.

[38] C.M. Oakley and R.H. Cannon, "Equations of motion for an experimental planar two-link flexible manipulator," *Proc. 1989 ASME Winter Annual Meet.*, San Francisco, CA, pp. 267–278, 1989.

[39] S. Panzieri and G. Ulivi, "Design and implementation of a state observer for a flexible robot," *Proc. 1993 IEEE Int. Conf. on Robotics and Automation*, Atlanta, GA, vol. 3, pp. 204–209, 1993.

[40] F. Pfeiffer, "A feedforward decoupling concept for the control of elastic robots," *J. of Robotic Systems*, vol. 6, pp. 407–416, 1989.

[41] H.R. Pota and T.E. Alberts, "Multivariable transfer functions for a slewing piezoelectric laminate beam," *ASME J. of Dynamic Systems, Measurement, and Control*, vol. 117, pp. 352–359, 1995.

[42] T. Ravichandran, G. Pang, and D. Wang, "Robust H_∞ control of a single flexible link," *Control — Theory and Advanced Technology*, vol. 9, pp. 887–908, 1993.

[43] K. Richter and F. Pfeiffer, "A flexible link manipulator as a force measuring and controlling unit," *Proc. 1991 IEEE Int. Conf. on Robotics and Automation*, Sacramento, CA, pp. 1214–1219, 1991.

[44] Y. Sakawa and Z.H. Luo, "Modeling and control of coupled bending and torsional vibrations of flexible beams," *IEEE Trans. on Automatic Control*, vol. 34, pp. 970–977, 1989.

[45] Y. Sakawa, F. Matsuno, and S. Fukushima, "Modelling and feedback control of a flexible arm," *J. of Robotic Systems*, vol. 2, pp. 453–472, 1985.

[46] B. Siciliano, "An inverse kinematics scheme for flexible manipulators," *Proc. 2nd IEEE Mediterranean Symp. on New Directions in Control & Automation*, Chania, GR, pp. 543–548, 1994.

[47] B. Siciliano and W.J. Book, "A singular perturbation approach to control of lightweight flexible manipulators," *Int. J. of Robotics Research*, vol. 7, no. 4, pp. 79–90, 1988.

[48] B. Siciliano, W.J. Book, and G. De Maria, "An integral manifold approach to control of a one link flexible arm," *Proc. 25th IEEE Conf. on Decision and Control*, Athina, GR, pp. 1131–1134, 1986.

[49] B. Siciliano, J.V.R. Prasad, and A.J. Calise, "Output feedback two-time scale control of multi-link flexible arms," *ASME J. of Dynamic Systems, Measurement, and Control*, vol. 114, pp. 70–77, 1992.

[50] S.N. Singh and A.A. Schy, "Control of elastic robotic systems by nonlinear inversion and modal damping," *ASME J. of Dynamic Systems, Measurement, and Control*, vol. 108, pp. 180–189, 1986.

[51] S.N. Singh and A.A. Schy, "Elastic robot control: Nonlinear inversion and linear stabilization," *IEEE Trans. on Aerospace and Electronic Systems*, vol. 22, pp. 340–348, 1986.

[52] W.H. Sunada and S. Dubowsky, "The application of finite element methods to the dynamic analysis of flexible linkage systems," *ASME J. of Mechanical Design*, vol. 103, pp. 643–651, 1983.

[53] P. Tomei and A. Tornambè, "Approximate modeling of robots having elastic links," *IEEE Trans. on Systems, Man, and Cybernetics*, vol. 18, pp. 831–840, 1988.

[54] M.W. Vandegrift, F.L. Lewis, and S.Q. Zhu, "Flexible-link robot arm control by a feedback linearization/singular perturbation approach," *J. of Robotic Systems*, vol. 11, pp. 591–603, 1994.

[55] D. Wang and M. Vidyasagar, "Transfer functions for a single flexible link," *Int. J. of Robotics Research*, vol. 10, pp. 540–549, 1991.

[56] F.-Y. Wang, "On the extremal fundamental frequencies of one-link flexible manipulators," *Int. J. of Robotics Research*, vol. 13, pp. 162–170, 1994.

[57] B.-S. Yuan, W.J. Book, and B. Siciliano, "Direct adaptive control of a one-link flexible arm with tracking," *J. of Robotic Systems*, vol. 6, pp. 663–680, 1989.

[58] S. Yurkovich, F.E. Pacheco, and A.P. Tzes, "On-line frequency domain identification for control of a flexible-link robot with varying payload," *IEEE Trans. on Automatic Control*, vol. 34, pp. 1300–1304, 1989.

[59] H. Zhao and D. Chen, "Exact and stable tip trajectory tracking for multi-link flexible manipulator," *Proc. 32nd IEEE Conf. on Decision and Control*, San Antonio, TX, pp. 1371–1376, 1993.

Part III

Mobile robots

Chapter 7

Modelling and structural properties

The third part of the book is concerned with modelling and control of *wheeled mobile robots*. A wheeled mobile robot is a wheeled vehicle which is capable of an autonomous motion (without external human driver) because it is equipped, for its motion, with actuators that are driven by an embarked computer.

The aim of this chapter is to give a general and unifying presentation of the *modelling* issues of wheeled mobile robots. Several examples of derivation of kinematic and/or dynamic models for wheeled mobile robots are available in the literature for particular prototypes of mobile robots. Here, a more general viewpoint is adopted and a general class of wheeled mobile robots with an arbitrary number of wheels of different types and actuation is considered. The purpose is to point out the *structural properties* of the kinematic and dynamic models, taking into account the restriction to robot mobility induced by the constraints. By introducing the concepts of *degree of mobility* and of *degree of steerability*, we show that, notwithstanding the variety of possible robot constructions and wheel configurations, the set of wheeled mobile robots can be partitioned in five classes.

We introduce four different kinds of state space models that are of interest for understanding the behaviour of wheeled mobile robots.

- The *posture kinematic model* is the simplest state space model able to give a global description of wheeled mobile robots. It is shown that within each of the five classes, this model has a particular generic structure which allows understanding the manoeuvrability properties

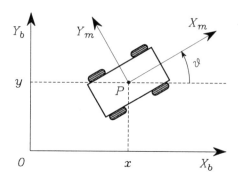

Figure 7.1: Posture coordinates.

of the robot. The reducibility, the controllability and the stabilizability of this model are also analyzed.

- The *configuration kinematic model* allows the behaviour of wheeled mobile robots to be analyzed within the framework of the theory of nonholonomic systems.

- The *configuration dynamic model* is the more general state space model. It gives a complete description of the dynamics of the system including the generalized forces provided by the actuators. In particular, the issue of the actuator configuration is addressed, and a criterion is proposed to check whether the motorization is sufficient to fully exploit the kinematic mobility.

- The *posture dynamic model* is feedback equivalent to the configuration dynamic model, and it is useful to analyze its reducibility, controllability and stabilizability properties.

7.1 Robot description

Without real loss of generality and to keep the mathematical derivation as simple as possible, we will assume that the mobile robots under study are made up of a rigid *cart* equipped with *non-deformable wheels* and that they are moving on a horizontal plane. The position of the robot in the plane is described as follows (Fig. 7.1). An arbitrary inertial base frame b is fixed in the plane of motion, while a frame m is attached to the mobile robot.

The robot posture can be described in terms of the two coordinates x, y of the origin P of the moving frame and by the orientation angle ϑ of the

7.1. ROBOT DESCRIPTION

moving frame, both with respect to the base frame with origin at O. Hence, the robot posture is given by the (3×1) vector

$$\xi = \begin{pmatrix} x \\ y \\ \vartheta \end{pmatrix}, \qquad (7.1)$$

and the rotation matrix expressing the orientation of the base frame with respect to the moving frame is

$$R(\vartheta) = \begin{pmatrix} \cos\vartheta & \sin\vartheta & 0 \\ -\sin\vartheta & \cos\vartheta & 0 \\ 0 & 0 & 1 \end{pmatrix}. \qquad (7.2)$$

We assume that, during motion, the plane of each wheel remains vertical and the wheel rotates about its (horizontal) axle whose orientation with respect to the cart can be fixed or varying. We distinguish between two basic classes of idealized wheels; namely, the *conventional* wheels and the *Swedish* wheels. In each case, it is assumed that the contact between the wheel and the ground is reduced to a single point of the plane.

For a *conventional wheel*, the contact between the wheel and the ground is supposed to satisfy both conditions of *pure rolling* and *non-slipping* along the motion. This means that the velocity of the contact point is equal to zero and implies that the two components, respectively parallel to the plane of the wheel and orthogonal to this plane, of this velocity are equal to zero.

For a *Swedish* wheel, only *one* component of the velocity of the contact point of the wheel with the ground is supposed to be equal to zero along the motion. The direction of this zero component of velocity is a priori arbitrary but is fixed with respect to the orientation of the wheel.

We now derive explicitly the expressions of the constraints for conventional and Swedish wheels.

7.1.1 Conventional wheels

Fixed wheel

The center of the *fixed wheel*, denoted by A, is a fixed point of the cart (Fig. 7.2). The position of A in the moving frame is characterized using polar coordinates, i.e., the distance l of A from P and the angle α. The orientation of the plane of the wheel with respect to l is represented by the constant angle β. The rotation angle of the wheel about its (horizontal) axle is denoted by φ and the radius of the wheel by r.

The position of the wheel is thus characterized by 4 constants: α, β, l, r, and its motion by a time-varying angle $\varphi(t)$. With this description, the

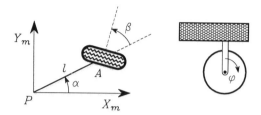

Figure 7.2: Fixed wheel or steering wheel.

components of the velocity of the contact point are easily computed and the 2 following constraints can be deduced:

- *on the wheel plane,*

$$(-\sin(\alpha+\beta)\quad \cos(\alpha+\beta)\quad l\cos\beta\,)\,R(\vartheta)\dot{\xi}+r\dot{\varphi}=0; \qquad (7.3)$$

- *orthogonal to the wheel plane,*

$$(\cos(\alpha+\beta)\quad \sin(\alpha+\beta)\quad l\sin\beta\,)\,R(\vartheta)\dot{\xi}=0. \qquad (7.4)$$

Steering wheel

A *steering wheel* is such that motion of the wheel plane with respect to the cart is a rotation about a vertical axle passing through the center of the wheel (Fig. 7.2). The description is the same as for a fixed wheel, except that now the angle β is not constant but time-varying. The position of the wheel is characterized by 3 constants: l, α, r, and its motion with respect to the cart by 2 time-varying angles $\beta(t)$ and $\varphi(t)$. The constraints have the same form as above, i.e.,

$$(-\sin(\alpha+\beta)\quad \cos(\alpha+\beta)\quad l\cos\beta\,)\,R(\vartheta)\dot{\xi}+r\dot{\varphi}=0 \qquad (7.5)$$
$$(\cos(\alpha+\beta)\quad \sin(\alpha+\beta)\quad l\sin\beta\,)\,R(\vartheta)\dot{\xi}=0. \qquad (7.6)$$

Castor wheel

A *castor wheel* is a wheel which is orientable with respect to the cart, but the rotation of the wheel plane is about a vertical axle which does *not* pass through the center of the wheel (Fig. 7.3). In this case, the description of the wheel configuration requires more parameters.

The center of the wheel is now denoted by B and is connected to the cart by a rigid rod from A to B of constant length d which can rotate about

7.2. RESTRICTIONS ON ROBOT MOBILITY

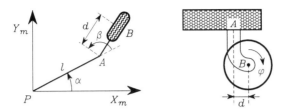

Figure 7.3: Castor wheel.

a fixed vertical axle at point A. This point A is itself a fixed point of the cart and its position is specified by the 2 polar coordinates l and α as above. The rotation of the rod with respect to the cart is represented by the angle β and the plane of the wheel is aligned with d.

The position of the wheel is described by 4 constants: α, l, r, d while its motion by 2 time-varying angles $\beta(t)$ and $\varphi(t)$. With these notations, the constraints have the following form:

$$(-\sin(\alpha+\beta) \quad \cos(\alpha+\beta) \quad l\cos\beta\,)\,R(\vartheta)\dot{\xi} + r\dot{\varphi} = 0 \qquad (7.7)$$

$$(\cos(\alpha+\beta) \quad \sin(\alpha+\beta) \quad d + l\sin\beta\,)\,R(\vartheta)\dot{\xi} + d\dot{\beta} = 0. \qquad (7.8)$$

7.1.2 Swedish wheel

The position of the *Swedish wheel* with respect to the cart is described, as for the fixed wheel, by 3 constant parameters: α, β, l. An additional parameter is required to characterize the direction, with respect to the wheel plane, of the zero component of the velocity of the contact point represented by the angle γ (Fig. 7.4).

The motion constraint is expressed as

$$(-\sin(\alpha+\beta+\gamma) \quad \cos(\alpha+\beta+\gamma) \quad l\cos(\beta+\gamma)\,)\,R(\vartheta)\dot{\xi} + r\cos\gamma\,\dot{\varphi} = 0 \qquad (7.9)$$

7.2 Restrictions on robot mobility

We now consider a general mobile robot, equipped with N wheels of the 4 above-described classes. We use the 4 following subscripts to identify quantities relative to these 4 classes: f for fixed wheels, s for steering wheels, c for castor wheels and sw for Swedish wheels. The numbers of wheels of each type are denoted by N_f, N_s, N_c, N_{sw} with $N_f + N_s + N_c + N_{sw} = N$.

CHAPTER 7. MODELLING AND STRUCTURAL PROPERTIES

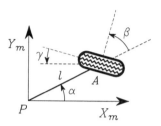

Figure 7.4: Swedish wheel.

The configuration of the robot is fully described by the following coordinate vectors, respectively:

- *posture coordinates* $\xi(t) = (\, x(t) \quad y(t) \quad \vartheta(t)\,)^T$ for the position in the plane;

- *orientation coordinates* $\beta(t) = (\, \beta_s^T(t) \quad \beta_c^T(t)\,)^T$ for the orientation angles of the steering and castor wheels, respectively;

- *rotation coordinates* $\varphi(t) = (\, \varphi_f(t) \quad \varphi_s(t) \quad \varphi_c(t) \quad \varphi_{sw}(t)\,)^T$ for the rotation angles of the wheels about their horizontal axle of rotation.

The whole set of posture, orientation and rotation coordinates $\xi, \beta_s, \beta_c, \varphi$ is termed the set of *configuration coordinates* in the sequel. The total number of the configuration coordinates is clearly $N_f + 2N_s + 2N_c + N_{sw} + 3$.

With these notations, the constraints can be written in the general matrix form

$$J_1(\beta_s, \beta_c)R(\vartheta)\dot{\xi} + J_2\dot{\varphi} = 0 \qquad (7.10)$$
$$C_1(\beta_s, \beta_c)R(\vartheta)\dot{\xi} + C_2\dot{\beta}_c = 0. \qquad (7.11)$$

In (7.10), it is

$$J_1(\beta_s, \beta_c) = \begin{pmatrix} J_{1f} \\ J_{1s}(\beta_s) \\ J_{1c}(\beta_c) \\ J_{1sw} \end{pmatrix}$$

where J_{1f}, J_{1s}, J_{1c}, and J_{1sw} are respectively $(N_f \times 3)$, $(N_s \times 3)$, $(N_c \times 3)$, and $(N_{sw} \times 3)$ matrices, whose forms derive directly from the constraints (7.3), (7.5), (7.7), and (7.9), respectively. In particular, J_{1f} and J_{1sw} are constant, while J_{1s} and J_{1c} are time-varying, respectively through $\beta_s(t)$

7.2. RESTRICTIONS ON ROBOT MOBILITY

and $\beta_c(t)$. Further, J_2 is a constant $(N \times N)$ matrix whose diagonal entries are the radii of the wheels, except for the radii of the Swedish wheels which are multiplied by $\cos\gamma$.

On the other hand, in (7.11), it is

$$C_1(\beta_s, \beta_c) = \begin{pmatrix} C_{1f} \\ C_{1s}(\beta_s) \\ C_{1c}(\beta_c) \end{pmatrix} \qquad C_2 = \begin{pmatrix} 0 \\ 0 \\ C_{2c} \end{pmatrix}$$

where C_{1f}, C_{1s}, and C_{1c} are 3 matrices respectively of dimensions $(N_f \times 3)$, $(N_s \times 3)$, and $(N_c \times 3)$, whose rows derive from the non-slipping constraints (7.4), (7.6), and (7.8), respectively. In particular, C_{1f} is constant while C_{1s} and C_{1c} are time-varying. Further, C_{2c} is a diagonal matrix whose diagonal entries are equal to $d\sin\gamma$ for the N_c castor wheels.

We introduce the following assumption concerning the configuration of the Swedish wheels.

Assumption 7.1 For each Swedish wheel it is $\gamma \neq \pi/2$.

□

The value $\gamma = \pi/2$ would correspond to the direction of the zero component of the velocity being orthogonal to the plane of the wheel. Such a wheel would be subject to a constraint identical to the non-slipping constraint of conventional wheels, hence loosing the benefit of implementing a Swedish wheel.

Consider now the first $(N_f + N_s)$ non-slipping constraints from (7.11) and written explicitly as

$$C_{1f} R(\vartheta) \dot{\xi} = 0 \qquad (7.12)$$
$$C_{1s}(\beta_s) R(\vartheta) \dot{\xi} = 0. \qquad (7.13)$$

These constraints imply that the vector $R(\vartheta)\dot{\xi} \in \mathcal{N}(C_1^*(\beta_s))$ where

$$C_1^*(\beta_s) = \begin{pmatrix} C_{1f} \\ C_{1s}(\beta_s) \end{pmatrix}. \qquad (7.14)$$

Obviously, it is $\text{rank}(C_1^*(\beta_s)) \leq 3$. If $\text{rank}(C_1^*(\beta_s)) = 3$, then $R(\vartheta)\dot{\xi} = 0$ and any motion in the plane is impossible! More generally, restrictions on robot mobility are related to the rank of C_1^*. This point will be discussed in detail hereafter.

Before that, it is worth noticing that conditions (7.12) and (7.13) have an interesting geometrical interpretation. At each time instant, the motion of the robot can be viewed as an instantaneous rotation about the

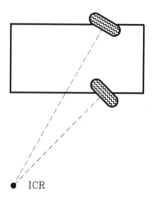

Figure 7.5: Instantaneous center of rotation.

instantaneous center of rotation (ICR) whose position with respect to the cart can be time-varying. Hence, at each time instant, the velocity vector of any point of the cart is orthogonal to the straight line joining this point and the ICR. In particular this is true for the centers of the fixed and steering wheels. This implies that, at each time instant, the horizontal rotation axles of all the fixed and steering wheels intersect at the ICR. This fact is illustrated in Fig. 7.5 and is equivalent to the condition that $\mathrm{rank}(C_1^*(\beta_s)) \leq 2$.

Clearly, the rank of matrix $C_1^*(\beta_s)$ depends on the design of the mobile robot. We define the *degree of mobility* δ_m of a mobile robot as

$$\delta_m = \dim(\mathcal{N}(C_1^*(\beta_s))) = 3 - \mathrm{rank}(C_1^*(\beta_s)).$$

Let us now examine the case $\mathrm{rank}(C_{1f}) = 2$ which implies that the robot has at least 2 fixed wheels and, if there are more than 2, that their axles intersect at the ICR whose position with respect to the cart is *fixed*. In such a case, it is clear that the only possible motion is a rotation of the robot about a fixed ICR. Obviously, this limitation is not acceptable in practice and thus we assume that $\mathrm{rank}(C_{1f}) \leq 1$. We assume moreover that the robot structure is *nondegenerate* in the following sense.

Assumption 7.2 A mobile robot is nondegenerate if

$$\mathrm{rank}(C_{1f}) \leq 1 \qquad \mathrm{rank}(C_1^*(\beta_s)) = \mathrm{rank}(C_{1f}) + \mathrm{rank}(C_{1s}(\beta_s)) \leq 2.$$

□

7.2. RESTRICTIONS ON ROBOT MOBILITY

This assumption is equivalent to the following conditions.

- If the robot has more than one fixed wheel ($N_f > 1$), then they are all on a single common axle.

- The centers of the steering wheels do not belong to this common axle of the fixed wheels.

- The number $\text{rank}(C_{1s}(\beta_s)) \leq 2$ is the number of steering wheels that can be oriented independently in order to steer the robot. We call this number the *degree of steerability*

$$\delta_s = \text{rank}(C_{1s}(\beta_s)).$$

The number of steering wheels N_s and their type are obviously a privilege of the robot designer. If a mobile robot is equipped with more than δ_s steering wheels ($N_s > \delta_s$), the motion of the extra wheels must be coordinated to guarantee the existence of the ICR at each time instant.

It follows that only *five* nonsingular structures are of practical interest, which can be inferred by the following conditions.

- The degree of mobility δ_m satisfies the inequality

$$1 \leq \delta_m \leq 3; \tag{7.15}$$

the upper bound is obvious, while the lower bound means that we consider only the case where a motion is possible, i.e., $\delta_m \neq 0$.

- The degree of steerability δ_s satisfies the inequality

$$0 \leq \delta_s \leq 2; \tag{7.16}$$

the upper bound can be reached only for robots without fixed wheels ($N_f = 0$), while the lower bound corresponds to robots without steering wheels ($N_s = 0$).

- The following inequality is satisfied:

$$2 \leq \delta_m + \delta_s \leq 3; \tag{7.17}$$

the case $\delta_m + \delta_s = 1$ is not acceptable because it corresponds to the rotation of the robot about a *fixed* ICR as we have seen above. The cases $\delta_m \geq 2$ and $\delta_s = 2$ are excluded because, according to Assumption 7.2, $\delta_s = 2$ implies $\delta_m = 1$.

δ_m	3	2	2	1	1
δ_s	0	0	1	1	2

Table 7.1: Degree of mobility and degree of steerability for possible wheeled mobile robots.

Therefore, there exist only five types of wheeled mobile robots, corresponding to the five pairs of values of δ_m and δ_s that satisfy inequalities (7.15), (7.16) and (7.17) according to Tab. 7.1.

In the sequel, each type of structure will be designated by using a denomination of the form "Type (δ_m, δ_s) robot." The main design characteristics of each type of mobile robot are now briefly presented.

Type (3,0) robot

In this case it is

$$\delta_m = \dim(\mathcal{N}(C_1^*(\beta_s))) = 3 \qquad \delta_s = 0.$$

These robots have *no* fixed wheels ($N_f = 0$) and *no* steering wheels ($N_s = 0$). They have only castor and/or Swedish wheels. Such robots are called *omnidirectional* because they have a full mobility in the plane which means that they can move at each time instant in any direction without any reorientation. In contrast, the other four types of robots have a restricted mobility (degree of mobility less than 3).

Type (2,0) robot

In this case it is

$$\delta_m = \dim(\mathcal{N}(C_1^*(\beta_s))) = \dim(\mathcal{N}(C_{1f})) = 2 \qquad \delta_s = 0.$$

These robots have *no* steering wheels ($N_s = 0$). They have either one or several fixed wheels but with a single common axle, otherwise rank(C_{1f}) would be greater than 1. The mobility of the robot is restricted in the sense that, for any admissible trajectory $\xi(t)$, the velocity $\dot{\xi}(t)$ is constrained to belong to the 2-dimensional distribution spanned by the vector fields $R^T(\vartheta)s_1$ and $R^T(\vartheta)s_2$, where s_1 and s_2 are two *constant* vectors spanning $\mathcal{N}(C_{1f})$.

7.2. RESTRICTIONS ON ROBOT MOBILITY

Type (2,1) robot

In this case it is

$$\delta_m = \dim(\mathcal{N}(C_1^*(\beta_s))) = \dim(\mathcal{N}(C_{1s}(\beta_s))) = 2 \qquad \delta_s = 1.$$

These robots have *no* fixed wheels ($N_f = 0$) and at least *one* steering wheel ($N_s \geq 1$). If there are more than one steering wheel, their orientations must be coordinated in such a way that $\text{rank}(C_{1s}(\beta_s)) = \delta_s = 1$. The velocity $\dot{\xi}(t)$ is constrained to belong to the 2-dimensional distribution spanned by the vector fields $R^T(\vartheta)s_1(\beta_s)$ and $R^T(\vartheta)s_2(\beta_s)$ where $s_1(\beta_s)$ and $s_2(\beta_s)$ are two vectors spanning $\mathcal{N}(C_{1s}(\beta_s))$ and parameterized by the angle β_s of one arbitrarily chosen steering wheel.

Type (1,1) robot

In this case it is

$$\delta_m = \dim(\mathcal{N}(C_1^*(\beta_s))) = 1 \qquad \delta_s = 1.$$

These robots have one or several fixed wheels with a single common axle. They also have one or several steering wheels, with the condition that the center of one of them is *not* located on the axle of the fixed wheels — otherwise the structure would be singular, see Assumption 7.2— and that their orientations are coordinated in such a way that $\text{rank}(C_{1s}(\beta_s)) = \delta_s = 1$. The velocity $\dot{\xi}(t)$ is constrained to belong to a 1-dimensional distribution parameterized by the orientation angle of one arbitrarily chosen steering wheel. Mobile robots that are built on the model of a conventional car (often called *car-like* robots) belong to this class.

Type (1,2) robot

In this case it is

$$\delta_m = \dim(\mathcal{N}(C_1^*(\beta_s))) = \dim \mathcal{N}(C_{1s}(\beta_s)) = 1 \qquad \delta_s = 2.$$

These robots have *no* fixed wheels ($N_f = 0$). They have at least *two* steering wheels ($N_s \geq 2$). If there are more than 2 steering wheels, their orientations must be coordinated in such a way that $\text{rank}(C_{1s}(\beta_s)) = \delta_s = 2$. The velocity $\dot{\xi}(t)$ is constrained to belong to a 1-dimensional distribution parameterized by the orientation angles of two arbitrarily chosen steering wheels of the robot.

In order to avoid useless notational complications, we will assume from now on that the degree of steerability is precisely equal to the number of

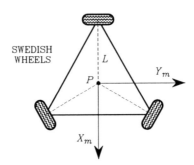

Figure 7.6: Type (3,0) robot with Swedish wheels.

steering wheels, i.e.,
$$\delta_s = N_s.$$
This is certainly a big restriction from a robot design viewpoint. However, for the mathematical analysis of the behaviour of mobile robots, there is no loss of generality in this assumption, although it considerably simplifies the technical derivation. Indeed, for robots having an excess of steering wheels ($\delta_s < N_s$), it is always possible by appropriate (but possibly tedious) manipulation to reduce the constraints (7.13) to a minimal subset of exactly δ_s independent constraints that correspond to the δ_s wheels that have been selected as the master steering wheels of the robot (see comment after Assumption 7.2) and to ignore the other slave steering wheels in the analysis.

7.3 Three-wheel robots

We present in this section six practical examples of mobile robots to illustrate the *five* types of structures that have been presented above. We restrict our attention to robots with *three* wheels.

As we have shown in Section 7.1, the wheels of a mobile robot are described by (at most) six characteristic constants; namely, three angles α, β, γ, and three lengths l, r, d. For each example, we give successively a table with the numerical values of these characteristic constants and a presentation of the various matrices J and C involved in the mathematical expressions (7.10) and (7.11) of the constraints.

However, we will assume that the radii r and the distances d are identical for all the wheels of all the examples. Hence, we will specify only the values of α, β, γ, l.

7.3.1 Type (3,0) robot with Swedish wheels

The robot has 3 Swedish wheels located at the vertices of the cart that has the form of an equilateral triangle (Fig. 7.6). The characteristic constants are specified in Tab. 7.2.

Wheels	α	β	γ	l
$1sw$	$\pi/3$	0	0	L
$2sw$	π	0	0	L
$3sw$	$5\pi/3$	0	0	L

Table 7.2: Characteristic constants of Type (3,0) robot with Swedish wheels.

The constraints have the form (7.10) where:

$$J_1 = J_{1sw} = \begin{pmatrix} -\sqrt{3}/2 & 1/2 & L \\ 0 & -1 & L \\ \sqrt{3}/2 & 1/2 & L \end{pmatrix}$$

$$J_2 = \begin{pmatrix} r & 0 & 0 \\ 0 & r & 0 \\ 0 & 0 & r \end{pmatrix}.$$

7.3.2 Type (3,0) robot with castor wheels

The robot has 3 castor wheels (Fig. 7.7). The characteristic constants are specified in Tab. 7.3.

Wheels	α	β	l
$1c$	0	–	L
$2c$	π	–	L
$3c$	$3\pi/2$	–	L

Table 7.3: Characteristic constants of Type (3,0) robot with castor wheels.

The constraints have the form (7.10) and (7.11) where:

$$J_1 = J_{1c}(\beta_c) = \begin{pmatrix} -\sin\beta_{c1} & \cos\beta_{c1} & L\cos\beta_{c1} \\ \sin\beta_{c2} & -\cos\beta_{c2} & L\cos\beta_{c2} \\ \cos\beta_{c3} & \sin\beta_{c3} & L\cos\beta_{c3} \end{pmatrix}$$

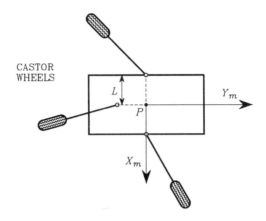

Figure 7.7: Type (3,0) robot with castor wheels.

$$J_2 = \begin{pmatrix} r & 0 & 0 \\ 0 & r & 0 \\ 0 & 0 & r \end{pmatrix}$$

$$C_1 = C_{1c}(\beta_c) = \begin{pmatrix} \cos\beta_{c1} & \sin\beta_{c1} & d + L\sin\beta_{c1} \\ -\cos\beta_{c2} & -\sin\beta_{c2} & d + L\sin\beta_{c2} \\ \sin\beta_{c3} & -\cos\beta_{c3} & d + L\sin\beta_{c3} \end{pmatrix}$$

$$C_2 = C_{2c} = \begin{pmatrix} d & 0 & 0 \\ 0 & d & 0 \\ 0 & 0 & d \end{pmatrix}.$$

7.3.3 Type (2,0) robot

The robot has 2 fixed wheels on the same axle and 1 castor wheel (Fig. 7.8), and it is typically referred to as the *unicycle robot*. The characteristic constants are specified in Tab. 7.4.

Wheels	α	β	l
1f	0	0	L
2f	π	0	L
3c	$3\pi/2$	–	L

Table 7.4: Characteristic constants of Type (2,0) robot.

7.3. THREE-WHEEL ROBOTS

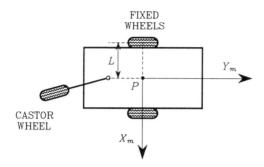

Figure 7.8: Type (2,0) robot.

The constraints have the form (7.10) and (7.11), where:

$$J_1 = \begin{pmatrix} J_{1f} \\ J_{1c}(\beta_{c3}) \end{pmatrix} = \begin{pmatrix} 0 & 1 & L \\ 0 & -1 & L \\ \cos\beta_{c3} & \sin\beta_{c3} & L\cos\beta_{c3} \end{pmatrix}$$

$$J_2 = \begin{pmatrix} r & 0 & 0 \\ 0 & r & 0 \\ 0 & 0 & r \end{pmatrix}$$

$$C_1 = \begin{pmatrix} C_{1f} \\ C_{1c}(\beta_{c3}) \end{pmatrix} = \begin{pmatrix} 1 & 0 & 0 \\ -1 & 0 & 0 \\ \sin\beta_{c3} & -\cos\beta_{c3} & d+L\sin\beta_{c3} \end{pmatrix}$$

$$C_2 = \begin{pmatrix} 0 \\ C_{2c} \end{pmatrix} = \begin{pmatrix} 0 \\ 0 \\ d \end{pmatrix}.$$

We note that the non-slipping constraints of the 2 fixed wheels are equivalent (see the first 2 rows of C_1); hence, the matrix C_1^* has rank equal to 1 as expected.

7.3.4 Type (2,1) robot

The robot has 1 steering wheel and 2 castor wheels (Fig. 7.9). The characteristic constants are specified in Tab. 7.5.

The constraints have the form (7.10) and (7.11), where:

$$J_1 = \begin{pmatrix} J_{1s}(\beta_{s1}) \\ J_{1c}(\beta_{c2}, \beta_{c3}) \end{pmatrix} = \begin{pmatrix} -\sin\beta_{s1} & \cos\beta_{s1} & 0 \\ -\cos\beta_{c2} & -\sin\beta_{c2} & \sqrt{2}L\cos\beta_{c2} \\ \sin\beta_{c3} & \cos\beta_{c3} & \sqrt{2}L\cos\beta_{c3} \end{pmatrix}$$

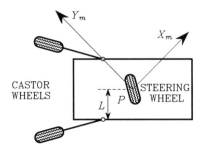

Figure 7.9: Type (2,1) robot.

Wheels	α	β	l
$1s$	0	–	0
$2c$	$\pi/2$	–	$L\sqrt{2}$
$3c$	π	–	$L\sqrt{2}$

Table 7.5: Characteristic constants of Type (2,1) robot.

$$J_2 = \begin{pmatrix} r & 0 & 0 \\ 0 & r & 0 \\ 0 & 0 & r \end{pmatrix}$$

$$C_1 = \begin{pmatrix} C_{1s}(\beta_{s1}) \\ C_{1c}(\beta_{c2}, \beta_{c3}) \end{pmatrix} = \begin{pmatrix} \cos\beta_{s1} & \sin\beta_{s1} & 0 \\ -\sin\beta_{c2} & \cos\beta_{c2} & d + \sqrt{2}L\sin\beta_{c2} \\ -\cos\beta_{c3} & -\sin\beta_{c3} & d + \sqrt{2}L\sin\beta_{c3} \end{pmatrix}$$

$$C_2 = \begin{pmatrix} 0 \\ C_{2c} \end{pmatrix} = \begin{pmatrix} 0 & 0 \\ d & 0 \\ 0 & d \end{pmatrix}.$$

7.3.5 Type (1,1) robot

The robot has 2 fixed wheels on the same axle and 1 steering wheel, like the children tricycles (Fig. 7.10). The characteristic constants are specified in Tab. 7.6.

The constraints have the form (7.10) and (7.11), where:

$$J_1 = \begin{pmatrix} J_{1f} \\ J_{1s}(\beta_{s3}) \end{pmatrix} = \begin{pmatrix} 0 & 1 & L \\ 0 & -1 & L \\ \cos\beta_{s3} & \sin\beta_{s3} & L\cos\beta_{s3} \end{pmatrix}$$

7.3. THREE-WHEEL ROBOTS

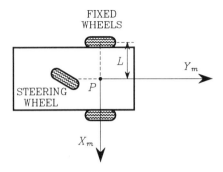

Figure 7.10: Type (1,1) robot.

Wheels	α	β	l
$1f$	0	0	L
$2f$	π	0	L
$3s$	$3\pi/2$	–	L

Table 7.6: Characteristic constants of Type (1,1) robot.

$$J_2 = \begin{pmatrix} r & 0 & 0 \\ 0 & r & 0 \\ 0 & 0 & r \end{pmatrix}$$

$$C_1 = \begin{pmatrix} C_{1f} \\ C_{1s}(\beta_{s3}) \end{pmatrix} = \begin{pmatrix} 1 & 0 & 0 \\ -1 & 0 & 0 \\ \sin\beta_{s3} & -\cos\beta_{s3} & L\sin\beta_{s3} \end{pmatrix}$$

$$C_2 = 0.$$

7.3.6 Type (1,2) robot

The robot has 2 steering wheels and 1 castor wheel (Fig. 7.11). The characteristic constants are specified in Tab. 7.7.

The constraints have the form (7.10) and (7.11), where:

$$J_1 = \begin{pmatrix} J_{1s}(\beta_{s1},\beta_{s2}) \\ J_{1c}(\beta_{c3}) \end{pmatrix} = \begin{pmatrix} -\sin\beta_{s1} & \cos\beta_{s1} & L\cos\beta_{s1} \\ \sin\beta_{s2} & -\cos\beta_{s2} & L\cos\beta_{s2} \\ \cos\beta_{c3} & \sin\beta_{c3} & L\cos\beta_{c3} \end{pmatrix}$$

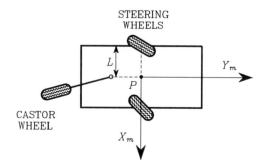

Figure 7.11: Type (1,2) robot.

Wheels	α	β	l
1s	0	—	L
2s	π	—	L
3c	$3\pi/2$	—	L

Table 7.7: Characteristic constants of Type (1,2) robot.

$$J_2 = \begin{pmatrix} r & 0 & 0 \\ 0 & r & 0 \\ 0 & 0 & r \end{pmatrix}$$

$$C_1 = \begin{pmatrix} C_{1s}(\beta_{s1}, \beta_{s2}) \\ C_{1c}(\beta_{c3}) \end{pmatrix} = \begin{pmatrix} \cos\beta_{s1} & \sin\beta_{s1} & L\sin\beta_{s1} \\ -\cos\beta_{s2} & -\sin\beta_{s2} & L\sin\beta_{s2} \\ \sin\beta_{c3} & -\cos\beta_{c3} & d + L\sin\beta_{c3} \end{pmatrix}$$

$$C_2 = \begin{pmatrix} 0 \\ C_{2c} \end{pmatrix} = \begin{pmatrix} 0 \\ 0 \\ d \end{pmatrix}.$$

7.4 Posture kinematic model

In this section, the analysis of mobility, as discussed in Section 7.2, is reformulated into a state space form which will be useful for our subsequent developments.

We have shown that, whatever the type of mobile robot, the velocity $\dot{\xi}(t)$ is restricted to belong to a distribution Δ_c defined as

$$\dot{\xi}(t) \in \Delta_c = \text{span}\{\text{col}(R^T(\vartheta)\Sigma(\beta_s))\} \qquad \forall t$$

7.4. POSTURE KINEMATIC MODEL

where the columns of the matrix $\Sigma(\beta_s)$ form a basis of $\mathcal{N}(C_1^*(\beta_s))$, i.e.,

$$\mathcal{N}(C_1^*(\beta_s)) = \mathrm{span}\{\mathrm{col}(\Sigma(\beta_s))\}.$$

This is equivalent to the following statement: for all t, there exists a time-varying vector $\eta(t)$ such that

$$\dot{\xi} = R^T(\vartheta)\Sigma(\beta_s)\eta. \qquad (7.18)$$

The dimension of the distribution Δ_c and, hence, of the vector $\eta(t)$ is the degree of mobility δ_m of the robot. Obviously, in the case where the robot has no steering wheels ($\delta_s = 0$), the matrix Σ is constant and the expression (7.18) reduces to

$$\dot{\xi} = R^T(\vartheta)\Sigma\eta. \qquad (7.19)$$

In the opposite case ($\delta_s \geq 1$), the matrix Σ explicitly depends on the orientation coordinates β_s and the expression (7.18) can be augmented as follows:

$$\dot{\xi} = R^T(\vartheta)\Sigma(\beta_s)\eta \qquad (7.20)$$
$$\dot{\beta}_s = \zeta. \qquad (7.21)$$

The representation (7.19) (or (7.20) and (7.21)) can be regarded as a state space representation of the system, termed the *posture kinematic model*, with the posture coordinates ξ and (possibly) the orientation coordinates β_s as state variables while η and ζ —which are homogeneous to velocities— can be interpreted as control inputs entering the model linearly. Nevertheless, this interpretation shall be taken with some care, since the true physical control inputs of a mobile robot are the torques provided by the embarked actuators; the kinematic state space model is in fact only a subsystem of the general dynamic model that will be presented in Section 7.6.

7.4.1 Generic models of wheeled robots

In Section 7.2, we have introduced a classification of all nondegenerate wheeled mobile robots according to the values of degree of mobility δ_m and degree of steerability δ_s. It is easy to determine to which class a particular robot belongs just by inspecting the number and the configuration of the fixed and steering wheels. Various examples have been given above.

In this section we wish to emphasize that for any particular wheeled mobile robot, whatever its constructive features, it is always possible to select the origin and the orientation of the moving frame (see Fig. 7.1) such that the posture kinematic model of the robot takes a *generic* form which is unique for each class and is completely determined by the two characteristic numbers δ_m and δ_s. In other terms, all robots of a given type can be described by the same posture kinematic model.

The five generic posture kinematic models are now presented.

Type (3,0) robot

The point P and axes X_m and Y_m can be selected arbitrarily. The matrix Σ can always be chosen as a (3×3) identity matrix. The posture kinematic model reduces to

$$\begin{pmatrix} \dot{x} \\ \dot{y} \\ \dot{\vartheta} \end{pmatrix} = \begin{pmatrix} \cos \vartheta & -\sin \vartheta & 0 \\ \sin \vartheta & \cos \vartheta & 0 \\ 0 & 0 & 1 \end{pmatrix} \begin{pmatrix} \eta_1 \\ \eta_2 \\ \eta_3 \end{pmatrix} \qquad (7.22)$$

where η_1 and η_2 are the robot velocity components along X_m and Y_m, respectively, and η_3 is the angular velocity.

Type (2,0) robot

The point P can be arbitrarily chosen along the axle of the fixed wheels, while axis X_m is aligned with this axle (see Fig. 7.8). The matrix Σ is selected as

$$\Sigma = \begin{pmatrix} 0 & 0 \\ 1 & 0 \\ 0 & 1 \end{pmatrix}.$$

The posture kinematic model (7.19) reduces to

$$\begin{pmatrix} \dot{x} \\ \dot{y} \\ \dot{\vartheta} \end{pmatrix} = \begin{pmatrix} -\sin \vartheta & 0 \\ \cos \vartheta & 0 \\ 0 & 1 \end{pmatrix} \begin{pmatrix} \eta_1 \\ \eta_2 \end{pmatrix} \qquad (7.23)$$

where η_1 is the robot velocity component along Y_m and η_2 is the angular velocity.

Type (2,1) robot

The point P is the center of the steering wheel of the robot. The orientation of the moving frame can be arbitrarily chosen; let us refer to the choice in Fig. 7.9. The matrix $\Sigma(\beta_s)$ is selected as

$$\Sigma(\beta_s) = \begin{pmatrix} -\sin \beta_{s1} & 0 \\ \cos \beta_{s1} & 0 \\ 0 & 1 \end{pmatrix}.$$

7.4. POSTURE KINEMATIC MODEL

The posture kinematic model (7.20) and (7.21) reduces to

$$\begin{pmatrix} \dot{x} \\ \dot{y} \\ \dot{\vartheta} \end{pmatrix} = \begin{pmatrix} -\sin(\vartheta + \beta_{s1}) & 0 \\ \cos(\vartheta + \beta_{s1}) & 0 \\ 0 & 1 \end{pmatrix} \begin{pmatrix} \eta_1 \\ \eta_2 \end{pmatrix} \quad (7.24)$$

$$\dot{\beta}_{s1} = \zeta_1. \quad (7.25)$$

Type (1,1) robot

The point P must be located on the axle of the fixed wheels, at the intersection with the normal passing through the center of the steering wheel; the axle Y_m is aligned with this normal (see Fig. 7.10). The matrix $\Sigma(\beta_s)$ is selected as

$$\Sigma(\beta_s) = \begin{pmatrix} 0 \\ L \sin \beta_{s3} \\ \cos \beta_{s3} \end{pmatrix}.$$

The posture kinematic model (7.20) and (7.21) reduces to

$$\begin{pmatrix} \dot{x} \\ \dot{y} \\ \dot{\vartheta} \end{pmatrix} = \begin{pmatrix} -L \sin \vartheta \sin \beta_{s3} \\ L \cos \vartheta \sin \beta_{s3} \\ \cos \beta_{s3} \end{pmatrix} \eta_1 \quad (7.26)$$

$$\dot{\beta}_{s3} = \zeta_1. \quad (7.27)$$

Type (1,2) robot

The point P is the midpoint of the segment joining the centers of the two steering wheels; axis X_m is aligned with this segment (see Fig. 7.11). The matrix $\Sigma(\beta_s)$ is selected as

$$\Sigma(\beta_s) = \begin{pmatrix} -2L \sin \beta_{s1} \sin \beta_{s2} \\ L \sin(\beta_{s1} + \beta_{s2}) \\ \sin(\beta_{s2} - \beta_{s1}) \end{pmatrix}.$$

The posture kinematic model (7.20) and (7.21) reduces to

$$\begin{pmatrix} \dot{x} \\ \dot{y} \\ \dot{\vartheta} \end{pmatrix} = \begin{pmatrix} -L\big(\sin \beta_{s1} \sin(\vartheta + \beta_{s2}) + \sin \beta_{s2} \sin(\vartheta + \beta_{s1})\big) \\ L\big(\sin \beta_{s1} \cos(\vartheta + \beta_{s2}) + \sin \beta_{s2} \cos(\vartheta + \beta_{s1})\big) \\ \sin(\beta_{s2} - \beta_{s1}) \end{pmatrix} \eta_1 \quad (7.28)$$

$$\dot{\beta}_{s1} = \zeta_1 \quad (7.29)$$

$$\dot{\beta}_{s2} = \zeta_2. \quad (7.30)$$

7.4.2 Mobility, steerability and manoeuvrability

It is convenient from now on to rewrite the posture kinematic model of mobile robots in the following compact form

$$\dot{z} = B(z)u \qquad (7.31)$$

with either $(\delta_s = 0)$

$$z = \xi \qquad B(z) = R^T(\vartheta)\Sigma \qquad u = \eta$$

or $(\delta_s > 0)$

$$z = \begin{pmatrix} \xi \\ \beta_s \end{pmatrix} \qquad B(z) = \begin{pmatrix} R^T(\vartheta)\Sigma(\beta_s) & 0 \\ 0 & I \end{pmatrix} \qquad u = \begin{pmatrix} \eta \\ \zeta \end{pmatrix}.$$

This posture kinematic model allows us to discuss further the manoeuvrability of wheeled mobile robots. The *degree of mobility* δ_m is a first index of manoeuvrability; it is equal to the number of degrees of freedom that can be *directly* manipulated from the inputs η, without reorientation of the steering wheels. Intuitively it corresponds to how many "degrees of freedom" the robot could have instantaneously from its current configuration, without steering any of its wheels. This number δ_m is not equal to the overall number of "degrees of freedom" of the robot that can be manipulated from the inputs η and ζ. In fact this number is equal to the sum $\delta_M = \delta_m + \delta_s$ that we could call *degree of manoeuvrability*. It includes the δ_s additional degrees of freedom that are accessible from the inputs ζ. But the action of ζ on the posture coordinates ξ is indirect, since it is achieved only through the coordinates β_s, that are related to ζ by an integral action. This reflects the fact that the modification of the orientation of a steering wheel cannot be achieved instantaneously.

The manoeuvrability of a wheeled mobile robot depends not only on δ_M, but also on the way these δ_M degrees of freedom are partitioned into δ_m and δ_s. Therefore, two indices are needed to characterize manoeuvrability: δ_M and δ_m, or, equivalently, δ_m and δ_s, which are the two indices identifying the five classes of robots illustrated above.

Two robots with the same value of δ_M, but different δ_m, are not equivalent. For robots with $\delta_M = 3$, it is possible to freely assign the position of the ICR, either directly from η, for Type (3,0) robots, or by orientation of 1 or 2 steering wheels for Type (2,1) and Type (1,2) robots. For robots with $\delta_M = 2$, the ICR is constrained to belong to a straight line (the axle of the fixed wheel). Its position on this line is assigned either directly for Type (2,0) robots, or by the orientation of a steering wheel for Type (1,1) robots.

7.4. POSTURE KINEMATIC MODEL

Similarly, two wheeled mobile robots with the same value of δ_m, but different δ_M, are not equivalent; the robot with the largest δ_M is more manoeuvrable. Compare, for instance, a Type (1,1) robot and a Type (1,2) robot with $\delta_m = 1$ and, respectively, $\delta_M = 2$ and $\delta_M = 3$. The position of the ICR for a Type (1,2) robot can be assigned freely in the plane, just by orienting 2 steering wheels, while for a Type (1,1) robot, the ICR is constrained to belong to the axle of the fixed wheels, its position on this axle being specified by the orientation of the steering wheel. Since the steering directions of the steering wheels can usually be changed very quickly, especially for small indoors robots, it follows, from a practical viewpoint, that a Type (1,2) robot is more manoeuvrable than a Type (1,1) robot.

Obviously, the ideal situation is that of omnidirectional robots where $\delta_m = \delta_M = 3$.

7.4.3 Irreducibility

In this section, we address the question of reducibility of the kinematic state space model (7.31). A state space model is *reducible* if there exists a change of coordinates such that some of the new coordinates are identically zero along the motion of the system. For a nonlinear dynamical system without drift like (7.31), reducibility is related to the involutive closure $\bar{\Delta}$ of the following distribution Δ, expressed in local coordinates as

$$\Delta(z) = \text{span}\{\text{col}(B(z))\}.$$

A well-known consequence of Frobenius theorem is that the system is reducible only if $\dim(\bar{\Delta}) < \dim(z)$.

In this section, we will prove that the posture kinematic model of nondegenerate mobile robots (see Assumption 7.3) is always *irreducible*. To establish this result, we proceed by first analyzing in detail the particular case of a Type (1,1) robot, whose posture kinematic model is

$$\dot{z} = \begin{pmatrix} \dot{x} \\ \dot{y} \\ \dot{\vartheta} \\ \dot{\beta}_{s3} \end{pmatrix} = \begin{pmatrix} -L \sin \vartheta \sin \beta_{s3} & 0 \\ L \cos \vartheta \sin \beta_{s3} & 0 \\ \cos \beta_{s3} & 0 \\ 0 & 1 \end{pmatrix} \begin{pmatrix} \eta_1 \\ \zeta_1 \end{pmatrix} = B(z)u. \qquad (7.32)$$

In this particular case, a basis of $\bar{\Delta}(z)$ is

$$\bar{\Delta}(z) = \text{span}\{b_1(z), b_2(z), b_3(z), b_4(z)\},$$

where the columns $b_1(z)$ and $b_2(z)$ of $B(z)$ are

$$b_1(z) = \begin{pmatrix} -L\sin\vartheta\sin\beta_{s3} \\ L\cos\vartheta\sin\beta_{s3} \\ \cos\beta_{s3} \\ 0 \end{pmatrix} \qquad b_2(z) = \begin{pmatrix} 0 \\ 0 \\ 0 \\ 1 \end{pmatrix}$$

and

$$b_3(z) = [b_1(z), b_2(z)] = \begin{pmatrix} L\sin\vartheta\cos\beta_{s3} \\ -L\cos\vartheta\cos\beta_{s3} \\ \sin\beta_{s3} \\ 0 \end{pmatrix}$$

$$b_4(z) = [b_1(z), b_3(z)] = \begin{pmatrix} L\cos\vartheta \\ L\sin\vartheta \\ 0 \\ 0 \end{pmatrix}.$$

We see that $\operatorname{rank}(B(z)) = \delta_m + \delta_s = 2$ and $\dim(\bar{\Delta}(z)) = \dim(z) = 4$ everywhere in the state space. It follows that the kinematic state space model of a Type (1,1) robot is irreducible.

The same line of reasoning that has been followed so far for a Type (1,1) robot can be followed easily for the other robots described above. It can be concluded that each posture kinematic model is irreducible. We summarize this analysis in a Property.

Property 7.1 For the posture kinematic model $\dot{z} = B(z)u$ of a wheeled mobile robot, the input matrix $B(z)$ has full rank, i.e.,

$$\operatorname{rank}(B(z)) = \delta_m + \delta_s \qquad \forall z,$$

and the involutive distribution $\bar{\Delta}(z)$ has constant maximal dimension, i.e.,

$$\dim(\bar{\Delta}(z)) = 3 + \delta_s \qquad \forall z.$$

As a consequence, the posture kinematic model of a wheeled mobile robot is irreducible. This is a coordinate-free property.

□

7.4.4 Controllability and stabilizability

In this subsection, we analyze the main *controllability* and feedback *stabilizability* properties of the posture kinematic model of wheeled mobile robots. We first examine the linear approximation around an arbitrary equilibrium configuration $\bar{z} = (\bar{\xi}^T \ \bar{\beta}_s^T)^T$. Equilibrium means that the robot is at rest somewhere, with a given constant posture $\bar{\xi}$ and a given constant orientation $\bar{\beta}_s$ of the steering wheels. Obviously, velocities are zero, i.e., $\bar{u} = 0$.

7.4. POSTURE KINEMATIC MODEL

Property 7.2 The controllability rank of the linear approximation of the posture kinematic model $\dot{z} = B(z)u$ around an equilibrium configuration is $\delta_m + \delta_s$.

□

This property follows from the fact that the linear approximation around $(\bar{z} = 0, \bar{u} = 0)$ can be written as

$$\frac{d}{dt}(z - \bar{z}) = B(\bar{z})u.$$

It follows that the controllability matrix reduces to $B(\bar{z})$ whose rank is $\delta_m + \delta_s$ for all $\bar{z}$ by Property 7.1.

This implies that the linear approximation of the posture kinematic model of *omnidirectional* robots (Type (3,0) with $\delta_m = 3$ and $\delta_s = 0$) is completely controllable since δ_m is precisely the state dimension in this case, whereas it is *not* controllable for *restricted mobility* robots (Type (2,1), Type (2,0), Type (1,1), and Type (1,2) with $\delta_m \leq 2$) since $\delta_m + \delta_s < 3 + \delta_s = \dim(z)$.

This property, however, does not prevent restricted mobility robots from being controllable, in accordance with physical intuition.

Property 7.3 The posture kinematic model $\dot{z} = B(z)u$ of a wheeled mobile robot is controllable.

□

Indeed, for a nonlinear dynamical system without drift of the form $\dot{z} = B(z)u$, the strong accessibility algebra coincides with the involutive distribution $\bar{\Delta}(z) = \text{inv span}\{\text{col}(B(z))\}$ that we have introduced in Section 7.4.3. It follows from Property 7.1, that the strong accessibility rank condition is satisfied for all z in the configuration space and, therefore, the system is strongly accessible from any configuration. For a system without drift, strong accessibility implies controllability.

Practically, this property means that a mobile robot can always be driven from any initial posture ξ_0 to any final one ξ_f, in a finite time, by manipulating the velocity control input $u = (\eta^T \ \zeta^T)^T$.

Let us now consider the question of the existence of a feedback control $u(z)$ able to stabilize a mobile robot at a particular configuration z^*. For omnidirectional robots, the answer to that question is obvious. For instance,

$$u(z) = B^{-1}(z)A(z - z^*),$$

with A an arbitrary Hurwitz matrix, is clearly a *linearizing* smooth feedback control law that drives the robot exponentially to z^*. Indeed, the closed

loop is described by the freely assignable linear dynamics
$$\frac{d}{dt}(z - z^*) = A(z - z^*).$$
Hence, omnidirectional mobile robots are full state feedback linearizable and therefore they are quite similar to fully actuated robot manipulators.

For restricted mobility robots, the situation is less favourable, since from Property 7.1 they are certainly not full state feedback linearizable; the controllability of the linear approximation is necessary for that.

Property 7.4 For restricted mobility robots the posture kinematic model $\dot{z} = B(z)u$ is not stabilizable by a continuous static time-invariant state feedback $u(z)$, but is stabilizable by a continuous *time-varying* static state feedback $u(z,t)$.

□

Indeed, the so-called Brockett's necessary condition is not satisfied by a continuous static time-invariant state feedback, since the map $(z, u) \to B(z)u$ is not onto on a neighbourhood of the equilibrium $\bar{z} = (\bar{\xi}^T \quad \bar{\beta}_s^T)^T$, $\bar{u} = 0$. Stabilizability by a continuous time-varying static state feedback is a special case of a general stabilizability result for driftless systems. A systematic procedure for the design of such stabilizing time-varying feedback controllers can be adopted, which is applicable to all the posture kinematic models because in each case one column of the matrix $B(z)$ is of the form $(0 \quad \ldots \quad 0 \quad 1)^T$.

7.5 Configuration kinematic model

So far, we have used only a subset of the constraints (7.10) and (7.11); namely, that part of the constraints which is relative to the fixed and steering wheels, expressed by (7.12) and (7.13). The remaining constraints are now used to derive the equations of the evolution of the orientation and rotation velocities $\dot{\beta}_c$ and $\dot{\varphi}$ not involved in the posture kinematic model (7.20) and (7.21).

From (7.11) and (7.10) it follows directly that
$$\dot{\beta}_c = -C_{2c}^{-1} C_{1c}(\beta_c) R(\vartheta) \dot{\xi} \tag{7.33}$$
$$\dot{\varphi} = -J_2^{-1} J_1(\beta_s, \beta_c) R(\vartheta) \dot{\xi}. \tag{7.34}$$

By combining these equations with the posture kinematic model (7.20), the state equations for β_c and φ become
$$\dot{\beta}_c = D(\beta_c) \Sigma(\beta_s) \eta \tag{7.35}$$
$$\dot{\varphi} = E(\beta_s, \beta_c) \Sigma(\beta_s) \eta \tag{7.36}$$

7.5. CONFIGURATION KINEMATIC MODEL

with the following definitions of $D(\beta_c)$ and $E(\beta_s, \beta_c)$:

$$D(\beta_c) = -C_{2c}^{-1} C_{1c}(\beta_c)$$
$$E(\beta_s, \beta_c) = -J_2^{-1} J_1(\beta_s, \beta_c).$$

We note also that these matrices satisfy the equations

$$J_1(\beta_s, \beta_c) + J_2 E(\beta_s, \beta_c) = 0 \qquad (7.37)$$
$$C_{1c}(\beta_c) + C_{2c} D(\beta_c) = 0. \qquad (7.38)$$

Defining q as the vector of configuration coordinates, i.e.,

$$q = \begin{pmatrix} \xi \\ \beta_s \\ \beta_c \\ \varphi \end{pmatrix}, \qquad (7.39)$$

the evolution of the configuration coordinates can be described by the following compact equation, resulting from (7.20), (7.21), (7.35) and (7.36), called the *configuration kinematic model*

$$\dot{q} = S(q) u \qquad (7.40)$$

where

$$S(q) = \begin{pmatrix} R^T(\vartheta) \Sigma(\beta_s) & 0 \\ 0 & I \\ D(\beta_c) \Sigma(\beta_s) & 0 \\ E(\beta_s, \beta_c) \Sigma(\beta_s) & 0 \end{pmatrix} \quad u = \begin{pmatrix} \eta \\ \zeta \end{pmatrix}. \qquad (7.41)$$

Equation (7.40) has the standard form of the kinematic model of a system subject to independent velocity constraints. We now connect this formulation with the standard theory of nonholonomic mechanical systems.

Reducibility of (7.40) is directly related to the dimension of the involutive closure of the distribution Δ_1 spanned in local coordinates q by the columns of the matrix $S(q)$, i.e.,

$$\Delta_1(q) = \text{span}\{\text{col}(S(q))\}.$$

It follows immediately that

$$\delta_m + N_s = \dim(\Delta_1) \leq \dim(\text{inv}(\Delta_1)) \leq \dim(q) = 3 + N + N_c + N_s.$$

We define the *degree of nonholonomy* M of a mobile robot as:

$$M = \dim(\text{inv}(\Delta_1)) - (\delta_m + N_s).$$

This number M represents the number of velocity constraints that are not integrable and therefore cannot be eliminated, whatever the choice of the generalized coordinates. It must be pointed out that this number depends on the particular structure of the robot, and thus it has not the same value for all the robots belonging to a given class.

On the other hand, for a particular choice of generalized coordinates, the number of coordinates that can be eliminated by integration of the constraints is equal to the difference between $\dim(q)$ and $\dim(\text{inv}(\Delta_1))$.

Property 7.5 The configuration kinematic model $\dot{q} = S(q)u$ of all types of wheeled mobile robot is nonholonomic, i.e., $M > 0$, but is reducible, i.e., $\dim(q) > \dim(\text{inv}(\Delta_1))$.

□

This property is not contradictory with irreducibility of the posture kinematic state space model (7.20) and (7.21), as discussed in Section 7.4.3; reducibility of (7.40) means that there exists at least one smooth function of $\xi, \beta_c, \varphi, \beta_s$, involving explicitly at least one of the variables β_c, φ that is constant along the trajectories of the system compatible with all the constraints (7.10) and (7.11).

This discussion is illustrated by two examples.

Type (3,0) robot with Swedish wheels

For this robot (Fig. 7.6), $\delta_m = 3$ and the configuration coordinates are

$$q = (x \quad y \quad \vartheta \quad \varphi_1 \quad \varphi_2 \quad \varphi_3)^T.$$

The configuration model is characterized by

$$S(q) = \begin{pmatrix} \cos\vartheta & -\sin\vartheta & 0 \\ \sin\vartheta & \cos\vartheta & 0 \\ 0 & 0 & 1 \\ \dfrac{\sqrt{3}}{2r} & -\dfrac{1}{2r} & -\dfrac{L}{r} \\ 0 & \dfrac{1}{r} & -\dfrac{L}{r} \\ -\dfrac{\sqrt{3}}{2r} & -\dfrac{1}{2r} & -\dfrac{L}{r} \end{pmatrix}.$$

It is easy to check that

$$\dim(\Delta_1) = 3 \qquad \dim(\text{inv}(\Delta_1)) = 5.$$

7.5. CONFIGURATION KINEMATIC MODEL

It follows that the degree of nonholonomy is equal to $5 - 3 = 2$, while the number of coordinates that can be eliminated is equal to $6 - 5 = 1$. In fact, the structure of the configuration model implies that

$$\dot{\varphi}_1 + \dot{\varphi}_2 + \dot{\varphi}_3 = -\frac{3L}{r}\dot{\vartheta}.$$

This means that $(\varphi_1 + \varphi_2 + \varphi_3 + 3L\vartheta/r)$ is constant along any trajectory compatible with the constraints. It is then possible to eliminate one of the four variables $\varphi_1, \varphi_2, \varphi_3, \vartheta$.

Type (2,0) robot

For this robot (Fig. 7.8), $\delta_m = 2$ and the configuration coordinates are

$$q = (\,x \quad y \quad \vartheta \quad \beta_{c3} \quad \varphi_1 \quad \varphi_2 \quad \varphi_3\,)^T.$$

The configuration model is characterized by

$$S(q) = \begin{pmatrix} -\sin\vartheta & 0 \\ \cos\vartheta & 0 \\ 0 & 1 \\ \frac{1}{d}\cos\beta_{c3} & -\frac{1}{d}(d + L\sin\beta_{c3}) \\ -\frac{1}{r} & -\frac{L}{r} \\ \frac{1}{r} & -\frac{L}{r} \\ -\frac{1}{r}\sin\beta_{c3} & -\frac{L}{r}\cos\beta_{c3} \end{pmatrix}.$$

It can be checked that

$$\dim(\Delta_1) = 2 \qquad \dim(\text{inv}(\Delta_1)) = 6.$$

It follows that the degree of nonholonomy is equal to $6 - 2 = 4$, and the number of coordinates that can be eliminated is equal to $7 - 6 = 1$. From the configuration model it is

$$\dot{\varphi}_1 + \dot{\varphi}_2 = -\frac{2L}{r}\dot{\vartheta}.$$

This means that the variable $(\varphi_1 + \varphi_2 + 2L\vartheta/r)$ has a constant value along any trajectory compatible with the constraints.

7.6 Configuration dynamic model

The aim of this section is the derivation of a general dynamic state space model of wheeled mobile robots describing the dynamic relations between the configuration coordinates ξ, β, φ and the torques developed by the embarked actuators.

This general state space model is termed *configuration dynamic model* and is made up of six kinds of state equations; namely, three for the coordinates ξ, β, φ and three for the internal coordinates η, ζ that were introduced in Section 7.4. The state equations for ξ, β, φ have been derived in Section 7.5 under the form of the configuration kinematic model. The state equation for η and ζ will be established in Section 7.6.1 using the Lagrange formalism. The actuator configuration will be discussed in Section 7.6.2.

7.6.1 Model derivation

We assume that the robot is equipped with actuators that can force either the orientation of the steering and castor wheels (orientation coordinates β_s and β_c) or the rotation of the wheels (rotation coordinates φ). The torques provided by the actuators are denoted by τ_φ for the rotation of the wheels, τ_c for the orientation of the castor wheels, and τ_s for the orientation of the steering wheels, respectively.

Using the *Lagrange formulation*, the dynamics of wheeled mobile robots is described by the following $(3 + N_c + N + N_s)$ Lagrange's equations:

$$\frac{d}{dt}\left(\frac{\partial T}{\partial \dot{\xi}}\right)^T - \left(\frac{\partial T}{\partial \xi}\right)^T = R^T(\vartheta) J_1^T(\beta_s, \beta_c)\lambda + R^T(\vartheta) C_1^T(\beta_s, \beta_c)\mu \quad (7.42)$$

$$\frac{d}{dt}\left(\frac{\partial T}{\partial \dot{\beta_c}}\right)^T - \left(\frac{\partial T}{\partial \beta_c}\right)^T = C_2^T \mu + \tau_c \quad (7.43)$$

$$\frac{d}{dt}\left(\frac{\partial T}{\partial \dot{\varphi}}\right)^T - \left(\frac{\partial T}{\partial \varphi}\right)^T = J_2^T \lambda + \tau_\varphi \quad (7.44)$$

$$\frac{d}{dt}\left(\frac{\partial T}{\partial \dot{\beta_s}}\right)^T - \left(\frac{\partial T}{\partial \beta_s}\right)^T = \tau_s \quad (7.45)$$

where T represents the kinetic energy and λ, μ are the Lagrange multipliers associated with the constraints (7.10) and (7.11) respectively.

In order to eliminate the Lagrange multipliers, we proceed as follows. The first three Lagrange's equations (7.42), (7.43), and (7.44) are premultiplied by the matrices $\Sigma^T(\beta_s)R(\vartheta)$, $\Sigma^T(\beta_s)D^T(\beta_c)$, and $\Sigma^T(\beta_s)E^T(\beta_s, \beta_c)$, respectively, and then summed up. This leads to the two following equa-

7.6. CONFIGURATION DYNAMIC MODEL

tions, from which the Lagrange multipliers have disappeared owing to equalities (7.37) and (7.38):

$$\Sigma^T(\beta_s)\left(R(\vartheta)[T]_\xi + D^T(\beta_c)[T]_{\beta_c} + E^T(\beta_s,\beta_c)[T]_\varphi\right) =$$
$$\Sigma^T(\beta_s)\left(D^T(\beta_c)\tau_c + E^T(\beta_s,\beta_c)\tau_\varphi\right) \qquad (7.46)$$
$$[T]_{\beta_s} = \tau_s \qquad (7.47)$$

where the compact notation

$$[T]_\psi = \frac{d}{dt}\left(\frac{\partial T}{\partial \dot\psi}\right)^T - \left(\frac{\partial T}{\partial \psi}\right)^T$$

has been used. The kinetic energy of wheeled mobile robots can be expressed as follows:

$$T = \dot\xi^T R^T(\vartheta)\left(M(\beta_c)R(\vartheta)\dot\xi + 2V(\beta_c)\dot\beta_c + 2W\dot\beta_s\right)$$
$$+ \dot\beta_c^T I_c \dot\beta_c + \dot\varphi^T I_\varphi \dot\varphi + \dot\beta_s^T I_s \dot\beta_s$$

with appropriate definitions of the matrices $M(\beta_c)$, $V(\beta_c)$, W, I_c, I_φ, and I_s which depend on the mass distribution and the inertia moments of the various rigid bodies (cart and wheels) that constitute the robot. The state equations for η and ζ are then obtained (after rather lengthy calculation) by substituting this expression of T in the dynamic equations (7.46) and (7.47), and by eliminating the velocities $\dot\xi$, $\dot\beta_c$, $\dot\varphi$, and $\dot\beta_s$, and the accelerations $\ddot\xi$, $\ddot\beta_c$, $\ddot\varphi$, and $\ddot\beta_s$ with the aid of the kinematic equations (7.20), (7.21), (7.35), (7.36), and their derivatives.

Therefore, the *configuration dynamic model* of wheeled mobile robots in the state space takes on the following general form:

$$\dot\xi = R^T(\vartheta)\Sigma(\beta_s)\eta \qquad (7.48)$$
$$\dot\beta_s = \zeta \qquad (7.49)$$
$$\dot\beta_c = D(\beta_c)\Sigma(\beta_s)\eta \qquad (7.50)$$
$$H_1(\beta_s,\beta_c)\dot\eta + \Sigma^T(\beta_s)V(\beta_c)\dot\zeta + f_1(\beta_s,\beta_c,\eta,\zeta)$$
$$= \Sigma^T(\beta_s)\left(D^T(\beta_c)\tau_c + E^T(\beta_s,\beta_c)\tau_\varphi\right) \qquad (7.51)$$
$$V^T(\beta_c)\Sigma(\beta_s)\dot\eta + I_s\dot\zeta + f_2(\beta_s,\beta_c,\eta,\zeta) = \tau_s \qquad (7.52)$$
$$\dot\varphi = E(\beta_c,\beta_s)\Sigma(\beta_s)\eta \qquad (7.53)$$

where

$$H_1(\beta_s,\beta_c) = \Sigma^T(\beta_s)\left(M(\beta_c) + D^T(\beta_c)V^T(\beta_c) + V(\beta_c)D(\beta_c)\right.$$
$$\left. + D^T(\beta_c)I_c D(\beta_c) + E^T(\beta_s,\beta_c)I_\varphi E(\beta_s,\beta_c)\right)\Sigma(\beta_s).$$

7.6.2 Actuator configuration

In the general configuration dynamic model (7.48)–(7.53), the vectors $\tau_\varphi, \tau_c, \tau_s$ represent all the torques that can be *potentially* applied for the rotation and orientation of robot wheels. In practice, however, only a limited number of actuators will be used, which means that many components of $\tau_\varphi, \tau_c, \tau_s$ are identically zero.

Our concern in this section is to characterize the actuator configurations that allow a full manoeuvrability of the robot while requiring a number of actuators as limited as possible.

First, it is clear that all the steering wheels must be provided with an actuator for their orientation; otherwise, these wheels would just play the role of fixed wheels.

Moreover, to ensure a full robot mobility, N_m additional actuators (with $N_m \geq \delta_m$) must be implemented for either the rotation of some wheels or the orientation of some castor wheels. The vector of the torques developed by these actuators is denoted by τ_m, and thus we have

$$\begin{pmatrix} \tau_c \\ \tau_\varphi \end{pmatrix} = P\tau_m \quad (7.54)$$

where P is an $((N_c + N) \times N_m)$ elementary matrix which selects the components of τ_c and τ_φ that are effectively used as control inputs.

Using (7.54), we can recognize that eq. (7.51) of the general dynamic model becomes

$$H_1(\beta_s, \beta_c)\dot{\eta} + \Sigma^T(\beta_s)V(\beta_c)\dot{\zeta} + f_1(\beta_s, \beta_c, \eta, \zeta) = B(\beta_s, \beta_c)P\tau_m \quad (7.55)$$

with

$$B(\beta_s, \beta_c) = \Sigma^T(\beta_s)(D^T(\beta_c) \quad E^T(\beta_s, \beta_c)).$$

We introduce the following assumption.

Assumption 7.3 The actuator configuration is such that the matrix $B(\beta_s, \beta_c)P$ has full rank for all $(\beta_s, \beta_c) \in R^{N_s+N_c}$.

□

We now present the minimal admissible actuator configurations for the various types of mobile robots that have been described in Section 7.3.

Type (3,0) robot with Swedish wheels

For this robot (Fig. 7.6), the matrix B is constant and reduces to

$$B = \Sigma^T E^T = -J_2^{-1}J_1 = -\frac{1}{r}\begin{pmatrix} -\sqrt{3}/2 & 1/2 & L \\ 0 & -1 & L \\ \sqrt{3}/2 & 1/2 & L \end{pmatrix}$$

7.6. CONFIGURATION DYNAMIC MODEL

which is nonsingular. We conclude that the only admissible configuration is to equip each wheel with an actuator.

Type (3,0) robot with castor wheels

For this robot (Fig. 7.7), the matrix $B(\beta_c)$ is

$$B(\beta_c) = \Sigma^T \left(D^T(\beta_c) \quad E^T(\beta_c) \right)$$

with

$$\Sigma^T D^T(\beta_c) = -\frac{1}{d} \begin{pmatrix} \cos\beta_{c1} & -\cos\beta_{c2} & \sin\beta_{c3} \\ \sin\beta_{c1} & -\sin\beta_{c2} & -\cos\beta_{c3} \\ d+L\sin\beta_{c1} & d+L\sin\beta_{c2} & d+L\sin\beta_{c3} \end{pmatrix}$$

$$\Sigma^T E^T(\beta_c) = -\frac{1}{r} \begin{pmatrix} -\sin\beta_{c1} & \sin\beta_{c2} & \cos\beta_{c3} \\ \cos\beta_{c1} & -\cos\beta_{c2} & \sin\beta_{c3} \\ L\cos\beta_{c1} & L\cos\beta_{c2} & L\cos\beta_{c3} \end{pmatrix}.$$

It appears that there is no set of 3 columns of $B(\beta_c)$ which are independent for any $\beta_{c1}, \beta_{c2}, \beta_{c3}$. It is therefore necessary to use (at least) $N_m = 4$ actuators. An admissible configuration is as follows: 2 actuators (one for the orientation and one for the rotation) on 2 of the 3 wheels (the third one being not actuated and hence self-aligning). For instance, if wheels 1 and 2 are actuated in this way, the selection matrix P is

$$P = \begin{pmatrix} 1 & 0 & 0 & 0 \\ 0 & 1 & 0 & 0 \\ 0 & 0 & 0 & 0 \\ 0 & 0 & 1 & 0 \\ 0 & 0 & 0 & 1 \\ 0 & 0 & 0 & 0 \end{pmatrix}$$

and the matrix $B(\beta_c)P$ has full rank ($= 3$) for any configuration of the robot.

Type (2,0) robot

For this robot (Fig. 7.8), the matrix $B(\beta_c)$ is

$$B(\beta_{c3}) = \begin{pmatrix} \frac{1}{d}\cos\beta_{c3} & -\frac{1}{r} & \frac{1}{r} & -\frac{1}{r}\sin\beta_{c3} \\ -\frac{1}{d}(d+L\sin\beta_{c3}) & -\frac{L}{r} & -\frac{L}{r} & -\frac{L}{r}\cos\beta_{c3} \end{pmatrix}.$$

Several configurations with 2 actuators are admissible.

- 2 rotation actuators on wheels 1 and 2; in this case it is

$$P = \begin{pmatrix} 0 & 0 \\ 1 & 0 \\ 0 & 1 \\ 0 & 0 \end{pmatrix}.$$

- 1 actuator for the orientation of wheel 3 and 1 actuator for the rotation of wheel 2 (or 3), provided that $d > L$; in this case it is

$$P = \begin{pmatrix} 1 & 0 \\ 0 & 1 \\ 0 & 0 \\ 0 & 0 \end{pmatrix}.$$

- 2 actuators (orientation and rotation) on castor wheel 3, provided that $d < L$; in this case it is

$$P = \begin{pmatrix} 1 & 0 \\ 0 & 0 \\ 0 & 0 \\ 0 & 1 \end{pmatrix}.$$

Type (2,1) robot

For this robot (Fig. 7.9), we first need an orientation actuator for the steering wheel. The matrix $B(\beta_s, \beta_c)$ is then

$$B(\beta_s, \beta_c) = \Sigma^T(\beta_s) \left(D^T(\beta_c) \quad E^T(\beta_s, \beta_c) \right)$$

with

$$\Sigma^T(\beta_s) = \begin{pmatrix} -\sin\beta_{s1} & \cos\beta_{s1} & 0 \\ 0 & 0 & 1 \end{pmatrix}$$

$$D^T(\beta_c) = -\frac{1}{d} \begin{pmatrix} -\sin\beta_{c2} & -\cos\beta_{c3} \\ \cos\beta_{c2} & -\sin\beta_{c3} \\ d + \sqrt{2}L\sin\beta_{c2} & d + \sqrt{2}L\sin\beta_{c3} \end{pmatrix}$$

$$E^T(\beta_s, \beta_c) = -\frac{1}{r} \begin{pmatrix} \cos\beta_{s1} & -\sin\beta_{c2} & -\cos\beta_{c3} \\ \sin\beta_{s1} & \cos\beta_{c2} & -\sin\beta_{c3} \\ 0 & d + \sqrt{2}L\sin\beta_{c2} & d + \sqrt{2}L\sin\beta_{c3} \end{pmatrix}.$$

Columns 1 and 3 (or 1 and 2) of $B(\beta_s, \beta_c)$ are independent provided $d > L\sqrt{2}$. Hence two admissible actuator configurations are obtained by using

7.6. CONFIGURATION DYNAMIC MODEL

a second actuator for the rotation of the steering wheel (number 1) and a third actuator for the orientation of either wheel 2 or wheel 3. The two corresponding matrices P are:

$$P = \begin{pmatrix} 1 & 0 \\ 0 & 1 \\ 0 & 0 \\ 0 & 0 \\ 0 & 0 \end{pmatrix} \qquad P = \begin{pmatrix} 1 & 0 \\ 0 & 0 \\ 0 & 1 \\ 0 & 0 \\ 0 & 0 \end{pmatrix}.$$

Type (1,1) robot

For this robot (Fig. 7.10), a first orientation actuator for the steering wheel is needed. The matrix $B(\beta_s)$ reduces to the vector

$$B = -\frac{L}{r}\begin{pmatrix} \sin\beta_{s3} + \cos\beta_{s3} & -\sin\beta_{s3} + \cos\beta_{s3} & 1 \end{pmatrix}.$$

Since $\delta_m = 1$, Assumption 7.3 will be satisfied if a second actuator is provided for the rotation of the third wheel. The matrix P is then

$$P = \begin{pmatrix} 0 \\ 0 \\ 1 \end{pmatrix}.$$

Type (1,2) robot

For this robot (Fig. 7.11), 2 actuators are required for the orientation of the 2 steering wheels. The matrix $B(\beta_s, \beta_c)$ is then

$$B(\beta_s, \beta_c) = \Sigma^T(\beta_s)\begin{pmatrix} D^T(\beta_c) & E^T(\beta_s, \beta_c) \end{pmatrix}$$

with

$$\Sigma^T(\beta_s) = \begin{pmatrix} -2L\sin\beta_{s1}\sin\beta_{s2} & L\sin(\beta_{s1}+\beta_{s2}) & \sin(\beta_{s2}-\beta_{s1}) \end{pmatrix}$$

$$D^T(\beta_c) = \begin{pmatrix} -\frac{1}{d}\sin\beta_{c3} \\ \frac{1}{d}\cos\beta_{c3} \\ -\frac{1}{d}(d+L\sin\beta_{c3}) \end{pmatrix}$$

$$E^T(\beta_s, \beta_c) = -\frac{1}{r}\begin{pmatrix} -\sin\beta_{s1} & \sin\beta_{s2} & \cos\beta_{c3} \\ \cos\beta_{s1} & -\cos\beta_{s2} & \sin\beta_{c3} \\ L\cos\beta_{s1} & L\cos\beta_{s2} & L\cos\beta_{c3} \end{pmatrix}.$$

Since $\delta_m = 1$, it would be sufficient to have one column of $B(\beta_s, \beta_c)$ being nonzero for all possible configurations. However, there is no such a column. It is therefore necessary to use 2 additional actuators, for instance for the rotation of wheels 1 and 2 giving the matrix P as

$$P = \begin{pmatrix} 0 & 0 \\ 1 & 0 \\ 0 & 1 \\ 0 & 0 \end{pmatrix}.$$

Finally, Tab. 7.8 summarizes the results.

Type (δ_m, δ_s)	Number of actuators $N_m + N_s$
Type (3,0)	3 or 4
Type (2,0)	2
Type (2,1)	3
Type (1,1)	2
Type (1,2)	4

Table 7.8: Number of actuators for different Type robots.

7.7 Posture dynamic model

The configuration dynamic model of wheeled mobile robots can be rewritten in the more compact form

$$\dot{q} = S(q)u \qquad (7.56)$$
$$H(\beta)\dot{u} + f(\beta, u) = F(\beta)\tau_0 \qquad (7.57)$$

with the following definitions:

$$\beta = \begin{pmatrix} \beta_s \\ \beta_c \end{pmatrix}$$

$$q = \begin{pmatrix} \xi \\ \beta \\ \varphi \end{pmatrix}$$

$$u = \begin{pmatrix} \eta \\ \zeta \end{pmatrix}$$

7.7. POSTURE DYNAMIC MODEL

$$H(\beta) = \begin{pmatrix} H_1(\beta_s, \beta_c) & \Sigma^T(\beta_s)V(\beta_c) \\ V^T(\beta_c)\Sigma(\beta_s) & I_s \end{pmatrix}$$

$$f(\beta, u) = \begin{pmatrix} f_1(\beta_s, \beta_c, \eta, \zeta) \\ f_2(\beta_s, \beta_c, \eta, \zeta) \end{pmatrix}$$

$$F(\beta) = \begin{pmatrix} B(\beta_s, \beta_c)P & 0 \\ 0 & I \end{pmatrix}$$

$$\tau_0 = \begin{pmatrix} \tau_m \\ \tau_s \end{pmatrix}.$$

It follows from Assumption 7.3 that the matrix $F(\beta)$ has full rank for all β. This property is important to analyze the behaviour of wheeled mobile robots and the design of feedback controllers. It is first used to transform the general state space model into a simpler and more convenient form, by a smooth static state feedback.

Property 7.6 The configuration dynamic model of a wheeled mobile robot (7.56) and (7.57) is feedback equivalent (by a smooth static time-invariant state feedback) to the following system:

$$\dot{q} = S(q)u \qquad (7.58)$$
$$\dot{u} = v \qquad (7.59)$$

where v represents a set of δ_m auxiliary control inputs.

□

Indeed, it follows directly from Property 7.1 that the following smooth static time-invariant state feedback is well defined everywhere in the state space, i.e.,

$$\tau_0 = F^\dagger(\beta)\big(H(\beta)\dot{u} - f(\beta, u)\big) \qquad (7.60)$$

where $F^\dagger$ denotes an arbitrary left inverse of $F(\beta)$.

We would like to emphasize that a further simplification is of interest from an operational viewpoint. In a context of trajectory planning or feedback control design, it is clear that we will be essentially concerned with controlling the posture of the robot (namely the coordinate $\xi(t)$) by using the control input v. We observe that this implies that we can ignore deliberately the coordinates β_c and φ and restrict our attention to the following *posture dynamic model*:

$$\dot{z} = B(z)u \qquad (7.61)$$
$$\dot{u} = v \qquad (7.62)$$

where we recall that $z = (\,\xi^T \quad \beta_s^T\,)^T$ and $u = (\,\eta^T \quad \zeta^T\,)^T$.

This posture dynamic model fully describes the system dynamics between the control input v and the posture ξ. The coordinates β_c and φ have apparently disappeared but it is important to notice that they are in fact hidden in the feedback (7.60).

The difference with the posture kinematic model is that now the variables u are part of the state vector. This implies the existence of a drift term and the fact that the input vector fields are constant.

The posture dynamic model inherits the structural properties of the posture kinematic model discussed in Section 7.4.3.

Property 7.7 The posture dynamic model is generic and irreducible, and small-time-locally-controllable; further, for restricted mobility robots, it is not stabilizable by a continuous static time-invariant state feedback, but is stabilizable by a time-varying static state feedback.

□

7.8 Further reading

This chapter on the kinematic and dynamic modelling of wheeled mobile robots has been adapted from [14]. Other examples of derivation of general kinematic models of wheeled mobile robots can be found in [3, 26]. Many designs of omnidirectional robots with Swedish wheels (also called directional sliding wheels [2]) have been reported in the literature. Typical examples are the UCL robot with three Swedish wheels [7], the URANUS robot with four Swedish wheels [25], the holonomic wheeled platform in [20], the ball wheel robot in [35] and more recently the ACTRESS robot [4]. Unicycle robots belong to the class of Type (2,0) robots. A well-known example is the HILARE robot [23]. Mobile robots that are built on the model of a conventional car (often called *car-like* robots) belong to the class of Type (1,1) robots. Examples from the literature are the AVATAR robot [5] and the HERO 1 robot [17]. Many commercial mobile robots are furnished with several steering wheels and therefore belong to the class of Type (1,2) robots. Typical example are the KLUDGE robot [18], the HERMIES-III robot [31, 32], and the robot described in [24]. The manoeuvrability of Type (1,2) robots can be improved by using compliant linkage [9]. The general approach followed in this chapter for the modelling of wheeled mobile robots can also be extended to other types of nonholonomic vehicles like cars with trailers [11], cars with front and rear wheel steering [1] or submarines [28, 33].

In this chapter, various structural properties regarding reducibility, controllability, stabilizability and nonholonomy of the kinematic and dynamic state space models of wheeled mobile robots have been given. For the sake of simplicity and readability, however, these properties have been stated in a rather intuitive way without a formal technical proof. The interested reader can find additional insight on the way these properties can be established in the following bibliographical notes.

Properties 7.1, 7.2 and 7.3 on reducibility and controllability stem from a fairly straightforward application of differential geometry theory for nonlinear control systems, as it is presented in some recent textbooks [19, 29, 34]. In particular, the irreducibility of the posture kinematic model is a direct consequence of the Frobenius theorem, e.g., [19, 34]. The concept of strong accessibility Lie algebra and its use in the analysis of the local controllability of nonlinear system is exposed, e.g., in [29] where it is shown also that strong accessibility implies controllability for driftless systems and hence for the posture kinematic model of a mobile robot.

Property 7.4 deals with state feedback stabilizability. The posture kinematic model of a restricted mobility robot is not stabilizable at an equilibrium point by a smooth time-invariant state feedback because it does not satisfy Brockett's necessary condition [10]. Nevertheless, it is stabilizable by a smooth time-varying state feedback; this is a consequence of a general stabilizability result for controllable driftless systems proved in [15]. A systematic procedure for the design of such time-varying control laws is proposed in [30]. This procedure is applicable to all types of mobile robots.

Property 7.5 emphasizes the nonholonomic nature of wheeled mobile robots which is also discussed in, e.g., [6, 21, 22, 27] in the context of motion planning. A treatment of general nonholonomic mechanical systems can be found in, e.g., [8, 12, 13]. It may arise also that, due for instance to slipping effects, the nonholonomic constraints be temporarily violated along the motion of the robot. A modelling approach of this situation is proposed in [16].

References

[1] J. Ackerman, "Robust yaw damping of cars with front and rear wheel steering," *Proc. 31st IEEE Conf. on Decision and Control*, Tucson, AZ, pp. 2586–2590, 1992.

[2] J. Agullo, S. Cardona, and J. Vivancos, "Kinematics of vehicles with directional sliding wheels," *Mechanism and Machine Theory*, vol. 22, pp. 295–301, 1987.

[3] J.C. Alexander and J.H. Maddocks, "On the kinematics of wheeled mobile robots," *Int. J. of Robotics Research*, vol. 8, no. 5, pp. 15–27, 1989.

[4] H. Asama, M. Sato, L. Bogoni, H. Kaetsu, A. Matsumoto, and I. Endo, "Development of an omni-directional mobile robot with 3 dof decoupling drive mechanism," *Proc. 1995 IEEE Int. Conf. on Robotics and Automation*, Nagoya, pp. 1925–1930, 1995.

[5] C. Balmer, "Avatar: a home built robot," *Robotics Age*, vol. 4, no 1, pp. 20–25, 1988.

[6] J. Barraquand and J.-C. Latombe, "On non-holonomic mobile robots and optimal manoeuvring", *Revue d'Intelligence Artificielle*, vol. 3, no. 2, pp. 77–103, 1989.

[7] G. Bastin and G. Campion, "On adaptive linearizing control of omnidirectional mobile robots," in *Progress in Systems and Control Theory 4*, Birkhäuser, Boston, MA, vol. 2, pp. 531–538, 1989.

[8] A.M. Bloch, N.H. McClamroch, and M. Reyhanoglu, "Control and stabilization of nonholonomic dynamical systems," *IEEE Trans. on Automatic Control*, vol. 37, pp. 1746–1757, 1992.

[9] J. Borenstein, "Control and kinematic design of multi-degree-of-freedom mobile robots with compliant linkage," *IEEE Trans. on Robotics and Automation*, vol. 11, pp. 21–35, 1995.

[10] R.W. Brockett, "Asymptotic stability and feedback stabilization," in *Differential Geometric Control Theory*, R.W. Brockett, R.S. Millman and H.J. Sussman (Eds.), Birkhäuser, Boston, MA, pp. 181–208, 1983.

[11] L. Bushnell, D. Tilbury, and S. Sastry, "Steering three-input nonholonomic systems: The fire truck example," *Int. J. of Robotic Research*, vol. 14, pp. 366–381, 1995.

[12] G. Campion, B. d'Andrea-Novel, and G. Bastin, "Controllability and state feedback stabilizability of nonholonomic mechanical systems," in *Advanced Robot Control*, Lecture Notes in Control and Information Science, vol. 162, C. Canudas de Wit (Ed.), Springer-Verlag, Berlin, D, pp. 106–124, 1991.

[13] G. Campion, B. d'Andréa-Novel, and G. Bastin, "Modelling and state feedback control of nonholonomic mechanical systems," *Proc. 30th IEEE Conf. on Decision and Control*, Brighton, UK, pp. 1184–1189, 1991.

[14] G. Campion, G. Bastin, and B. d'Andréa Novel, "Structural properties and classification of kinematic and dynamic models of wheeled mobile robots," *IEEE Trans. on Robotics and Automation*, vol. 12, pp. 47–62, 1996.

[15] J.M. Coron, "Global asymptotic stabilization for controllable systems without drift," *Mathematics of Control, Signals, and Systems*, vol. 5, pp. 295–312, 1992.

[16] B. d'Andréa-Novel, G. Campion, and G. Bastin, "Control of wheeled mobile robots not satisfying ideal velocity constraints: A singular perturbation approach," *Int. J. of Robust and Nonlinear Control*, vol. 5, pp. 243–267, 1995.

[17] C. Helmers, "Ein Hendenleben (or a hero's life)," *Robotics Age*, vol. 5, no. 2, pp. 7–16, 1983.

[18] J.M. Holland, "Rethinking robot mobility," *Robotics Age*, vol. 7, no. 1, pp. 26–30, 1988.

[19] A. Isidori, *Nonlinear Control Systems*, (3rd ed.), Springer-Verlag, London, UK, 1995.

[20] S.M. Killough and F.G. Pin, "Design of an omnidirectional and holonomic wheeled platform design," *Proc. 1992 IEEE Int. Conf. on Robotics and Automation*, Nice, F, pp. 84–90, 1992.

[21] J.-C. Latombe, *Robot Motion Planning*, Kluwer Academic Publishers, Boston, MA, 1991.

[22] J.-P. Laumond, "Finding collision-free smooth trajectories for a nonholonomic system," *Proc. 10th Int. Joint Conf. on Artificial Intelligence*, Milano, I, pp. 1120–1123, 1987.

[23] J.-P. Laumond, "Controllability of a multibody mobile robot," *IEEE Trans. on Robotics and Automation*, vol. 9, pp. 755–763, 1993.

[24] M.G. Mehrabi, R.M.H. Cheng, and A. Henami, "Control of a wheeled mobile robot with double steering," *Proc. IEEE/RSJ Int. Work. on Intelligent Robots and Systems*, Osaka, J, pp. 806–810, 1991.

[25] P.F. Muir and C.P. Neuman, "Kinematic modelling for feedback control of an omnidirectional mobile robot," *Proc. 1987 IEEE Int. Conf. on Robotics and Automation*, Raleigh, NC, pp. 1172–1178, 1987.

[26] P.F. Muir and C.P. Neuman, "Kinematic modeling of wheeled mobile robots," *J. of Robotic Systems*, vol. 4, pp. 281–329, 1987.

[27] R.M. Murray and S.S. Sastry, "Nonholonomic motion planning: Steering with sinusoids," *IEEE Trans. on Automatic Control*, vol. 38, pp. 700–716, 1993.

[28] Y. Nakamura and S. Savant, "Nonlinear tracking control of autonomous underwater vehicles," *Proc. 1992 IEEE Int. Conf. on Robotics and Automation*, Nice, F, pp. A4–A9, 1992.

[29] H. Nijmeijer and A.J. van der Schaft, *Nonlinear Dynamical Control Systems*, Springer-Verlag, New York, 1990.

[30] J.-B. Pomet, "Explicit design of time-varying stabilizing control laws for a class of controllable systems without drift," *Systems & Control Lett.*, vol. 18, pp. 147–158, 1992.

[31] D.B. Reister, J.P. Jones, P.L. Butler, M. Beckerman, and F.J. Sweeney, "DEMO89 – The initial experiment with the HERMIES-III robot," *Proc. 1991 IEEE Int. Conf. on Robotics and Automation*, Sacramento, CA, pp. 2562–2567, 1991.

[32] D.B. Reister and M.A. Unseren, "Position and constraint force control of a vehicle with two or more steerable drive wheels," *IEEE Trans. on Robotics and Automation*, vol. 9, pp. 723–731, 1993.

[33] O.J. Sørdalen, M. Dalsmo, and O. Egeland, "An exponentially convergent control law for a nonholonomic underwater vehicle," *Proc. 1993 IEEE Int. Conf. on Robotics and Automation*, Atlanta, GA, vol. 3, pp. 790–795, 1993.

[34] M. Vidyasagar, *Nonlinear Systems Analysis*, 2nd Ed., Prentice-Hall, Englewood Cliffs, NJ, 1993.

[35] M. West and H. Asada, "Design and control of ball wheel omnidirectional vehicles," *Proc. 1995 IEEE Int. Conf. on Robotics and Automation*, Nagoya, J, pp. 1931–1938, 1995.

Chapter 8

Feedback linearization

The last two chapters of the book are concerned with state feedback control of wheeled mobile robots.

The most challenging issue from a theoretical viewpoint is to find feedback control laws that can stabilize the robot about an equilibrium point. The reason is that a nonholonomic mobile robot cannot be stabilized by a smooth state feedback, as we have anticipated in Section 7.4.4. It is therefore necessary to find more clever solutions involving nonstationary (time varying) and/or singular feedback controls. This issue will be further discussed in the next chapter.

In this chapter, we will not be concerned with the stabilization problem about an equilibrium point (which is a *regulation* problem) but with another control problem, possibly more important in practice; namely, stable *tracking* of a reference motion (this is called also stabilization about a trajectory). Interestingly enough, the tracking problem is easier to solve than the regulation problem for wheeled mobile robots.

Our purpose in the present chapter is to investigate the solvability of tracking problems for mobile robots by means of smooth static and dynamic *feedback linearization*. In particular, we will elucidate the connection between the intrinsic structural mobility of the robots and their feedback linearizability.

We will address two basic tracking problems; namely, *point tracking* and *posture tracking* that will be described in the next section. By means of *static* feedback linearization, it is shown how to solve the *point and posture* tracking problems for omnidirectional robots and the *point* tracking problem only for robots having a restricted mobility. Then, it is shown how the *posture* tracking problem can be solved for all types of robots by *dynamic* feedback linearization, albeit with minor singularities. Design specifications that guarantee the avoidance of singularities are also given.

8.1 Feedback control problems

Let us now formulate the two main feedback control problems that will be considered in this chapter; namely, *posture tracking* and *point tracking*.

8.1.1 Posture tracking

The problem is to find a state feedback controller that can achieve tracking, with stability, of a given reference moving posture $\xi_r(t)$ which will be assumed to be twice differentiable. More precisely, the objective is to find a state feedback control law v such that:

- the tracking error $\tilde{\xi}(t) = \xi(t) - \xi_r(t)$ and the control v are bounded for all t;

- the tracking error asymptotically converges to zero, i.e., $\lim_{t \to \infty}(\xi(t) - \xi_r(t)) = 0$;

- if $\xi(0) = \xi_r(0)$, then $\xi(t) = \xi_r(t)$ for all t.

In other words, this can be seen as the problem of tracking the posture $\xi_r(t)$ of a virtual reference robot of the same type.

For omnidirectional robots, this problem can be solved by a smooth *static* linearizing state feedback as we have seen in Section 7.4.4. For restricted mobility robots we will show that this problem can be solved by a smooth *dynamic* linearizing state feedback, as long as the robot is moving, i.e., $\dot{\xi}(t) \neq 0$ for all t.

8.1.2 Point tracking

In a number of instances, full control of the robot posture is not required while it is sufficient (or even desirable) to control only the position of a *fixed* point P' on the cart of the robot (see Fig. 8.1). The polar coordinates of this point in the moving frame are denoted by (e, δ). The Cartesian coordinates of P' in the base frame are then expressed by

$$\begin{pmatrix} x' \\ y' \end{pmatrix} = \begin{pmatrix} x + e\cos(\vartheta + \delta) \\ y + e\sin(\vartheta + \delta) \end{pmatrix}. \tag{8.1}$$

The *point tracking problem* is to find a state feedback controller that can achieve tracking, with stability, of a given reference moving position $x'_r(t)$, $y'_r(t)$ which is assumed to be twice differentiable.

8.1. FEEDBACK CONTROL PROBLEMS

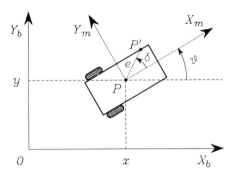

Figure 8.1: Point coordinates.

More precisely, the objective is to find a state feedback control law v such that:

- the tracking errors $\tilde{x}'(t) = x'(t) - x'_r(t)$, $\tilde{y}'(t) = y'(t) - y'_r(t)$ and the control $v(t)$ are bounded for all t,

- $\lim_{t \to \infty}(x'(t) - x'_r(t)) = 0$ and $\lim_{t \to \infty}(y'(t) - y'_r(t)) = 0$,

- if $x'(0) = x'_r(0)$ and $y'(0) = y'_r(0)$, then $x'(t) = x'_r(t)$ and $y'(t) = y'_r(t)$ for all t.

Although we will see that restricted mobility robots are not full state feedback linearizable via static state feedback, a partial feedback linearization can help us solve the point tracking problem.

8.1.3 Velocity and torque control

The control problems, as they have been formulated above, consist of finding a feedback control law to achieve tracking of a reference trajectory, for the posture dynamic model (see Section 7.7)

$$\dot{z} = B(z)u \qquad (8.2)$$
$$\dot{u} = v \qquad (8.3)$$

where we recall that $z = (\,\xi^T \quad \beta_s^T\,)^T$ and $u = (\,\eta^T \quad \zeta^T\,)^T$. This is called feedback *torque control* because the control action v is homogeneous to a torque.

The same posture and point tracking problems can also be considered for the posture kinematic model only, i.e.,

$$\dot{z} = B(z)u. \tag{8.4}$$

This is called feedback *velocity control* because the control action u is homogeneous to a velocity.

8.2 Static state feedback

In this section, we use static feedback linearization to solve the above mentioned feedback control problems. We first have the following property about the maximal subsystem that can be linearized by static state feedback.

Property 8.1 The largest subsystem of the posture kinematic model (8.4) linearizable by a smooth static state feedback has dimension $(\delta_m + \delta_s)$. The largest linearizable subsystem of the dynamic model (8.2) and (8.3) has dimension $2(\delta_m + \delta_s)$.

□

Indeed, it can be checked that the largest linearizable subsystem of the posture kinematic state space model (8.4) is obtained by selecting $(\delta_m + \delta_s)$ linearizing output functions depending on z. A vector of $(3 - \delta_m)$ coordinates remains non linearized. Furthermore, the largest linearizable subsystem of the posture dynamic state space model (8.2) and (8.3) has dimension $2(\delta_m + \delta_s)$ with exactly the same linearizing output functions which will be made explicit in Lemma 8.1 and Tab. 8.1.

For omnidirectional robots ($\delta_m = 3$ and $\delta_s = 0$) Property 8.1 shows that the posture kinematic (dynamic) model is fully linearizable by static state feedback. In contrast, restricted mobility robots are only partially feedback linearizable. Hence, it will be shown in this section that, with a *smooth static state feedback* the posture tracking problem can be solved for omnidirectional robots, whereas only the point tracking problem can be considered for restricted mobility robots.

8.2.1 Omnidirectional robots

We know from Chapter 7 that the posture dynamic model of omnidirectional robots has the following form:

$$\dot{\xi} = R^T(\vartheta)\eta \tag{8.5}$$
$$\dot{\eta} = v. \tag{8.6}$$

8.2. STATIC STATE FEEDBACK

Differentiating the first equation gives

$$\ddot{\xi} = R^T(\vartheta)\dot{\eta} + \dot{R}^T(\vartheta)\eta. \tag{8.7}$$

Since matrix $R^T(\vartheta)$ is invertible for all ϑ, it is clear that we can use the control $v = \dot{\eta}$ to assign any arbitrary dynamics to the posture $\xi(t)$. More precisely, we have:

Lemma 8.1 *The posture tracking problem is solved for omnidirectional robots by choosing the state feedback torque control*

$$v = R^{-T}(\vartheta)\left(-\dot{R}^T(\vartheta)\eta + \ddot{\xi}_r - (\Lambda_1 + \Lambda_2)\dot{\tilde{\xi}} - \Lambda_1\Lambda_2\tilde{\xi}\right) \tag{8.8}$$

where $\tilde{\xi} = \xi - \xi_r$ and the (3 × 3) matrices Λ_1 and Λ_2 are positive and diagonal.

◇ ◇ ◇

Proof. Substituting (8.8) in (8.6), and then in (8.7) gives that the dynamics of the tracking error $\tilde{\xi} = \xi - \xi_r$ is governed by the stable linear differential equation

$$\ddot{\tilde{\xi}} + (\Lambda_1 + \Lambda_2)\dot{\tilde{\xi}} + \Lambda_1\Lambda_2\tilde{\xi} = 0 \tag{8.9}$$

where the (3 × 3) matrices Λ_1 and Λ_2 are positive and diagonal.

◇

Remarks

- Eq. (8.9) can be interpreted as a *reference model* for the tracking error, that is a model of how we want the tracking error to decrease. We observe that this control law is quite similar to the *inverse dynamics control* that has been classically derived for rigid link manipulators (see Chapter 2).

- It is worth writing the reference model (8.9) in state space form

$$\begin{aligned} \dot{\tilde{\xi}} &= -\Lambda_1\tilde{\xi} + \tilde{\sigma} \\ \dot{\tilde{\sigma}} &= -\Lambda_2\tilde{\sigma} \end{aligned} \tag{8.10}$$

with the new state $\tilde{\sigma} = \Lambda_1\tilde{\xi} - \dot{\xi}_r + R^T(\vartheta)\eta$. It is easily checked that the new state $\tilde{\sigma}$ can also be written as

$$\tilde{\sigma} = R^T(\vartheta)(\eta - \eta_r) \tag{8.11}$$

with $\eta_r = R(\vartheta)(\dot{\xi}_r - \Lambda_1 \tilde{\xi})$, being the linearizing velocity control of the kinematic state space model. We can therefore interpret (8.8) as a torque control law that performs tracking of the reference linearizing velocity control signal $\eta_r(t)$ together with the reference posture $\xi_r(t)$.

8.2.2 Restricted mobility robots

As we have seen in Chapter 7, the posture dynamic model (8.2) and (8.3) can be written in the following form for restricted mobility robots:

$$\dot{\xi} = R^T(\vartheta)\Sigma(\beta_s)\eta \qquad (8.12)$$
$$\dot{\beta}_s = \zeta \qquad (8.13)$$
$$\dot{\eta} = v_1 \qquad (8.14)$$
$$\dot{\zeta} = v_2. \qquad (8.15)$$

We know from Property 8.1 that this model (8.12)–(8.15) is *not* full state linearizable by a smooth static time-invariant state feedback and that the largest linearizable subsystem has dimension $2(\delta_m + \delta_s)$. The purpose of this section is to examine how this property can be used to solve the point tracking problem by static state feedback linearization. Before that, we discuss the partial feedback linearization of the dynamic model (8.12)–(8.15) in a more detailed way.

Lemma 8.2 *The general dynamic model (8.12)–(8.15) can be generically transformed by state feedback and diffeomorphism into a controllable linear subsystem of dimension $2(\delta_m + \delta_s)$ and a nonlinear subsystem of dimension $3 - \delta_m$ of the form:*

$$\begin{aligned} \dot{z}_1 &= z_2 \\ \dot{z}_2 &= w \\ \dot{z}_3 &= \bar{Q}(z_1, z_3)z_2, \end{aligned} \qquad (8.16)$$

where both z_1 and z_2 have dimensions $\delta_m + \delta_s$, z_3 has dimension $3 - \delta_m$, and w is an auxiliary torque control input.

◇ ◇ ◇

Proof. As a consequence of Property 8.1, there exists a linearizing output vector function

$$z_1 = h(\xi, \beta_s) = h(z) \qquad (8.17)$$

of dimension $(\delta_m + \delta_s)$ which depends on the posture coordinates ξ and the angular coordinates β_s only, but *not* on the states η and ζ, such that

8.2. STATIC STATE FEEDBACK

the largest linearizable subsystem is obtained by differentiating twice z_1 as follows:

$$\dot{z}_1 = \frac{\partial h}{\partial z} B(z) u = K(z) u \qquad (8.18)$$
$$\ddot{z}_1 = K(z) v + g(z, u),$$

with the decoupling matrix

$$K(z) = \left(\frac{\partial h}{\partial \xi} R^T(\vartheta) \Sigma(\beta_s) \quad \frac{\partial h}{\partial \beta_s} \right) \qquad (8.19)$$

and

$$g(z,u) = \frac{\partial}{\partial \xi} \left(K(\xi^T \ \beta_s^T) \begin{pmatrix} \eta \\ \zeta \end{pmatrix} \right) R^T(\vartheta) \Sigma(\beta_s) \eta + \frac{\partial}{\partial \beta_s} \left(K(\xi^T \ \beta_s^T) \begin{pmatrix} \eta \\ \zeta \end{pmatrix} \right). \qquad (8.20)$$

A change of coordinates can be defined as

$$z_1 = h(z)$$
$$z_2 = K(z) u$$
$$z_3 = k(z)$$

where $k(z)$ is selected such that the mapping

$$\begin{pmatrix} \xi \\ \beta_s \end{pmatrix} \longrightarrow \begin{pmatrix} h(z) \\ k(z) \end{pmatrix} \qquad (8.21)$$

is a diffeomorphism on $R^{\delta_s + 3}$. In the new coordinates, the system dynamics becomes

$$\dot{z}_1 = z_2$$
$$\dot{z}_2 = \bar{g}(z_1, z_2, z_3) + \bar{K}(z_1, z_3) \begin{pmatrix} v_1 \\ v_2 \end{pmatrix} \qquad (8.22)$$
$$\dot{z}_3 = \bar{Q}(z_1, z_3) z_2$$

where $\bar{g}$, $\bar{K}$, and $\bar{Q}$ are representative of g, K, and Q, respectively expressed in the new coordinates, with

$$Q(z) = \left(\frac{\partial k}{\partial \xi} R^T(\vartheta) \Sigma(\beta_s) \quad \frac{\partial k}{\partial \beta_s} \right) K^{-1}(z). \qquad (8.23)$$

Since the decoupling matrix $K(z)$ is generically invertible, by applying the control

$$\begin{pmatrix} v_1 \\ v_2 \end{pmatrix} = \bar{K}^{-1}(z_1, z_3)(w - \bar{g}(z_1, z_2, z_3)) \qquad (8.24)$$

to (8.22), we obtain the expected partially linear system (8.16) and this concludes the proof.

◇

Once we have obtained the partially linearized structure (8.22), the auxiliary input w can be used to freely assign the dynamics of z_1 as the dynamics of a stable second-order linear system. But, in order to perform tracking with stability of some suitable reference trajectory, we have to ensure at least boundedness of the nonlinear part z_3.

Lemma 8.3 *Let $(z_{1d}, \dot{z}_{1d}, \ddot{z}_{1d})$ be a reference smooth trajectory such that $\|z_{1d}(t)\|$, $\|\dot{z}_{1d}(t)\|$, $\|\ddot{z}_{1d}(t)\|$ are bounded for all t and such that $\dot{z}_{1d}(t)$ is $\mathcal{L}_1$. If the matrix $\bar{Q}(z_1, z_3)$ defined in (8.22) is bounded for all z_1, z_3 then, the auxiliary control law*

$$w = \ddot{z}_{1d}(t) - (\Lambda_1 + \Lambda_2)\dot{\tilde{z}}_1 - \Lambda_1\Lambda_2\tilde{z}_1 \tag{8.25}$$

where $\tilde{z}_1 = z_1 - z_{1d}$, Λ_1 and Λ_2 are arbitrary $((\delta_m + \delta_s) \times (\delta_m + \delta_s))$ positive diagonal matrices, generically ensures that $\tilde{z}_1$ and $\dot{\tilde{z}}_1$ exponentially converge to zero, and $z_3(t)$ is bounded for all t.

◇ ◇ ◇

(δ_m, δ_s)	$z_1 = h(\xi, \beta_s)$	$z_3 = k(\xi, \beta_s)$	$\det(K(\xi, \beta_s)) \neq 0$
(2,0)	$\begin{pmatrix} x + e\cos(\vartheta + \delta) \\ y + e\sin(\vartheta + \delta) \end{pmatrix}$	ϑ	$e \neq 0$ $\delta \neq 2k\pi$
(2,1)	$\begin{pmatrix} x + e\cos(\vartheta + \delta) \\ y + e\sin(\vartheta + \delta) \\ \beta_s \end{pmatrix}$	ϑ	$e \neq 0$ $\delta \neq \beta_s + 2k\pi$
(1,1)	$\begin{pmatrix} x + L\sin\vartheta + e\cos(\vartheta + \beta_s) \\ y - L\cos\vartheta + e\sin(\vartheta + \beta_s) \end{pmatrix}$	$\begin{pmatrix} \vartheta \\ \beta_s \end{pmatrix}$	$e \neq 0$
(1,2)	$\begin{pmatrix} x + L\cos\vartheta - e\sin(\vartheta + \beta_{s1}) \\ y + L\sin\vartheta + e\cos(\vartheta + \beta_{s1}) \\ \beta_{s2} \end{pmatrix}$	$\begin{pmatrix} \vartheta \\ \beta_{s1} \end{pmatrix}$	$e\sin\beta_{s2} \neq 0$

Table 8.1: Linearizing outputs, nonlinear coordinates, and regularity conditions for different Type robots.

Proof. Substituting (8.25) in (8.16) leads to the following stable linear decoupled second-order equation in $\tilde{z}_1$:

$$\ddot{\tilde{z}}_1 + (\Lambda_1 + \Lambda_2)\dot{\tilde{z}}_1 + \Lambda_1\Lambda_2\tilde{z}_1 = 0,$$

8.2. STATIC STATE FEEDBACK

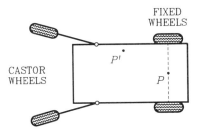

Figure 8.2: Type (2,0) robot with two fixed wheels and two castor wheels.

and thus there exist positive constants c and α such that

$$\|\tilde{z}_2(t)\| \leq c \exp(-\alpha t)\|\tilde{z}_2(0)\|. \tag{8.26}$$

Moreover, since by assumption $\bar{Q}(z_1, z_3)$ defined in (8.22) is bounded for every z_1, z_3, say by a positive constant K_1, we have

$$\|z_3(t)\| \leq \|z_3(0)\| + K_1 \left(\int_0^t \|z_{2d}(\tau)\| d\tau + \frac{C}{\alpha}(1 - \exp(-\alpha t))\|\tilde{z}_2(0)\| \right).$$

Since the reference trajectory is supposed to be such that $\dot{z}_{1d}(t) = z_{2d}(t)$ is $\mathcal{L}_1$ we can conclude that z_3 remains bounded for all t.

◇

We now examine the possibility of using the above linearizing control approach to solve the point tracking problem for restricted mobility robots. For each type of robots, we apply Lemmas 8.2 and 8.3 and we give admissible choices of output functions z_1 containing the Cartesian coordinates of a point P' on the cart of the robot (see Section 8.1.2), and of the non-linearized coordinates z_3 in Tab. 8.1. The last column of this table gives also in each case, the conditions for the decoupling matrix $K(\xi, \beta_s)$ given by (8.19) to be nonsingular.

From this table, we see that the point tracking problem is solvable by static feedback linearization for Type (2,0) and Type (2,1) robots, because the Cartesian coordinates x', y' of a fixed point P' on the cart that we wish to track, are components of the vector $z_1 = h(z)$ of linearizing output functions, in both cases.

However, the regularity conditions may imply that some particular choices of the controlled point P' must be excluded. Let us examine this issue on two concrete examples.

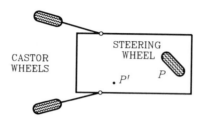

Figure 8.3: Type (2,1) robot.

Type (2,0) robot

The robot of Fig. 8.2 has two fixed wheels and two castor wheels. The regularity condition means that P' must not be located on the axle of the fixed wheels.

In fact, as we have seen in Section 7.2, any Type (2,0) robot has no steering wheel while they have either one or several fixed wheels, but with a single common axle. Therefore, the regularity condition will always mean that P' must not be located on this axle.

Type (2,1) robot

The robot of Fig. 8.3 has one steering wheel and two castor wheels. The point P must be the center of the steering wheel. In fact, any Type (2,1) robot has only one steering wheel (see Section 7.2). The regularity conditions always mean that the point P' should not be the center of the steering wheel (e must not be zero) and that $\sin(\delta - \beta_s)$ must not be zero. Due to the block-triangular form of the corresponding decoupling matrix $K(\vartheta, \beta_s)$, this singular configuration characterized by $\delta = \beta_s + 2k\pi$ ($k \in Z$) can be avoided at each time instant by a proper choice of the control input v_2 whereas v_1 ensures linearization of dynamics of the tracking error vector $\tilde{z}_1$.

In contrast, for Type (1,1) and Type (1,2) robots, the point tracking problem for a fixed point P' of the cart defined by (8.1) *cannot* be solved by feedback linearization, the corresponding decoupling matrix being generically singular. It is worth noticing, however, that for these robots there are linearizing outputs which can be interpreted from a "physical" viewpoint.

Type (1,1) robot

The robot of Fig. 8.4, classically referred to as the *front-wheel drive robot* (car), has two steering wheels with coordinated orientations, and two fixed

8.2. STATIC STATE FEEDBACK   317

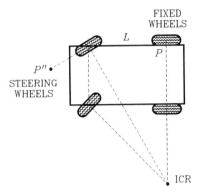

Figure 8.4: Type (1,1) front-wheel drive robot.

wheels. The axles of the two steering wheels converge to the instantaneous center of rotation placed on the axle of the fixed wheels. One of the steering wheels is selected (arbitrarily) as the "master" wheel, with an orientation characterized by the angle β_s, while the orientation of the other is constrained so as to satisfy the above condition. The origin of the moving frame P is located on the axle of the fixed wheels, at the intersection with the normal passing through the center of the master steering wheel. Then, the constant L is the distance between P and the center of the master wheel.

The linearizing outputs (see Tab. 8.1) correspond to the Cartesian coordinates of a material point P'' of the robot attached to the plane of one of the steering wheels, except the center of the wheel (belonging to the cart), for which $e = 0$. In fact, Type (1,1) robots have one or several fixed wheels with a single common axle. They have also one or several steering wheels but with one independent steering orientation ($\delta_s = 1$), and the center of one of these steering wheels is not located on the axle of the fixed wheels. Consequently, for any Type (1,1) robot, the linearizing outputs can be chosen as the Cartesian coordinates of a point P'' of the robot attached to the plane of one of the steering wheels, except the center of the wheel.

Type (1,2) robot

The robot of Fig. 8.5 has two independent steering wheels and two castor wheels. The point P is the center of the axle joining the centers of the two steering wheels. Moreover, the constant L is half the length of this axle. There is again a "physical" choice of linearizing outputs (see Tab. 8.1) that

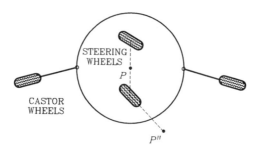

Figure 8.5: Type (1,2) robot.

allows solving the point tracking problem for a material point P'' attached to the plane of one of the two independent steering wheels, except the center of the corresponding wheel. The regularity conditions also mean that $\beta_{s2} \neq 2k\pi$. As for the Type (2,1) case, due to the block-triangular form of the corresponding decoupling matrix $K(\vartheta, \beta_s)$, this singular configuration can be avoided at each instant by an appropriate choice of the control input v_2 whereas v_1 ensures linearization of dynamics of the tracking error vector $\tilde{z}_1$. Type (1,2) robots have no fixed wheel and two or more steering wheels, two of them being oriented independently. Therefore, for any Type (1,2) robot, a possible choice of linearizing outputs are the Cartesian coordinates of a point attached to the plane of one of the steering wheels except the center of the corresponding wheel.

For each type of restricted mobility robots, it is easy to check from (8.19) that the decoupling matrix $K(\xi, \beta_s)$ depends only on sine and cosine functions of the angles ϑ and β_s, and from Tab. 8.1 we see that $z_3 = k(\xi, \beta_s)$ only depends linearly on ϑ and β_s. Hence, we easily deduce that the matrix $\bar{Q}$ in (8.16) defined via (8.23) is bounded for every ϑ and β_s such that $\det(K(\vartheta, \beta_s)) > \epsilon > 0$. Therefore, in view of Lemma 8.3, if the reference trajectory is such that $z_{1d}(t)$, $\dot{z}_{1d}(t)$, $\ddot{z}_{1d}(t)$ are bounded for all t and $\dot{z}_{1d}(t)$ is $\mathcal{L}_1$, we can conclude that the point tracking problem is solvable for any restricted mobility robot, with a fixed point P' on the cart when $\delta_m = 2$ and a fixed point P'' attached to one steering wheel when $\delta_m = 1$.

8.3 Dynamic state feedback

The objective of this section is to use dynamic feedback linearization to solve the posture tracking problem for restricted mobility robots. Design

8.3. DYNAMIC STATE FEEDBACK

specifications that guarantee the avoidance of singularities are also given.

8.3.1 Dynamic extension algorithm

We consider a dynamical system given in general state space form

$$\dot{z} = f(z) + \sum_{i=1}^{m} g_i(z) u_i \qquad (8.27)$$

where the state $z \in R^n$, the input $u \in R^m$, and the vector fields f and g_i are smooth.

When the system is not completely linearizable by diffeomorphism and static state feedback (as for restricted mobility robots, see the previous section), full linearization can nevertheless possibly be achieved by considering more general dynamic feedback laws of the form

$$\begin{aligned} u &= \alpha(z, \chi, w) \\ \dot{\chi} &= a(z, \chi, w) \end{aligned} \qquad (8.28)$$

where w is an auxiliary control input. Such a dynamic feedback is obtained through the choice of m suitable linearizing output functions

$$y_i = h_i(z) \qquad i = 1, \cdots, m \qquad (8.29)$$

leading to a singular decoupling matrix. We apply the so-called *dynamic extension algorithm* to system (8.27) and (8.29). The idea of this algorithm is to delay some "combinations of inputs" simultaneously affecting several outputs, via the addition of integrators, in order to enable other inputs to act in the meanwhile and therefore hopefully to obtain an extended decoupled system of the form

$$y_k^{(r_k)} = w_k \qquad k = 1, \cdots, m \qquad (8.30)$$

where $y_k^{(i)}$ denotes the i-th derivative of y_k with respect to time, r_k is the relative degree of y_k, and w_k's denote the new auxiliary inputs. Moreover, in order to get *full* linearization, we shall have for the n_e-dimensional extended system

$$\sum_{i=1}^{m} r_i = n_e, \qquad (8.31)$$

where n_e is the dimension of the extended state vector $z_e = (\, z^T \quad \chi^T \,)^T$ and if (8.31) is satisfied, then

$$\zeta = \Psi(z_e) = \left(y_1 \quad \cdots \quad y_1^{(r_1-1)} \quad \cdots \quad y_m \quad \cdots \quad y_m^{(r_m-1)} \right)^T \qquad (8.32)$$

is a local diffeomorphism.

The dynamic extension algorithm is illustrated below with an example.

Type (2,0) robots

In order to apply the dynamic extension algorithm, we have to choose linearizing output functions which correspond to a singular decoupling matrix when considering static state feedback laws. From Tab. 8.1, a possible choice consists of taking as output functions the coordinates of a point P' located *on* the axle of the fixed wheels ($\delta = 2k\pi$), i.e.,

$$h_1 = x + e\cos\vartheta$$
$$h_2 = y + e\sin\vartheta. \qquad (8.33)$$

We first consider the posture *kinematic* state space model of Type (2,0) robots

$$\dot{x} = -\sin\vartheta\,\eta_1$$
$$\dot{y} = \cos\vartheta\,\eta_1 \qquad (8.34)$$
$$\dot{\vartheta} = \eta_2.$$

We easily check that the only combination of inputs appearing in $\dot{h}_1$ and $\dot{h}_2$ is $\chi_1 = \eta_1 + e\eta_2$, χ_1 being the linear velocity of P'. Hence, applying the dynamic extension algorithm, we can delay χ_1, i.e., by introducing a new input w_1 such that

$$\dot\chi_1 = w_1. \qquad (8.35)$$

Then, by computing $\ddot{h}_1$ and $\ddot{h}_2$, the extended system with extended state vector

$$z_e = (\,x\quad y\quad \vartheta\quad \chi_1\,)^T$$

is linearizable by static state feedback with new inputs w_1 and η_2 and diffeomorphism

$$\Psi = (\,h_1\quad h_2\quad \dot{h}_1\quad \dot{h}_2\,)^T. \qquad (8.36)$$

Moreover, since

$$\ddot{h}_1 = -\chi_1 \cos\vartheta\,\eta_2 - \sin\vartheta\,w_1$$
$$\ddot{h}_2 = -\chi_1 \sin\vartheta\,\eta_2 + \cos\vartheta\,w_1, \qquad (8.37)$$

we notice that the new decoupling matrix is singular only when the linear velocity χ_1 of P' is zero.

Let us now consider the posture *dynamic* state space model of Type (2,0) robots:

$$\dot{x} = -\eta_1 \sin\vartheta$$
$$\dot{y} = \eta_1 \cos\vartheta$$
$$\dot{\vartheta} = \eta_2 \qquad (8.38)$$
$$\dot\eta_1 = v_1$$
$$\dot\eta_2 = v_2.$$

8.3. DYNAMIC STATE FEEDBACK

We easily see that the only combination of inputs appearing in $\ddot{h}_1$ and $\ddot{h}_2$ is now $\chi_2 = v_1 + ev_2 = \dot{\chi}_1$. Hence, by applying the dynamic extension algorithm, we can delay χ_2, i.e., by introducing a new input w_2 such that

$$\dot{\chi}_2 = w_2. \tag{8.39}$$

Then, by computing $h_1^{(3)}$ and $h_2^{(3)}$, the extended system with state vector

$$z_e = \begin{pmatrix} x & y & \vartheta & \chi_1 & \eta_2 & \chi_2 \end{pmatrix}^T$$

is linearizable by static state feedback with new inputs w_2 and v_2 and diffeomorphism

$$\Psi = \begin{pmatrix} h_1 & h_2 & \dot{h}_1 & \dot{h}_2 & \ddot{h}_1 & \ddot{h}_2 \end{pmatrix}^T. \tag{8.40}$$

Moreover, since

$$\begin{aligned} h_1^{(3)} &= -w_2 \sin\vartheta - v_2 \cos\vartheta \chi_1 - 2\eta_2 \cos\vartheta \chi_2 + \eta_2^2 \sin\vartheta \chi_1 \\ h_2^{(3)} &= w_2 \cos\vartheta - v_2 \sin\vartheta \chi_1 - 2\eta_2 \sin\vartheta \chi_2 - \eta_2^2 \cos\vartheta \chi_1, \end{aligned} \tag{8.41}$$

we notice that the new decoupling matrix is also singular when the linear velocity χ_1 of P' is zero. We summarize this discussion in the following lemma.

Lemma 8.4 *By considering as output functions the Cartesian coordinates given by (8.1) of a point P' on the axle of the fixed wheels (corresponding to $\delta = 2k\pi$), Type (2,0) robots are generically fully dynamic feedback linearizable.*

⋄ ⋄ ⋄

We can see that the kinematic and dynamic state space models of Type (2,0) robots are both fully dynamic feedback linearizable by considering the same linearizing output functions. This result can be generalized to all types of robots as we discuss in the next section.

8.3.2 Differential flatness

A system which can be shown to be linearizable by using the dynamic extension algorithm is an example of a so-called *differentially flat system*. The linearizing outputs $y_i = h_i(z)$ are also called differentially flat outputs. Conversely, differential flatness refers by definition to a system which can be fully linearized by dynamic state feedback. In this chapter, the two following properties of differentially flat systems are of interest.

Property 8.2 Any controllable driftless system with m inputs and at most $m + 2$ states is a differentially flat system.

□

Property 8.3 If a nonlinear system $\dot{z} = f(z, u)$ is differentially flat, then the augmented system $\dot{z} = f(z, u)$, $\dot{u} = v$ is also differentially flat with the same flat output functions.

□

The following lemma then states that any mobile robot is a differentially flat system.

Lemma 8.5 *The posture kinematic model (8.4) and the posture dynamic model (8.2) and (8.3) is a differentially flat system and is therefore generically fully linearizable by dynamic state feedback.*

◇ ◇ ◇

Proof. We first consider the posture kinematic model (8.4) of wheeled mobile robots. We observe that it is a driftless system. From the analysis of Chapter 7, we also know that the posture kinematic model is a state space model with $\delta_m + \delta_s$ inputs and $3 + \delta_s$ state variables satisfying the following inequality:
$$3 + \delta_s \leq \delta_m + \delta_s + 2.$$
Then it follows from Property 8.2 that the posture kinematic model is a differentially flat system and consequently, from Property 8.3, that the posture dynamic model is also a differentially flat system.

◇

In order to use feedback linearization to solve the posture tracking problem, it remains obviously to characterize the linearizing output functions for each type of robot. This problem will be considered in Section 8.3.4. Before that, the issue of *singularities* is briefly discussed.

8.3.3 Avoiding singularities

The decoupling matrix $K(z_e)$ is the square matrix with entries

$$K_{ij}(z_e) = \frac{\partial y_i^{(r_i)}}{\partial u_j}. \tag{8.42}$$

Obviously, the linearizing control law becomes singular when the determinant of the decoupling matrix $K(z_e)$ is zero. Generically, this corresponds in the extended state space to an $(n_e - 1)$-dimensional submanifold. We

8.3. DYNAMIC STATE FEEDBACK

know that this submanifold contains the domain where the diffeomorphism $\Psi(z_e)$ is singular. Consequently, at any point where the decoupling matrix is not singular, the diffeomorphism $\Psi(z_e)$ exists.

The objective is to use dynamic feedback linearization to achieve tracking with stability of a reference trajectory. More precisely let $z_r(t)$ be a reference smooth trajectory; then, according to (8.29), we have

$$y_{ri} = h_i(z_r) \qquad i = 1, \ldots, m \qquad (8.43)$$

and according to (8.32) $\zeta_r(t)$ is defined as

$$\zeta_r = \begin{pmatrix} y_{r1} & \ldots & y_{r1}^{(r_1-1)} & \ldots & y_{rm} & \ldots & y_{rm}^{(r_m-1)} \end{pmatrix}^T. \qquad (8.44)$$

This implies that
$$z_{er} = \Psi^{-1}(\zeta_r). \qquad (8.45)$$

The following lemma shows how to select the auxiliary control w (see (8.30)) in order to achieve tracking with stability of $\zeta_r(t)$ and consequently of $z_r(t)$.

Lemma 8.6 *The auxiliary control laws w_i, for $i = 1, \ldots m$,*

$$w_i = y_{ri}^{(r_i)} + \sum_{j=0}^{r_i-1} \alpha_i^j (y_i^{(j)} - y_{ri}^{(j)}) \qquad (8.46)$$

generically ensure that the trajectory error vector $\tilde{\zeta} = \zeta - \zeta_r$ exponentially converges to zero, provided that the arbitrary constants α_i^j are chosen to place the poles of the linear closed-loop system in the left-half plane. Further, $\tilde{z} = z - z_r$ also asymptotically converges to zero.

◊ ◊ ◊

Proof. The exponential convergence of $\tilde{\zeta}$ to zero is evident. Moreover, Ψ being a diffeomorphism, this implies that z_e tends to z_{er} and therefore that z tends to z_r.

◊

The following lemma shows that if the reference trajectory is bounded away from singularities, then the same will hold for the closed-loop trajectory, provided that the initial conditions are suitably chosen.

Lemma 8.7 *Let system (8.27) be generically fully linearizable by dynamic state feedback and diffeomorphism*

$$\zeta = \Psi(z_e). \qquad (8.47)$$

If

- there exists an open simply connected domain D included in
$$S = \{\zeta \in R^{n_e} : \det(K(\Psi^{-1}(\zeta))) \neq 0\};$$

- the reference trajectory $\zeta_r(t)$ belongs to D for all t and moreover there exists a positive continuous function $M_1(t)$ such that
$$\|\tilde{\zeta}(t)\| < M_1(t) \implies \zeta(t) \in D \qquad (8.48)$$
where $\tilde{\zeta} = \zeta - \zeta_r$;

- the initial conditions satisfy
$$\|\tilde{\zeta}(0)\| \leq \min_t F(t) \qquad F(t) = \left(\frac{M_1(t) - \epsilon}{K_2(t)}\right) \qquad (8.49)$$
for some positive constant ϵ and where $K_2(t) = \exp(-r_1 t)K_1 > 0$ characterizes the closed-loop exponentially stable dynamics (with $K_1 > 0$, $r_1 > 0$):
$$\|\tilde{\zeta}(t)\| \leq K_2(t)\|\tilde{\zeta}(0)\| \qquad \forall t; \qquad (8.50)$$

then
$$\zeta(t) \in D \qquad \forall t. \qquad (8.51)$$

◇ ◇ ◇

Proof. To show (8.51) it is sufficient to show that
$$\|\tilde{\zeta}(t)\| < M_1(t) \qquad \forall t. \qquad (8.52)$$
By contradiction, let us suppose that there exists a positive time T such that
$$\begin{aligned}\|\tilde{\zeta}(T)\| &= M_1(T) \\ \|\tilde{\zeta}(t)\| &< M_1(t) \qquad \forall t \in [0, T).\end{aligned} \qquad (8.53)$$
In $[0, T)$ we have from (8.50) and (8.49) that
$$\|\tilde{\zeta}(t)\| \leq M_1(t) - \epsilon. \qquad (8.54)$$
By continuity of $\tilde{\zeta}(t)$ and $M_1(t)$, eq. (8.54) is in contradiction with (8.53), and this concludes the proof.

◇

We can easily show that the following choice of $M_1(t)$ satisfies (8.48):
$$M_1(t) = \min_\zeta \|\tilde{\zeta}(t)\| \qquad (8.55)$$
when $\det(K(\Psi^{-1}(\zeta))) = 0$.

8.3.4 Solving the posture tracking problem

By applying the dynamic extension algorithm to each of the four posture kinematic models of restricted mobility robots, we have obtained the results summarized in Tab. 8.2 where we give in each case the linearizing output functions $h_i(z)$, the extended state vector z_e, the diffeomorphism $\Psi(z_e)$, and the regularity conditions ensuring that the decoupling matrix $K(z_e)$ of the extended system is not singular.

Type	$y_i = h_i(z)$	z_e	$\zeta = \Psi(z_e)$	$\det(K(z_e)) \neq 0$
(2,0)	$\begin{pmatrix} x + e\cos\vartheta \\ y + e\sin\vartheta \end{pmatrix}$	$\begin{pmatrix} x \\ y \\ \vartheta \\ \chi_1 \end{pmatrix}$	$\begin{pmatrix} h_1 \\ h_2 \\ \dot{h}_1 \\ \dot{h}_2 \end{pmatrix}$	$\chi_1 \neq 0$
(2,1)	$\begin{pmatrix} x \\ y \\ \vartheta \end{pmatrix}$	$\begin{pmatrix} x \\ y \\ \vartheta \\ \beta_s \\ \eta_1 \\ \eta_2 \end{pmatrix}$	$\begin{pmatrix} h_1 \\ h_2 \\ h_3 \\ \dot{h}_1 \\ \dot{h}_2 \\ \dot{h}_3 \end{pmatrix}$	$\eta_1 \neq 0$
(1,1)	$\begin{pmatrix} x + e\cos\vartheta \\ y + e\sin\vartheta \end{pmatrix}$	$\begin{pmatrix} x \\ y \\ \vartheta \\ \beta_s \\ \eta_1 \\ \chi_2 \end{pmatrix}$	$\begin{pmatrix} h_1 \\ h_2 \\ \dot{h}_1 \\ \dot{h}_2 \\ \ddot{h}_1 \\ \ddot{h}_2 \end{pmatrix}$	$\eta_1 \neq 0$
(1,2)	$\begin{pmatrix} x + e\cos(\vartheta + \delta) \\ y + e\sin(\vartheta + \delta) \\ \vartheta \end{pmatrix}$	$\begin{pmatrix} x \\ y \\ \vartheta \\ \beta_{s1} \\ \beta_{s2} \\ \eta_1 \end{pmatrix}$	$\begin{pmatrix} h_1 \\ h_2 \\ h_3 \\ \dot{h}_1 \\ \dot{h}_2 \\ \dot{h}_3 \end{pmatrix}$	$L\eta_1 \sin\beta_{s1} \sin\beta_{s2} \neq 0$

Table 8.2: Linearizing outputs, extended state vector, diffeomorphism, and regularity conditions for different Type robots.

In this table, for Type (2,0) robots the notation χ_1 means

$$\chi_1 = \eta_1 + e\eta_2,$$

whereas for Type (1,1) robots the notation χ_2 means

$$\chi_2 = u_1(L\sin\beta_s + e\cos\beta_s) + \eta_1\zeta_1(L\cos\beta_s - e\sin\beta_s).$$

From Tab. 8.2 we can see that, for any type of robot, the posture coordinate vector $\xi = (\begin{array}{ccc} x & y & \vartheta \end{array})^T$ is a part of z_e. Then, it follows from Lemma 8.6 that the posture tracking problem is generically solvable by dynamic feedback linearization.

We have already seen that for Type (2,0) robots, the linearizing output functions are the coordinates of a point P' located on the axle of the fixed wheels and the posture tracking problem is solvable as long as the linear velocity χ_1 of P' is not zero.

For Type (2,1) robots, the linearizing output functions are the coordinates of the center P' of the steering wheel completed by the orientation ϑ, and the posture tracking problem is solvable as long as the linear velocity η_1 of P' is not zero.

For Type (1,1) robots, the linearizing output functions are the Cartesian coordinates of a point P' on the axle of the fixed wheels, and the posture tracking problem is solvable as long as the linear velocity η_1 of P' is not zero.

For Type (1,2) robots, the linearizing output functions are the Cartesian coordinates given by (8.1) of a point P' of the cart completed by the orientation ϑ, and the posture tracking problem is solvable as long as η_1 and $\sin\beta_{s1}$, $\sin\beta_{s2}$ are not zero.

8.3.5 Avoiding singularities for Type (2,0) robots

Finally, the following lemma characterizes the function $M_1(t)$ of Lemma 8.7 for Type (2,0) robots. Let us recall from Tab. 8.2 that $z_e = (\begin{array}{cc} \xi^T & \chi_1 \end{array})^T$ for these robots.

Lemma 8.8 *For Type (2,0) robots, the function $M_1(t)$ satisfying (8.55) is*

$$M_1(t) = |\chi_{1r}(t)|, \qquad (8.56)$$

$\chi_{1r}(t)$ *being the reference trajectory of the linear velocity of P' given by (8.33). It follows that, if condition (8.49) is satisfied with (8.56), then the robot will follow the reference trajectory without singularity.*

⋄ ⋄ ⋄

Proof. From (8.55) and (8.36) we easily obtain

$$M_1(t) = \min |h_1^2 + h_2^2 + \dot{h}_1^2 + \dot{h}_2^2|^{\frac{1}{2}}$$

8.3. DYNAMIC STATE FEEDBACK

when $\chi_1 = 0$, i.e.,
$$M_1(t) = |\chi_{1r}(t)|.$$
Moreover, the diffeomorphism Ψ given by (8.47) is singular when $\chi_1 = 0$. Let us introduce:
$$S^+ = \{z_e : \chi_1 > 0\}$$
and similarly
$$S^- = \{z_e : \chi_1 < 0\}.$$
Then, Ψ can be inverted respectively on $\Psi(S^+)$ and $\Psi(S^-)$ as follows:
$$z_e = \Psi^{-1}(\zeta) = \begin{pmatrix} h_1 - e\cos\left(-\arctan(\dot{h}_1/\dot{h}_2) + f(\dot{h}_2)\pi\right) \\ h_2 - e\sin\left(-\arctan(\dot{h}_1/\dot{h}_2) + f(\dot{h}_2)\pi\right) \\ -\arctan(\dot{h}_1/\dot{h}_2) + f(\dot{h}_2)\pi \\ \sqrt{\dot{h}_1^2 + \dot{h}_2^2} \end{pmatrix}$$
and
$$z_e = \Psi^{-1}(\zeta) = \begin{pmatrix} h_1 - e\cos\left(-\arctan(\dot{h}_1/\dot{h}_2) + (1 - f(\dot{h}_2))\pi\right) \\ h_2 - e\sin\left(-\arctan(\dot{h}_1/\dot{h}_2) + (1 - f(\dot{h}_2))\pi\right) \\ -\arctan(\dot{h}_1/\dot{h}_2) + (1 - f(\dot{h}_2))\pi \\ -\sqrt{\dot{h}_1^2 + \dot{h}_2^2} \end{pmatrix},$$
where
$$f(a) = \begin{cases} 0 & \text{if } a \geq 0 \\ 1 & \text{if } a < 0 \end{cases} \tag{8.57}$$
and
$$\arctan\left(\frac{a}{b}\right) = \text{sgn}(a)\frac{\pi}{2} \tag{8.58}$$
when $b = 0$. Therefore, Ψ^{-1} is well defined as soon as the sign of $\chi_1(0)$ is defined. The conclusion then holds by applying Lemma 8.7.

◇

Remark

- For a reference trajectory not to be singular, $\chi_{1r}(t)$ must always have the same sign and the same holds for the actual closed-loop trajectory. It follows that $\chi_1(0)$ and $\chi_{1r}(0)$ must have the same sign. Provided that condition (8.49) is satisfied, if $\chi_{1r}(0) > 0$ ($\chi_{1r}(0) < 0$) the robot will follow the reference trajectory with the castor wheel backward (forward).

8.4 Further reading

In this chapter, adapted from [3], we have investigated how point and tracking problems for wheeled mobile robots can be solved by state feedback linearization.

Linearization by static state feedback and change of coordinates has been a major research topic in the theory of nonlinear control systems for the last twenty years. Thorough treatments of this theory can be found in many recent textbooks on nonlinear control systems, e.g., [7, 9]. An algorithm to determine the largest linearizable subsystem of a given nonlinear system is described in [8]. Property 8.1 follows from a straightforward application of this algorithm.

Full linearization by dynamic state feedback is a more recent topic in the literature. This problem is stated and addressed in [1, 2]. The dynamic extension algorithm is described in [4, 14]. The dynamic feedback linearization of Type (2,0) robots was previously described in [15].

Differential flatness [5, 6] is an important structural property for nonlinear control systems which is helpful to solve motion planning, path following or, as in this chapter, feedback linearization problems [10]. Checking whether a given nonlinear system is differentially flat is still an open and challenging problem. Property 8.2 is proved in [11, 12] and brings a partial solution to the problem which is fortunately sufficient for wheeled mobile robots. Property 8.3 is established in [3] and readily follows from the definition of differential flatness.

We have briefly discussed the avoidance of singularities resulting from the application of feedback linearization to wheeled mobile robots. There is however another kind of structural singularity appearing in Type (1,2) robots with multiple steering wheels. The issue is thoroughly treated in [13, 16].

Finally, we mention that there are obviously other approaches for solving posture tracking problems with stability for mobile robots. For instance, a solution based on a linear approximation of the system around sufficiently exciting trajectories is described and analyzed in [17].

References

[1] B. Charlet, J. Lévine, and R. Marino, "On dynamic feedback linearization," *Systems & Control Lett.*, vol. 13, pp. 143–151, 1989.

[2] B. Charlet, J. Lévine, and R. Marino, "Sufficient conditions for dynamic state feedback linearization," *SIAM J. of Control and Optimization*, vol. 29, pp. 38–57, 1991.

[3] B. d'Andréa-Novel, G. Campion, and G. Bastin, "Control of nonholonomic wheeled mobile robots by state feedback linearization," *Int. J. of Robotics Research*, vol. 14, pp. 543–559, 1995.

[4] J. Descusse and C.H. Moog, "Decoupling with dynamic compensation for strong invertible affine nonlinear systems," *Int. J. of Control*, vol. 42, pp. 1387–1398, 1985.

[5] M. Fliess, J. Lévine, P. Martin, and P. Rouchon, "On differentially flat nonlinear systems," *Prepr. 3rd IFAC Symp. on Nonlinear Control Systems Design*, Bordeaux, F, pp. 408–412, 1992.

[6] M. Fliess, J. Lévine, P. Martin, and P. Rouchon, "Flatness and defect of nonlinear systems: Introductory theory and applications," *Int. J. of Control*, vol. 61, pp. 1327–1361, 1995.

[7] A. Isidori, *Nonlinear Control Systems*, (3rd ed.), Springer-Verlag, London, UK, 1995.

[8] R. Marino, "On the largest feedback linearizable subsystem," *Systems & Control Lett.*, vol. 6, pp. 345–351, 1986.

[9] R. Marino and P. Tomei, *Nonlinear Control Design: Geometric, Adaptive and Robust*, Prentice-Hall, London, UK, 1995.

[10] P. Martin and P. Rouchon, "Feedback linearization and driftless systems," *Mathematics of Control, Signals, and Systems*, vol. 7, pp. 235–254, 1994.

[11] P. Martin and P. Rouchon, "Any controllable driftless system with 3 inputs and 5 states is flat," *Systems & Control Lett.*, vol. 25, pp. 167–175, 1995.

[12] P. Martin and P. Rouchon, "Any controllable driftless system with m inputs and $m + 2$ states is flat," *Proc. 34th IEEE Conf. on Decision and Control*, New Orleans, LA, pp. 167–175, 1995.

[13] A. Micaelli, B. d'Andréa-Novel, and B. Thuilot, "Modelling and asymptotic stabilisation of mobile robots with two or more steering wheels," *Proc. Int. Conf. on Advanced Robotics and Computer Vision*, Singapore, vol. 13, pp. 5.1–5.5, 1992.

[14] H. Nijmeijer and A.J. van der Schaft, *Nonlinear Dynamical Control Systems*, Springer-Verlag, New York, 1990.

[15] J.J. Slotine, W. Li, *Applied Nonlinear Control*, Prentice-Hall, Englewoods Cliff, NJ, 1991.

[16] B. Thuilot, B. d'Andréa-Novel, and A. Micaelli, "Modelling and state feedback control of mobile robots with several steering wheels," *IEEE Trans. on Robotics and Automation*, vol. 12, pp. 375–390, 1996.

[17] G. Walsh, D. Tilbury, S. Sastry, R. Murray, and J.-P. Laumond, "Stabilization of trajectories for systems with nonholonomic constraints," *IEEE Trans. on Automatic Control*, vol. 39, pp. 216–222, 1994.

Chapter 9

Nonlinear feedback control

The previous chapter has been devoted to solving point and posture tracking problems by state feedback linearization for the five generic types of wheeled mobile robots. However, as it has been already mentioned, feedback linearization through regular controllers has serious limitations for control of mobile robots. In particular, it does not allow a robot to be stabilized about a fixed point in the configuration space.

In this chapter we present several advanced *nonlinear feedback control* design methods allowing us to solve various control problem; namely, *posture tracking*, *path following*, and *posture stabilization*.

In order to keep the exposition as clear and pedagogical as possible, we will limit ourselves to *unicycle robots* of Type (2,0), as represented in Section 7.3.3. Indeed, this type of mobile robots is sufficient to capture the underlying nonholonomy property of restricted mobility robots which is the core of the difficulties involved in the control problems discussed in this chapter.

9.1 Unicycle robot

In order to make the notations more convenient and consistent with those which are most often used in the literature, the posture coordinates (x, y, θ) of a Type (2,0) robot are redefined according to Fig. 9.1. The posture kinematic model of a *unicycle robot* is then described by

$$\dot{x} = v \cos \theta$$

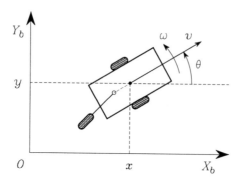

Figure 9.1: Redefinition of posture coordinates.

$$\dot{y} = v\sin\theta \qquad (9.1)$$
$$\dot{\theta} = \omega$$

where the two velocity control inputs are the linear velocity v and the angular velocity ω. Notice that, differently from the model in (7.23), θ indicates the angle between the direction of v and axis X_b.

We will sometimes use the compact notation

$$\dot{z} = G(z)u \qquad (9.2)$$

where

$$z = \begin{pmatrix} x \\ y \\ \theta \end{pmatrix} \qquad G(z) = \begin{pmatrix} \cos\theta & 0 \\ \sin\theta & 0 \\ 0 & 1 \end{pmatrix} \qquad u = \begin{pmatrix} v \\ \omega \end{pmatrix}.$$

9.1.1 Model transformations

Control design may in some cases be facilitated by a preliminary change of state coordinates which transforms the model equations of the robot into a simpler "canonical" form.

For instance, the following change of coordinates:

$$\begin{pmatrix} x_1 \\ x_2 \\ x_3 \end{pmatrix} = \begin{pmatrix} 0 & 0 & 1 \\ \cos\theta & \sin\theta & 0 \\ \sin\theta & -\cos\theta & 0 \end{pmatrix} \begin{pmatrix} x \\ y \\ \theta \end{pmatrix} \qquad (9.3)$$

together with the change of inputs

$$u_1 = \omega \qquad (9.4)$$
$$u_2 = v - \omega x_3$$

9.1. UNICYCLE ROBOT

transforms system (9.1) into

$$\begin{aligned} \dot{x}_1 &= u_1 \\ \dot{x}_2 &= u_2 \\ \dot{x}_3 &= x_2 u_1. \end{aligned} \quad (9.5)$$

This system belongs to the more general class of so-called *chained* systems characterized by equations in the form

$$\begin{aligned} \dot{x}_1 &= u_1 \\ \dot{x}_2 &= u_2 \\ \dot{x}_3 &= x_2 u_1. \\ &\vdots \\ \dot{x}_n &= x_{n-1} u_1. \end{aligned} \quad (9.6)$$

Chained systems are of particular interest in the field of mobile robotics because the modelling equations of several nonholonomic systems (e.g., unicycle-type and car-like vehicles pulling trailers) can locally be transformed into this form.

An alternative *local* change of coordinates is

$$\begin{aligned} x_1 &= x \\ x_2 &= \tan\theta \\ x_3 &= y \end{aligned} \quad (9.7)$$

associated with the change of control inputs

$$\begin{aligned} u_1 &= \cos\theta \, v \\ u_2 &= \frac{1}{\cos^2\theta} \omega. \end{aligned} \quad (9.8)$$

This yields the same system (9.5). However, the transformation presents in this case a singularity when $\cos\theta = 0$. It is thus only valid in domains where $\theta \in (-\pi/2 + k\pi, \pi/2 + k\pi)$ with $k \in Z$.

Depending on the transformation which is considered, a control law derived for the model (9.5) will yield a different transient behaviour of the trajectories for $x(t)$, $y(t)$, and $\theta(t)$.

9.1.2 Linear approximation

For many nonlinear systems, linear approximations can be used as a basis for a first simple control design. Linearization may also provide indications concerning the controllability and feedback stabilizability of the nonlinear system. More precisely, if the linear approximation is controllable, then the original nonlinear system is controllable and feedback stabilizable, at least locally. But the converse is not true. In our case, the linear approximation of system (9.1) about the equilibrium point $(z = 0, u = 0)$ gives

$$\dot{z} = G(0) \begin{pmatrix} v \\ \omega \end{pmatrix} = \begin{pmatrix} 1 & 0 \\ 0 & 0 \\ 0 & 1 \end{pmatrix} \begin{pmatrix} v \\ \omega \end{pmatrix}. \qquad (9.9)$$

This linear system is not controllable since the rank of the associated controllability matrix

$$C = G(0) = \begin{pmatrix} 1 & 0 \\ 0 & 0 \\ 0 & 1 \end{pmatrix} \qquad (9.10)$$

is only two. Information about the nonlinear model is thus lost when using the linearized model.

9.1.3 Smooth state feedback stabilization

The problem of smooth state feedback stabilization can be formulated as follows: find a feedback $u = k(z)$, where $k(z)$ is a smooth function of z, such that the closed-loop system

$$\dot{z} = G(z)k(z) = f(z) \qquad (9.11)$$

is asymptotically stable, i.e., the solutions $z(t)$ asymptotically converge to zero for any initial $z(0)$ in the neighbourhood of zero. As we have discussed in Chapter 7, a nonlinear system can be controllable without being feedback stabilizable, in the sense of the above definition. This is true in particular for the model (9.1). This negative result has motivated the search for control structures other than pure smooth state feedback laws. Such structures have the characteristic of being time-varying ($u = k(z,t)$, with $k(z,t)$ a smooth function) or piecewise continuous ($u = k(z)$, with $k(z)$ a piecewise continuous function). These control structures will be presented in the last section of this chapter when returning to the stabilization problem. Before this, the simpler problems of *posture tracking* and *path following* are addressed.

9.2 Posture tracking

Consider a *virtual reference* unicycle-type robot, whose equations are

$$\begin{aligned}\dot{x}_r &= v_r \cos \theta_r \\ \dot{y}_r &= v_r \sin \theta_r \\ \dot{\theta}_r &= \omega_r.\end{aligned} \qquad (9.12)$$

The subscript r stands for reference, and v_r and ω_r are assumed to be bounded and have bounded derivatives.

Assumption 9.1 The reference linear velocity v_r and angular velocity ω_r are so that

$$\lim_{t \to \infty} v_r(t) \neq 0 \qquad \lim_{t \to \infty} \omega_r(t) \neq 0. \qquad (9.13)$$

□

On the above assumption, the *tracking problem* is to find a feedback control law

$$\begin{pmatrix} v \\ \omega \end{pmatrix} = k(z, z_r, v_r, \omega_r)$$

such that

$$\lim_{t \to \infty} (z(t) - z_r(t)) = 0. \qquad (9.14)$$

The tracking problem is illustrated in Fig. 9.2. Note that Assumption 9.1 implies that the reference robot is not at rest all the time. Hence, stabilization to a fixed posture is not included in the above tracking problem definition.

The tracking problem so defined involves error equations which describe the time evolution of the difference $z - z_r$. The following change of coordinates can be used

$$\begin{pmatrix} e_1 \\ e_2 \\ e_3 \end{pmatrix} = \begin{pmatrix} \cos \theta & \sin \theta & 0 \\ -\sin \theta & \cos \theta & 0 \\ 0 & 0 & 1 \end{pmatrix} \begin{pmatrix} x_r - x \\ y_r - y \\ \theta_r - \theta \end{pmatrix} \qquad (9.15)$$

where e_1 and e_2 are the coordinates of the position error vector, and e_3 is the orientation error.

The associated tracking error equations are then obtained by differentiating (9.15) with respect to time. Introducing the change of inputs

$$\begin{aligned} u_1 &= -v + v_r \cos e_3 \\ u_2 &= \omega_r - \omega \end{aligned}$$

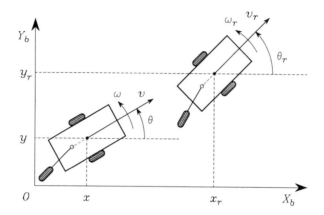

Figure 9.2: Illustration of the tracking problem.

gives, after simple calculation,

$$\dot{e} = \begin{pmatrix} 0 & \omega & 0 \\ -\omega & 0 & 0 \\ 0 & 0 & 0 \end{pmatrix} e + \begin{pmatrix} 0 \\ \sin e_3 \\ 0 \end{pmatrix} v_r + \begin{pmatrix} 1 & 0 \\ 0 & 0 \\ 0 & 1 \end{pmatrix} \begin{pmatrix} u_1 \\ u_2 \end{pmatrix}. \qquad (9.16)$$

9.2.1 Linear feedback control

Linearization of system (9.16) about the equilibrium point ($e = 0, u = 0$) yields the following linear time-varying system

$$\dot{e} = \begin{pmatrix} 0 & \omega_r(t) & 0 \\ -\omega_r(t) & 0 & v_r(t) \\ 0 & 0 & 0 \end{pmatrix} e + \begin{pmatrix} 1 & 0 \\ 0 & 0 \\ 0 & 1 \end{pmatrix} \begin{pmatrix} u_1 \\ u_2 \end{pmatrix}. \qquad (9.17)$$

If v_r and ω_r are constant, we fall upon a time-invariant linear system, whose controllability matrix is

$$C = \begin{pmatrix} B & AB & A^2B \end{pmatrix} = \begin{pmatrix} 1 & 0 & 0 & 0 & -\omega_r^2 & v_r\omega_r \\ 0 & 0 & -\omega_r & v_r & 0 & 0 \\ 0 & 1 & 0 & 0 & 0 & 0 \end{pmatrix}. \qquad (9.18)$$

In this case, it is simple to verify that the linearized model is controllable, provided that either v_r or ω_r is different from zero. When the reference robot is at rest ($v_r = \omega_r = 0$), controllability of the linearized model is lost. Due to this fact, we shall pay attention to the manner in which the linear control law is designed. For instance, a linear state feedback obtained by imposing closed-loop poles independent of v_r and ω_r may become ill-conditioned when $|v_r|$ and $|\omega_r|$ tend to become small.

In order to be more specific, let us consider the linear feedback law

$$\begin{aligned} u_1 &= -k_1 e_1 \\ u_2 &= -k_2 \mathrm{sgn}(v_r) e_2 - k_3 e_3, \end{aligned} \tag{9.19}$$

chosen so as to set the closed-loop system poles equal to the roots of the characteristic polynomial equation

$$(s + 2\xi a)(s^2 + 2\xi a s + a^2) = 0$$

where ξ and a are positive real numbers. The corresponding control gains are

$$\begin{aligned} k_1 &= 2\xi a \\ k_2 &= \frac{a^2 - \omega_r^2}{|v_r|} \\ k_3 &= 2\xi a. \end{aligned}$$

With a fixed pole placement strategy (a and ξ are constant), the control gain k_2 increases without bound when v_r tends to zero.

Regularization of the controller is possible by letting the closed-loop poles depend on the values of v_r and ω_r. This procedure is called *velocity scaling*. Choose, for example, $a = (\omega_r^2 + bv_r^2)^{\frac{1}{2}}$ with $b > 0$. The control gains then become

$$\begin{aligned} k_1 &= 2\xi(\omega_r^2 + bv_r^2)^{\frac{1}{2}} \\ k_2 &= b|v_r| \\ k_3 &= 2\xi(\omega_r^2 + bv_r^2)^{\frac{1}{2}} \end{aligned} \tag{9.20}$$

and the resulting control is now defined for all values of v_r and ω_r. The particular values $v_r = \omega_r = 0$ for which the system is not controllable simply yield zero control action, which makes sense intuitively.

9.2.2 Nonlinear feedback control

It is possible to design nonlinear feedback controllers for the nonlinear model (9.16) so as to enlarge the domain where asymptotic stability is granted and cover the case when v_r and ω_r are time-varying. One of such controllers is

$$\begin{aligned} u_1 &= -k_1(v_r, \omega_r) e_1 \\ u_2 &= -k_4 v_r \frac{\sin e_3}{e_3} e_2 - k_3(v_r, \omega_r) e_3 \end{aligned} \tag{9.21}$$

where k_4 is a positive constant and $k_1(\cdot)$ and $k_3(\cdot)$ are continuous functions, strictly positive on $R \times R - (0,0)$.

Notice the resemblance between this control and the linear control (9.19) and (9.20) derived in the previous section. It can be utilized in the choice of $k_1(\cdot)$ and $k_3(\cdot)$ to have the two controls behave in the same way near the origin $e = 0$.

Lemma 9.1 *On Assumption 9.1, control (9.21) globally asymptotically stabilizes the origin $e = 0$.*

⋄ ⋄ ⋄

Proof. Consider the Lyapunov function candidate

$$V(e) = \frac{k_4}{2}(e_1^2 + e_2^2) + \frac{e_3^2}{2}$$

which is nonincreasing along any system solution, since

$$\dot{V} = k_4 e_1 (u_1 + we_2) + k_4 e_2 (v_r \sin e_3 - we_1) + e_3 u_2$$
$$= -k_1 k_4 e_1^2 - k_3 e_3^2.$$

Along a system solution, $\|e(t)\|$, and thus $\|\dot{e}(t)\|$, are bounded. Since $v_r(t)$ and $\omega_r(t)$, and their time derivatives, are bounded (by assumption), $k_1(v_r(t), \omega_r(t))$ and $k_3(v_r(t), \omega_r(t))$ are uniformly continuous. As a consequence, $\dot{V}(t)$ is also uniformly continuous. Moreover, $V(t)$ does not increase and thus converges to some limit value, denoted by $\bar{V}$. From Barbalat's lemma, $\dot{V}(t)$ tends to zero. This in turn implies (omitting the time index from now on) that $k_1(v_r, \omega_r)e_1$ and $k_3(v_r, \omega_r)e_3$ tend to zero. Using the properties of $k_1(\cdot)$ and $k_3(\cdot)$, we deduce that both $(v_r^2 + \omega_r^2)e_1^2$ and $(v_r^2 + \omega_r^2)e_3^2$ tend to zero. In fact, e_1 and e_3 unconditionally converge to zero if $k_1(\cdot)$ and $k_3(\cdot)$ are chosen strictly positive on $R \times R$.

From the third system's equation, and using some of the above results gives

$$\dot{e}_3 = -k_4 v_r e_2 \frac{\sin(e_3)}{e_3} + o(t) \qquad \lim_{t \to \infty} o(t) = 0.$$

Hence, using the fact that $v_r e_3$ tends to zero leads to

$$\frac{d}{dt}(v_r^2 e_3) = -k_4 v_r^3 e_2 \frac{\sin(e_3)}{e_3} + o(t).$$

Note that $v_r^3 e_2 \sin(e_3)/e_3$ is uniformly continuous (since its time derivative is bounded). From Barbalat's lemma (slightly generalized), $d(v_r^2 e_3)/dt$ tends to zero. Therefore, $v_r^3 e_2 \sin(e_3)/e_3$ also tends to zero. Now, since $v_r e_3$

tends to zero, $v_r^2 e_2^2((\sin(e_3)/e_3)^2 + e_3^2)$ tends to zero. Since $((\sin(e_3)/e_3)^2 + e_3^2)$ is strictly larger than some positive number, $v_r e_2$ tends to zero.

From the first system equation, using the convergence of u_1 and u_2 to zero, it is
$$\dot{e}_1 = \omega_r e_2 + o(t).$$
Hence, using the fact that $\omega_r e_1$ tends to zero gives
$$\frac{d}{dt}(\omega_r^2 e_1) = \omega_r^3 e_2 + o(t),$$
i.e., $\omega_r^3 e_2$ is uniformly continuous, since its time derivative is bounded. Thus, from Barbalat's lemma, $d(\omega_r^2 e_1)/dt$ tends to zero. Therefore, $\omega_r^3 e_2$, and thus $\omega_r e_2$, tend to zero.

Finally, it has been shown that $(v_r^2 + \omega_r^2)e_i^2$ for $i = 1, 2, 3$ tends to zero. Hence, $(v_r^2 + \omega_r^2)V(e)$ tends to zero, with $V(e)$ converging to $\bar{V}$. Since $(v_r^2 + \omega_r^2)$ does not tend to zero (Assumption 9.1), $\bar{V}$ is necessarily equal to zero.

◊

Remark

- Comparing the expressions of the linear controller (9.19) and (9.20) and the nonlinear controller (9.21) suggests using the following functions:

$$k_1(v_r, \omega_r) = k_3(v_r, \omega_r) = 2\xi(\omega_r^2 + bv_r^2)^{\frac{1}{2}} \qquad k_4 = b.$$

9.3 Path following

The mobile robot and the path to be followed, denoted by $\mathcal{P}$, are represented in Fig. 9.3, where M is the orthogonal projection of the robot point P on $\mathcal{P}$, and x_t and x_n are the tangent and the normal unit vectors to the path at M, respectively. This point exists and is uniquely defined if $\mathcal{P}$ satisfies some conditions and the distance between the robot and $\mathcal{P}$ is not "too large". Also, l is the signed distance between M and P; s is the signed distance along the path between some arbitrary fixed point on the path and point M; $\theta_r(s)$ is the angle between axis X_b and x_t; and $c(s)$ is the path curvature at point M, assumed to be uniformly bounded and differentiable.

Let $\tilde{\theta} = \theta - \theta_r$ denote the orientation error. The variables s, l, and $\tilde{\theta}$ constitute a new set of state coordinates for the mobile robot. Note that they coincide with x, y, and θ in the particular case where the path $\mathcal{P}$

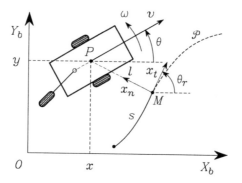

Figure 9.3: Illustration of the path following problem.

coincides with axis X_b. Using this parameterization, it is rather simple to verify that the following kinematic equations hold:

$$\begin{aligned} \dot{s} &= v\cos\tilde{\theta}\frac{1}{1-c(s)l} \\ \dot{l} &= v\sin\tilde{\theta} \\ \dot{\tilde{\theta}} &= \omega - v\cos\tilde{\theta}\frac{c(s)}{1-c(s)l}. \end{aligned} \qquad (9.22)$$

Given a path $\mathcal{P}$ in the xy-plane and the mobile robot linear velocity $v(t)$, assumed to be bounded together with its time derivative $\dot{v}(t)$, the *path following problem* consists of finding a (smooth) feedback control law

$$\omega = k(s, l, \tilde{\theta}, v(t))$$

such that

$$\lim_{t\to\infty} l(t) = 0 \qquad \lim_{t\to\infty} \tilde{\theta}(t) = 0.$$

Note that the path following problem, as stated above, is less stringent than the tracking problem, in the sense that only the stabilization of the coordinates l and $\tilde{\theta}$ is required. On the other hand, this objective shall be achieved by using only one control variable, namely the angular velocity ω.

By introducing the auxiliary control variable

$$u = \omega - v\cos\tilde{\theta}\frac{c(s)}{1-c(s)l}, \qquad (9.23)$$

9.3. PATH FOLLOWING

eqs. (9.22) become

$$\dot{s} = v\cos\tilde{\theta}\frac{1}{1-c(s)l}$$
$$\dot{l} = v\sin\tilde{\theta} \qquad (9.24)$$
$$\dot{\tilde{\theta}} = u.$$

9.3.1 Linear feedback control

In the neighbourhood of $(l = 0, \tilde{\theta} = 0)$, tangent linearization of the last two equations in (9.24) gives

$$\dot{l} = v\tilde{\theta}$$
$$\dot{\tilde{\theta}} = u. \qquad (9.25)$$

When v is constant and different from zero, this linear system is clearly controllable, and thus asymptotically stabilizable by using a linear state feedback. In fact, it is rather simple to verify that a stabilizing linear feedback is of the form

$$u = -k_2 v l - k_3 |v|\tilde{\theta} \qquad (9.26)$$

with $k_2 > 0$ and $k_3 > 0$. The closed-loop equation for the output l is then

$$\ddot{l} + k_3|v|\dot{l} + k_2 v^2 l = 0, \qquad (9.27)$$

which may also be written as

$$l'' + k_3 l' + k_2 l = 0 \qquad (9.28)$$

with $l' = \partial l/\partial \gamma$ and $\gamma = \int_0^t |v|d\tau$. At first approximation, γ represents the distance traveled by point M along the path. The transformation of eq. (9.27) into (9.28) is related to the velocity scaling procedure already evoked in Section 9.2, and the second-order linear equation (9.28) suggests selecting the control gains k_2 and k_3 as

$$k_2 = a^2$$
$$k_3 = 2\xi a \qquad (9.29)$$

where a must be chosen so as to specify the transient "rise distance" (the equivalent of the rise time in the case of time equations), and ξ is the damping coefficient (critical damping is obtained by setting $\xi = 1/\sqrt{2}$).

9.3.2 Nonlinear feedback control

Instead of the linear control (9.26) previously derived, consider the nonlinear control

$$u = -k_2 v l \frac{\sin \tilde{\theta}}{\tilde{\theta}} - k(v)\tilde{\theta}, \qquad (9.30)$$

where k_2 is a positive constant, and $k(v)$ is a continuous function strictly positive when $v \neq 0$.

In order to have the two controls behave similarly near $(l = 0, \tilde{\theta} = 0)$ we may choose, for instance, $k(v) = k_3|v|$ with k_2 and k_3 given by (9.29).

Assumption 9.2 *The linear velocity $v(t)$ is such that*

$$\lim_{t \to \infty} v(t) \neq 0.$$

□

Lemma 9.2 *On Assumption 9.2, control (9.30) asymptotically stabilizes $(l = 0, \tilde{\theta} = 0)$, provided that the robot initial configuration is such that*

$$l(0)^2 + \frac{1}{k_2}\tilde{\theta}(0)^2 < \frac{1}{\limsup(c^2(s))}. \qquad (9.31)$$

◇ ◇ ◇

Proof. Consider the Lyapunov function candidate

$$V = k_2 \frac{l^2}{2} + \frac{\tilde{\theta}^2}{2}.$$

Taking the time derivative of this function along a solution to the closed-loop system gives

$$\begin{aligned}\dot{V} &= k_2 l \dot{l} + \tilde{\theta}\dot{\tilde{\theta}} \\ &= k_2 l \sin\tilde{\theta} v + \tilde{\theta} u \\ &= -k(v)\tilde{\theta}^2 \leq 0.\end{aligned}$$

By invoking arguments similar to the ones used in the proof of Lemma 9.1 (boundedness of l and $\tilde{\theta}$, convergence of V to a limit value $\bar{V}$, and convergence of $\dot{V}$ to zero), we obtain that $k(v)\tilde{\theta}$ and $v\tilde{\theta}$ tend to zero. Then, in view of the control expression and the last system equation, it is

$$\dot{\tilde{\theta}} = -k_2 v l \frac{\sin\tilde{\theta}}{\tilde{\theta}} + o(t) \qquad \lim_{t \to \infty} o(t) = 0.$$

Hence, using the convergence of $v\tilde{\theta}$ to zero and the boundedness of $\dot{v}$ yields

$$\frac{d}{dt}(v^2 \tilde{\theta}) = -k_2 v^3 l \frac{\sin\tilde{\theta}}{\tilde{\theta}} + o(t).$$

Since $v^3 l \sin\tilde{\theta}/\tilde{\theta}$ is uniformly continuous (its time derivative is bounded), $d(v^2\tilde{\theta})/dt$ tends to zero by application of Barbalat's lemma. Thus $vl\sin\tilde{\theta}/\tilde{\theta}$ also tends to zero. This in turn implies that $v^2 l^2((\sin\tilde{\theta}/\tilde{\theta})^2 + \tilde{\theta}^2)$ tends to zero. Since $((\sin\tilde{\theta}/\tilde{\theta})^2 + \tilde{\theta}^2)$ is larger than some positive real number, vl tends to zero. Finally, from the convergence of $v\tilde{\theta}$ and vl to zero, we deduce that vV, and thus $v\bar{V}$, tend to zero. Now, since v does not tend to zero (see Assumption 9.2), $\bar{V}$ must be equal to zero.

◇

Remarks

- Condition (9.31) is needed in the proof to ensure that $(1 - c(s)l)$ remains positive (larger than some positive number) and avoid singularities due to the parameterization.

- The robot location along the path is characterized by the value of s (the signed distance traveled along the path) and thus depends on the linear velocity v which is not used as a control in the case of path following. This degree of freedom will be used later to stabilize s about a prescribed value s_r. This complementary problem clearly brings us back to the problem of stabilization about an arbitrary given posture.

9.4 Posture stabilization

Given an arbitrary reference posture z_r, the problem is to find a control law

$$\begin{pmatrix} v \\ \omega \end{pmatrix} = k(z - z_r, t)$$

which asymptotically stabilizes $z - z_r$ about zero, whatever the initial robot posture $z(0)$. Without loss of generality, we may take $z_r = 0$.

Recall that there is no smooth control law $k(z)$ that can solve the point-stabilization problem for the class of systems considered in this chapter. Three alternatives, the exploration of which is still the object of active research, are considered here; namely, *smooth* (differentiable) *time-varying control*, *piecewise continuous control*, and *time-varying piecewise continuous control*.

9.4.1 Smooth time-varying control

A nonholonomic mobile robot (with restricted mobility) can be stabilized about a desired posture by using smooth time-varying feedback control of the type

$$v = -k_1 x$$
$$\omega = -g(t)y - k_3\theta, \quad (9.32)$$

where x and y are the Cartesian coordinates of point P located on the axle of the actuated wheels, θ is the orientation of the robot cart, k_1 and k_3 are positive numbers, and $g(t)$ is a bounded function at least once-differentiable such that $\partial g(t)/\partial t$ does not tend to zero when t tends to infinity; for instance, $g(t) = \sin t$.

This control globally asymptotically stabilizes the origin ($x = 0, y = 0, \theta = 0$). However, it appears from practical cases that the asymptotic convergence rate for the state variables is not better than $1/\sqrt{t}$ for most initial configurations. In other words, instead of the usual exponential stability associated with stable linear systems and characterized by the inequality

$$||z(t)|| \leq a\exp(-bt)||z(0)||$$

for some positive numbers a, b, and c, we can only expect to have with this type of control:

$$||z(t)|| \leq c||z(0)||\sqrt{\frac{1}{t+1}}.$$

Nevertheless, it is important to realize that the asymptotic convergence rate is not, by itself, sufficient to properly evaluate the overall control performance. For instance, the time needed to reach a small neighbourhood of zero is not directly linked to the asymptotic convergence rate. This time can be reasonably short, while the final convergence phase is slow. In the case of control (9.32), for instance, the time required to reach a small neighbourhood to zero can be much reduced by replacing the term $g(t)y$ with another time-varying function $g(y,t)$.

Extension of the posture tracking problem

Consider the tracking problem described in Section 9.2, and assume that the virtual reference robot moves along a path which passes through the point $(x_r = 0, y_r = 0)$. Assume also that the tangent to the path at this point is aligned with axis X_b. The posture stabilization problem may then be treated as a tracking problem (convergence of the tracking errors to zero) with the additional requirement that the reference robot should itself be asymptotically stabilized about the desired configuration.

9.4. POSTURE STABILIZATION

To simplify, we will assume here, although this is not technically necessary, that the reference robot moves along the axis X_b. Hence, it is $y_r(t) = 0$ and $\theta_r(t) = 0$, $\forall t$ (and thus $\omega_r(t) = 0$, $\forall t$).

Consider now the nonlinear tracking control law proposed in Section 9.2, i.e.,

$$\begin{aligned} u_1 &= -k_1(v_r, \omega_r) e_1 \\ u_2 &= -k_2 v_r e_2 \frac{\sin(e_3)}{e_3} - k_3(v_r, \omega_r) e_3, \end{aligned} \quad (9.33)$$

which has been shown to be globally stabilizing when either $v_r(t)$ or $\omega_r(t)$ does not tend to zero.

The idea consists of using v_r (equal to $\dot{x}_r$ in this case) as a new control variable, whose selection shall be made so as to ensure both convergence of the tracking errors e_1, e_2, e_3 to zero and convergence of the reference robot coordinate x_r to zero, when control (9.33) is used.

A possible choice is the time-varying control

$$v_r = -k_4 x_r + g(e, t) \quad (9.34)$$

where $k_4 > 0$.

Assumption 9.3 *The function $g(e, t)$ in (9.34) is differentiable up to order $p + 1$ ($p \geq 1$) and uniformly bounded with respect to t, with partial derivatives up to order p also uniformly bounded with respect to t, and such that*

$$g(0, t) = 0 \quad \forall t.$$

□

Assumption 9.4 *There exists a diverging time sequence $\{t_i\}_{i \in N}$ and a continuous function $\alpha(\cdot)$ such that*

$$\|e\| > \epsilon > 0 \implies \sum_{j=1}^{p} \left(\frac{\partial^j g}{\partial t^j}(e, t_i) \right)^2 > \alpha(\epsilon) > 0 \quad \forall t_i.$$

□

Lemma 9.3 *The control law (9.33), with v_r calculated according to (9.34), globally asymptotically stabilizes the posture ($e = 0, x_r = 0$) and thus solves the posture stabilization problem.*

◇ ◇ ◇

Proof. The proof is performed by considering an arbitrary solution to the closed-loop system. All functions involved in the proof should thus be seen as time functions. However, in order to lighten the notations, the time index will be systematically omitted.

Since the Lyapunov function V used in the proof of Lemma 9.1 is nonincreasing, the tracking errors e_i ($i = 1, 2, 3$) are bounded. Therefore $g(e, t)$ —read $g(e(t), t)$— is bounded. Eq. (9.34) may then be interpreted as the equation of a stable linear system subject to the additive bounded perturbation $g(e, t)$. The state x_r associated with this equation remains bounded and, according to (9.34), v_r is also bounded.

By taking the time derivative of (9.34), it can be shown in the same way that $\dot{v}_r$ is bounded.

Since v_r and $\dot{v}_r$ are bounded, Lemma 9.1 applies to the present situation. In particular, if v_r does not tend to zero, then e must tend to zero. By uniform continuity, and by using Assumption 9.1, $g(e, t)$ tends to zero. Now, in view of (9.34), x_r and v_r must also tend to zero, yielding a contradiction. Hence v_r tends to zero.

By taking the time derivative of (9.34) and using the unconditional convergence of $\dot{e}$ to zero (as in the proof of Lemma 9.1), it is

$$\dot{v}_r = \frac{\partial g}{\partial t}(e, t) + o(t) \qquad \lim_{t \to \infty} o(t) = 0.$$

Since $\partial g(e, t)/\partial t$ is uniformly continuous (its time derivative is bounded), and since v_r tends to zero, $\dot{v}_r$ tends to zero, by application of Barbalat's lemma. Hence $\partial g(e, t)/\partial t$ tends to zero.

By taking the time derivative of $\partial g(e, t)/\partial t$, using the convergence of $\dot{e}$ to zero, and applying Barbalat's lemma, we obtain that $\partial^2 g(e, t)/\partial t^2$ tends to zero. Repeating the same procedure as many times as necessary, we obtain that $\partial^j g(e, t)/\partial t^j$ tends to zero, for $1 \leq j \leq p$.

Now, V is nonincreasing and converges to some limit value denoted by $\bar{V}$. If $\bar{V} \neq 0$, there exists a positive real number ϵ such that $||e(t)|| > \epsilon > 0$, $\forall t$. Hence, in view of Assumption 9.2,

$$\sum_{j=1}^{p} \left(\frac{\partial^j g}{\partial t^j}(e(t_i), t_i) \right)^2 > \alpha(\epsilon) > 0.$$

This contradicts the previously established fact that

$$\sum_{j=1}^{p} \left(\frac{\partial^j g}{\partial t^j}(e(t), t) \right)^2 \to 0.$$

The only alternative is $\bar{V} = 0$, which proves that e, and subsequently $g(e, t)$, tend to zero. Finally, we deduce from eq. (9.34) that also x_r tends to zero. ◇

Remarks

- The time-varying function $g(e,t)$ does not need to depend upon e_1 and e_3 when the functions $k_1(v_r, \omega_r)$ and $k_3(v_r, \omega_r)$ are chosen strictly positive on $R \times R$. The reason is that e_1 and e_3 unconditionally converge to zero in this case, due to the convergence of $\dot{V}$ to zero.

- A particular function $g(e,t)$ which satisfies Assumptions 9.1 and 9.2 is
$$g(e,t) = ||e||^2 \sin t.$$

- Another possibility, when the functions $k_1(\cdot)$ and $k_3(\cdot)$ are strictly positive, is the bounded function
$$g(e,t) = \frac{\exp(k_5 e_2) - 1}{\exp(k_5 e_2) + 1} \sin t \tag{9.35}$$

with $k_5 > 0$. By choosing a large constant k_5, $g(e,t)$ is approximately equal to $\sin t$ as long as $|e_2|$ is not small. Integration of (9.34) then gives
$$x_r(t) \approx \frac{k_4}{1 + k_4^2} \sin t - \frac{1}{1 + k_4^2} \cos t.$$

This relation shows that the reference robot maintains a periodic motion, whose amplitude is approximately equal to two (i.e., the amplitude of $\sin t$) as long as $|e_2|$ does not become small. This motion in turn produces a quick reduction of $||e||$. This fast transient is then followed by the final asymptotic convergence phase, which is slow as already pointed out.

- Going further in this direction, we may think of the following choice:
$$g(e,t) = \begin{cases} \sin t & \text{when } ||e|| \geq \epsilon > 0 \\ 0 & \text{otherwise.} \end{cases}$$

Note that in this case the control law is no longer smooth, and that asymptotic convergence to the desired posture is not granted. However, it is still possible to show that $||e||$ becomes, and stays, smaller than ϵ after a finite time, and that the reference robot exponentially converges to the desired posture. Thus, by choosing a small value for ϵ, the mobile robot will get very close to the desired posture (at a distance smaller than ϵ) after a finite and reasonably short time, while the control inputs will asymptotically exponentially converge to zero. From a practical viewpoint, this solution may be quite acceptable.

Extension of the path following problem

Consider the path following problem described in Section 9.3, and assume that the chosen path passes through the point $(x_r = 0, y_r = 0)$, and that the tangent to the path at this point is aligned with axis X_b. The path coordinate may arbitrarily be set equal to zero at this point ($s_r = 0$).

Stabilization of the mobile robot about the desired posture is then equivalent to having the variables l, $\tilde{\theta}$, and s asymptotically converge to zero.

In order to achieve this objective, we may again consider the nonlinear angular velocity control proposed in Section 9.2.2 for path following

$$u = -k(v)\tilde{\theta} - k_2 v l \frac{\sin(\tilde{\theta})}{\tilde{\theta}}. \tag{9.36}$$

It has been established in Lemma 9.2 that, if the robot linear velocity $v(t)$ and its time derivative are bounded, and if $v(t)$ does not asymptotically tend to zero, then l and $\tilde{\theta}$ asymptotically converge to zero provided that some initial conditions are satisfied.

The idea now consists of using the linear velocity v as a second control input, whose selection shall be made in order to also have s converge to zero. A possible choice, among others, is the following time-varying feedback

$$v = -k_1 \cos(\tilde{\theta}) \frac{\exp(k_3 s) - 1}{\exp(k_3 s) + 1} + g(l, \tilde{\theta}, t), \tag{9.37}$$

where k_1 and k_3 are positive real numbers.

Assumption 9.5 *The function $g(l, \tilde{\theta}, t)$ in (9.37) has the same properties as the function in Assumption 9.3, and in particular*

$$g(0, 0, t) = 0 \quad \forall t.$$

□

Assumption 9.6 *There exists a diverging time sequence $\{t_i\}_{i \in N}$ and a continuous function $\alpha(\cdot)$ such that*

$$(l^2 + \tilde{\theta}^2)^{\frac{1}{2}} > \epsilon > 0 \implies \sum_{j=1}^{j=p} \left(\frac{\partial^j g}{\partial t^j}(l, \tilde{\theta}, t_i) \right)^2 > \alpha(\epsilon) > 0 \quad \forall t_i.$$

□

Lemma 9.4 *The control (9.36) and (9.37) asymptotically stabilizes the point $(l = 0, \tilde{\theta} = 0, s = 0)$ provided that:*

$$l^2(0) + \frac{1}{k_2}\tilde{\theta}^2(0) < \frac{1}{\limsup(c^2(s))}.$$

◇ ◇ ◇

Proof. The proof is quite similar to the one of Lemma 9.3. It is first established that l, $\tilde{\theta}$, v, and $\dot{v}$ are bounded along any solution to the closed-loop system. Then, using Lemma 9.2, it is shown that v must tend to zero. We deduce from there that $\partial^j g(l, \tilde{\theta}, t)/\partial t^j$ tends to zero, for $1 \leq j \leq p$. This in turn implies, using (9.6) and the convergence of the Lyapunov function used in the proof of Lemma 9.2, that l and $\tilde{\theta}$ tend to zero. The convergence of $g(l, \tilde{\theta}, t)$ to zero then follows from (9.5). Finally the convergence of s to zero can be worked out from

$$\dot{s} = -k_1 \frac{\exp(k_3 s) - 1}{\exp(k_3 s) + 1} + o(t) \qquad \lim_{t \to \infty} o(t) = 0.$$

◇

The previous remarks concerning the choice of the time-varying function $g(\cdot, t)$ also hold in this case. In particular, the dependence upon the variable $\tilde{\theta}$ is not necessary when $k(v)$ is chosen strictly positive (unconditional convergence of $\tilde{\theta}$ to zero).

9.4.2 Piecewise continuous control

In this section we present two alternative approaches for stabilization of the kinematic model of a mobile robot. The first one consists of dividing the state space into disjoint subspaces. The manifold that divides these complementary subspaces defines nonattractive discontinuous surface. A piecewise continuous feedback control law is presented which makes the robot exponentially converge to the origin. This approach will be called the *coordinate projection* approach. The second approach is based on stabilization of nonlinear systems in chained forms. It relies on the concept of mixing piecewise *constant* feedback with piecewise *continuous* feedback control laws. This approach will be named *hybrid piecewise control* since it combines a discrete-time law with a continuous-time one.

Coordinate projection

The idea behind the control design based on a coordinate projection is the following. First, we introduce the circle family $\mathcal{P}$ as:

$$\mathcal{P} = \{(x, y) : x^2 + (y - r)^2 = r^2\} \tag{9.38}$$

as the set of circles with radius $r = r(x, y)$ passing through the origin and centered on axis Y_b with $\partial y / \partial x = 0$ at the origin. Associated with these circles, we can define θ_r as being the angle of the tangent of $\mathcal{P}$ at (x, y),

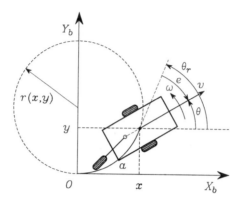

Figure 9.4: Coordinate projection.

i.e.,

$$\theta_r(x,y) = \begin{cases} 2\arctan(y/x) & (x,y) \neq (0,0) \\ 0 & (x,y) = (0,0), \end{cases} \qquad (9.39)$$

and

$$a(x,y) = r(x,y)\theta_r \qquad a(x,0) = x \qquad (9.40)$$
$$e(x,y,\theta) = \theta - \theta_r, \qquad (9.41)$$

where $a(x,y)$ defines the arc length from the origin to (x,y) along a circle which is centered on axis Y_b and passes through these two points; $a(x,y)$ may be positive or negative according to the sign of x. When $y = 0$, we define $a(x,0) = 0$ which makes $a(x,y)$ continuous with respect to y since $a(x,\varepsilon) \approx x$ when $\varepsilon \approx 0$. Discontinuities in $a(x,y)$ only take place in the set $d(z) = \{z : x = 0, y \neq 0\}$. The variable e is the orientation error. An illustration of these definitions is shown in Fig. 9.4.

The robot can then be stabilized by finding a feedback control law that orientates the angle θ according to the tangent of one of the members of the circle family $\mathcal{P}$ and then decreases the arc length of the associated circle. The design of such a control law can be easily understood by writing the open-loop equations in the projected coordinates a and e

$$\dot{a} = b_1(z)v \qquad (9.42)$$
$$\dot{e} = b_2(z)v + \omega \qquad (9.43)$$

where, as before, z is the original system state vector coordinates, i.e., $z = (x \ y \ \theta)^T$ and the functions $b_i(z)$ have the following properties.

9.4. POSTURE STABILIZATION

Property 9.1 $b_{\min} \leq b_1(z) \leq b_{\max}$.

□

Property 9.2 $b_1(e, x, y)$ is continuous in e.

□

Property 9.3 $\lim_{e \to 0} b_1(z) = 1$.

□

Property 9.4 $|b_2(z)a(z)| \leq N$ for some constant $N > 0$.

□

Here, $b_{\min}$ and $b_{\max}$ are bounded functions independent of e. These properties are useful for establishing exponential convergence of the closed-loop trajectories (but without asymptotic stability in a Lyapunov sense). Taking the following control law with $\gamma > 0$ and $k > 0$

$$v = -\gamma b_1 a \qquad (9.44)$$
$$\omega = -b_2 v - ke = \gamma b_1 b_2 a - ke. \qquad (9.45)$$

gives, away from the discontinuous surface $d(z)$, the closed-loop equations

$$\dot{a} = -\gamma b_1^2 a \qquad (9.46)$$
$$\dot{e} = -ke.$$

From (9.46) we can see that e exponentially converges to zero and from Property 9.3 we see that the function $b_1(z)$ tends to a positive constant and therefore the arc length a converges exponentially to zero. Note also that the boundedness of $b_2(z)$ indicated in Property 9.4 implies the boundedness of the control vector ω. Finally, it can also be proved that the surface $d(z)$ is not attractive and that the convergence rate away from $d(z)$ is not influenced by the crossing of the discontinuous surface $d(z)$.

It can also be proved that the discontinuous surface can only be crossed twice in finite time. Therefore it is always possible to find an exponentially decaying time function that gives an upper bound on the solution of the projected coordinates a and e. Finally, these properties can be transferred to the original coordinates z.

Hybrid piecewise control

Another alternative approach of piecewise control design consists of mixing constant and continuous piecewise control design. The main idea is presented below using the model transformation (9.7) and (9.8), which leads to the following 3-dimensional chained structure:

$$\dot{x}_1 = u_1$$
$$\dot{x}_2 = u_2 \qquad (9.47)$$
$$\dot{x}_3 = x_2 u_1.$$

The input $u_1(t)$ is piecewise constant for the time interval $I_k = [k\delta, (k+1)\delta]$, $\forall k \in \{0, 1, 2, 3, \ldots\}$ and some $\delta > 0$; namely,

$$u_1(t) = u_1(k\delta) \qquad \forall t \in I_k \qquad (9.48)$$

where $u_1(k\delta)$ describes a discrete-time control law for the subsystem $\dot{x}_1 = u_1$. Hence, system (9.47) can be written as

$$\dot{x}_1(t) = u_1(k\delta) \qquad (9.49)$$

$$\dot{z}(t) = \begin{pmatrix} 0 & 0 \\ u_1(k\delta) & 0 \end{pmatrix} z(t) + \begin{pmatrix} 1 \\ 0 \end{pmatrix} u_2(t) \qquad (9.50)$$

where $z = (x_2, x_3)$. This representation describes two subsystems, where one of them (x_1) has $u_1(k\delta)$ as a piecewise *constant* input, and the other (z) has $u_2(t)$ as a piecewise *continuous* input function. For this reason the approach can be understood as a hybrid piecewise control design. Notice that subsystem (9.49) describes a piecewise continuous controllable linear time-invariant system as long as $u_1(k\delta)$ does not vanish.

The following control law ensures the uniform ultimate stabilization of the x-coordinates to an arbitrarily small set containing the origin

$$u_1(k\delta) = k_1(k\delta) + \mathrm{sgn}(k_1(k\delta))\gamma(\|z(k\delta)\|) \qquad (9.51)$$
$$u_2(t) = -|u_1(k\delta)|k_2 x_2(t) - u_1(k\delta) k_3 x_3(t), \qquad (9.52)$$

where:

- $k_1(k\delta)$ is any discrete-time stabilizing feedback law for the subsystem $\dot{x}_1(t) = k_1(k\delta)$, i.e.,

$$k_1(k\delta) = \frac{(a-1)}{\delta} x_1(k\delta) \qquad 0 < |a| < 1; \qquad (9.53)$$

- $\gamma(\|z(k\delta)\|)$ is a positive definite function vanishing only when $\|z(k\delta)\| = 0$, i.e.,

$$\gamma(\|z(k\delta)\|) = \frac{\gamma_0(1-a)}{\delta} \frac{\exp\left(c\|z(k\delta)\|^2\right) - 1}{\exp\left(c\|z(k\delta)\|^2\right) + 1} \qquad (9.54)$$

where $\gamma_0 > 0$ and $c > 0$;

9.4. POSTURE STABILIZATION

- k_2 and k_3 are positive constants.

Asymptotic stability follows from the analysis of the closed-loop system

$$x_1(k) = ax_1(k-1) + s(k)\gamma(\|z(k\delta)\|) \qquad (9.55)$$
$$\dot{z}(t) = |u_1(k)|A(s(k))z(t), \qquad (9.56)$$

where $s(k) = \text{sgn}(k_1(k\delta))$ and $A(s(k))$ is defined as

$$A(s(k)) = \begin{pmatrix} -k_2 & -s(k)k_3 \\ s(k) & 0 \end{pmatrix}. \qquad (9.57)$$

Since suitable choices of k_2 and k_3 can make $A(s(k))$ stable with eigenvalues invariant with respect to $s(k)$, the solution to (9.56) in I_k can be written as

$$z(t) = \exp\bigl(|u_1(k)|A(s(k))(t - k\delta)\bigr)z(k\delta). \qquad (9.58)$$

Hence its norm is bounded as:

$$\begin{aligned}\|z(t)\| &\leq c\exp\bigl(-|u_1(k)|\lambda_0(t-k\delta)\bigr)\|z(k\delta)\| \\ &\leq c\exp\bigl(-\gamma(\|z(k\delta)\|)\lambda_0(t-k\delta)\bigr)\|z(k\delta)\|\end{aligned} \qquad (9.59)$$

where c is a constant and $\lambda_0 = |\lambda_{\min}(A)|$ and the last inequality is obtained from the definition of the control law $u_1(k)$. Then, for $t = (k+1)\delta$, we get

$$\|z((k+1)\delta)\| \leq c\exp\bigl(-\lambda_0\delta\gamma(\|z(k\delta)\|)\bigr)\|z(k\delta)\|, \qquad (9.60)$$

and thus $\|z(k\delta)\|$ decreases as long as $c\exp\bigl(-\lambda_0\delta\gamma(\|z(k\delta)\|)\bigr) < 1$, or equivalently as long as $\|z(k\delta)\| < \varepsilon$, where ε can be rendered arbitrarily small by a suitable choice of gains. Finally, eq. (9.55) describes a stable discrete-time system driven by a bounded and asymptotically ε-decaying input $(\gamma(\|z(k\delta)\|))$. Therefore, it is easy to conclude that $x_1(k)$ also tends asymptotically to an arbitrarily small set including the origin.

Decaying rates can be modified by a proper choice of the function $\gamma(\|z(k\delta)\|)$. The behaviour of the state trajectories is similar to the behaviour that has been obtained with the time-varying controllers but using piecewise continuous changes in the velocity inputs. Another way to introduce smoothness into the control design is to combine the *piecewise continuous* and periodic *time-varying* controllers. The advantage of doing this is that exponential convergence rates can be obtained. This idea is explained below.

9.4.3 Time-varying piecewise continuous control

In this section, a stabilizing time-varying feedback which depends smoothly on the state, except at periodic time instants, is presented. The control for the kinematic model is given by

$$\begin{aligned} \dot{x}_1 &= u_1 \\ \dot{x}_2 &= u_2 \\ \dot{x}_3 &= x_2 u_1. \end{aligned} \quad (9.61)$$

By using time-periodic functions in the control law, the inputs are smoothed out and become continuous with respect to time. By letting the feedback be nonsmooth with respect to the state at discrete instants of time, exponential convergence is obtained.

The idea behind the control law is to choose the input u_1 such that the subsystem

$$\begin{aligned} \dot{x}_2 &= u_2 \\ \dot{x}_3 &= x_2 u_1 \end{aligned} \quad (9.62)$$

becomes linear and time-varying in the time intervals $[k\delta, (k+1)\delta)$, $k \in \{0, 1, 2, \ldots\}$, where $k\delta, (k+1)\delta, \ldots$ are discrete time instants, as defined in the previous section. This is obtained by the following structure for u_1

$$u_1 = k_1(x(k\delta)) f(t) \quad (9.63)$$

where $k\delta$ denotes the last element in the sequence $(0, \delta, 2\delta, \ldots)$ such that $t \geq k\delta$, and $x = (\, x_1 \quad x_2 \quad x_3 \,)^T$. The function $f(t)$ is smooth and periodic. One possible choice of $f(t)$ is

$$f(t) = \frac{1 - \cos \omega t}{2} \qquad \omega = \frac{2\pi}{\delta}. \quad (9.64)$$

The other input u_2 will be chosen such that $\| (\, x_2(t) \quad x_3(t) \,)^T \|$ converges exponentially to zero.

The sign and magnitude of the parameter $k_1(x(k\delta))$ is chosen such that u_1 makes $x_1(t)$ converge exponentially to zero as $\| (\, x_2(t) \quad x_3(t) \,)^T \|$ converges to zero. Let $k_1(x(k\delta))$ be given by

$$k_1(x(k\delta)) = -\bigl(x_1(k\delta) + \operatorname{sgn}(x_1(k\delta)) \gamma(\|z(k\delta)\|)\bigr)\beta \quad (9.65)$$

where $z = (\, x_2 \quad x_3 \,)^T$ and

$$\gamma(\|z(k\delta)\|) = \kappa \|z(k\delta)\|^{\frac{1}{2}} = \kappa (x_2^2(k\delta) + x_3^2(k\delta))^{\frac{1}{4}} \quad (9.66)$$

$$\frac{1}{\beta} = \int_{k\delta}^{(k+1)\delta} f(\tau) d\tau \quad (9.67)$$

9.4. POSTURE STABILIZATION

where κ is a positive constant and $\text{sgn}(x_1(k\delta))$ is defined in the previous section. We see that $\gamma(\cdot)$ is a function of class $\mathcal{K}$.

In order to find the control law for u_2, we introduce the following auxiliary variable

$$x_{20} = -\frac{\lambda_3}{k_1(x(k\delta))} f^2(t) x_3. \qquad (9.68)$$

Time differentiation gives, together with (9.62) and (9.63),

$$\dot{x}_{20} = -\lambda_3 \left(2 f \dot{f} \frac{x_3}{k_1(x(k\delta))} + f^3 x_2 \right). \qquad (9.69)$$

The input u_2 can be chosen as

$$u_2 = -\lambda_2 (x_2 - x_{20}) + \dot{x}_{20} \qquad (9.70)$$

where the controller parameters λ_2 and λ_3 are positive constants.

The control law for system (9.61) is then given from (9.63), (9.65), (9.68), (9.69), and (9.70) as

$$u_1 = k_1(x(k\delta)) f(t) \qquad (9.71)$$

$$u_2 = -(\lambda_2 + \lambda_3 f^3) x_2 - \lambda_3 (\lambda_2 f^2 + 2 f \dot{f}) \frac{1}{k_1(x(k\delta))} x_3 \qquad (9.72)$$

where $f(t)$ and $k_1(x(k\delta))(x(k\delta))$ are given by (9.64) and (9.65).

To analyze the convergence and stability of the system, we introduce the variable

$$\tilde{x}_2 = x_2 - x_{20}.$$

The control law (9.71) and (9.72) with (9.62) implies that

$$\dot{\tilde{x}}_2 = -\lambda_2 \tilde{x}_2 \qquad (9.73)$$

$$\dot{x}_3 = -\lambda_3 f^3(t) x_3 + k_1(x(k\delta)) f(t) \tilde{x}_2. \qquad (9.74)$$

It can be shown that $\| (\tilde{x}_2(t) \;\; x_3(t))^T \|$ converges exponentially to zero. Hence, this can be used to show that $|k_1(x(k\delta))| \geq c |x_3(t)|^{\frac{1}{2}}$, $\forall t \geq 0$, where c is a positive constant. Since

$$x_2 = x_{20} + \tilde{x}_2 = -\lambda_3 f^2(t) \frac{x_3}{k_1(x(k\delta))} + \tilde{x}_2$$

and $\| (\tilde{x}_2(t) \;\; x_3(t))^T \|$ converges exponentially to zero, $x_2(t)$ converges exponentially to zero as well. From the control law for u_1 in (9.71) and from the system equations (9.61) we obtain

$$\|x_1((k+1)\delta)\| = \gamma(\|z(k\delta)\|) = \kappa (x_2^2(k\delta) + x_3^2(k\delta))^{\frac{1}{4}} \qquad \forall k \in \{0, 1, 2, \ldots\}.$$

This can be used to show that $x_1(t)$ converges exponentially to zero as $\|z(t)\| = \|(x_2(t) \quad x_3(t))^T\|$ converges to zero.

In conclusion, if the control law is given by (9.71) and (9.72), where $k_1(x(k\delta))$ is given by (9.65), then the origin of system (9.61) is globally asymptotically stable and the rate of convergence to the origin is exponential, i.e.,

$$\|x(t)\| \leq \exp(-\gamma t) h(\|x(0)\|) \qquad \forall t \geq 0. \qquad (9.75)$$

Here, $h(\cdot)$ is of class $\mathcal{K}$ and γ is a positive constant.

9.5 Further reading

Conversion of nonholonomic wheeled mobile robots kinematic equations into the so-called canonical chained form was introduced in [27, 28]. The class of wheeled mobile robots whose kinematic equations can be locally converted to the chained form, on the assumptions of *pure rolling* and *non-slipping*, include the restricted mobility vehicles described in Chapter 7 and also, on additional assumptions, vehicles pulling trailers [40, 7]. The impossibility, already pointed out in Chapter 8, of asymptotically stabilizing a given point for these systems by means of continuous pure-state feedback derives from violation of Brockett's necessary condition for stabilizability [5]. Early published works on the tracking and path following problems, as they are stated in this chapter, were based on the classical tangent (or pseudo) linearization technique.

The velocity scaling procedure is already present, although implicitly, in the work [13] on vision-based road following. Its importance in the control design has subsequently been made explicit in [33, 35]. The globally stabilizing nonlinear feedback proposed in Section 9.2.2 for robot tracking was given in [38]. The path planning results of Section 9.3 are taken from [35].

Although not mentioned in the third part of this book, path planning for nonholonomic systems has received a great deal of attention. Some of the methodologies employed for motion planning or steering, which indeed belong to the class of open-loop control design, are differential geometry and differential algebra [21, 14, 32], motion planning using geometric phases [20] or based on a parameterized input belonging to a given finite-dimensional family of functions [7], and design of optimal controllers [6]. Some of these aspects are also described in [22, 23, 26]. The survey [18] gives a clear view of the current research in this field.

The idea of using continuous time-varying feedback in order to circumvent the difficulty captured by Brockett's necessary condition in the case of point-stabilization was proposed in [34]. Subsequent general results establishing the existence of stabilizing continuous time-varying feedbacks

9.5. FURTHER READING

when the system to be controlled is small-time locally controllable, as in the case of a chained system, are due to [11, 12]. The first systematic approach for the design of explicit time-varying feedback laws which stabilize a general class of driftless nonlinear control systems is due to [30]. The slow (polynomial) asymptotic rate of convergence associated with smooth time-varying feedback was pointed out in [39] and then discussed by several authors (see [29, 14] for example). This limitation no longer holds when the time-varying feedback is continuous only.

The simple smooth time-varying feedback which introduces the method in Section 9.4.1 corresponds to an early scheme proposed in [39]. The other technique presented in that section, which consists of addressing the point-stabilization problem as an extension of the tracking and path following problems, was proposed in [35] and generalized in [37]. The possibility of achieving exponential convergence by using time-periodic feedbacks, which are not differentiable at the origin, has been pointed out in [27, 36, 31, 25] and is the subject of current investigation; the underlying control design and analysis is based on the use of the properties associated with the so-called homogeneous systems [16, 15].

Discontinuous feedback has been considered as another means to circumvent Brockett's necessary condition. In the case of nonholonomic mechanical systems, early works in this direction are those by [3, 4, 10]. The coordinate projection method described in Section 9.4.2 was presented in [10]. It was the pioneering result yielding exponential convergence of the closed-loop system solutions to zero. However, it is not asymptotically stable in the sense of Lyapunov, in contrast to the time-varying continuous methods. The time-varying piecewise continuous method presented in Section 9.4.3 is due to [41]. It is asymptotically stable in the sense of Lyapunov, in contrast to the time-varying continuous methods. The hybrid method presented in the same section is taken from [8]. Other discontinuous feedbacks yielding exponential rate of convergence are given in [9, 17, 1]. Sliding control has been studied by [2] and other forms of hybrid controllers have been studied by [19]. For an overview on the various existing methods for point stabilization see [18].

Time-varying feedbacks which are continuous everywhere and share the aforementioned properties (asymptotic stability in the sense of Lyapunov and exponential rate of convergence) have recently been proposed; see [24, 31]), for example. Stabilization of Hamiltonian systems with time-varying feedback has been addressed in [17]. Research is still very active in the domain of point-stabilization of nonholonomic systems, and several important issues related to the overall performance of the controlled system, including monitoring of the closed-loop system transient behaviour and robustness, remain to be explored.

References

[1] A. Astolfi, "Exponential stabilization of nonholonomic systems via discontinuous control," *Prepr. 4th IFAC Symp. on Nonlinear Control Systems Design*, Tahoe City, pp. 741–746, 1995.

[2] A.M. Bloch and D. Drakunov, "Tracking in nonholonomic systems via sliding modes," *Proc. 34th IEEE Conf. on Decision and Control*, New Orleans, LA, pp. 2103–2107, 1995.

[3] A.M. Bloch and N.H. McClamroch, "Control of mechanical systems with classical nonholonomic constraints," *Proc. 28th IEEE Conf. on Decision and Control*, Tampa, FL, pp. 201–205, 1989.

[4] A.M. Bloch, M. Reyhanoglu, and N.H. McClamroch, "Control and stabilization of nonholonomic dynamic systems," *IEEE Trans. on Automatic Control*, vol. 37, pp. 1746–1757, 1992.

[5] R.W. Brockett, "Asymptotic stability and feedback stabilization," in *Differential Geometric Control Theory*, R.W. Brockett, R.S. Millman and H.J. Sussman (Eds.), Birkhäuser, Boston, MA, pp. 181–208, 1983.

[6] R.W. Brockett and L. Dai "Non-holonomic kinematics and the role of elliptic functions in constructive controllability," in *Nonholonomic Motion Planning*, Z.X. Li and J. Canny (Eds.), Kluwer Academic Publishers, Boston, MA, pp. 1–21, 1993.

[7] L.G. Bushnell, D.M. Tilbury, and S.S. Sastry, "Steering three-input chained form nonholonomic systems using sinusoids. The fire truck example," *Proc. 2nd European Control Conf.*, Groningen, NL, pp. 1432–1437, 1993.

[8] C. Canudas de Wit, H. Berghuis, and H. Nijmeijer, "Hybrid stabilization of nonlinear systems in chained form," *Proc. 33rd IEEE Conf. on Decision and Control*, Lake Buena Vista, FL, pp. 3475–3480, 1994.

[9] C. Canudas de Wit and H. Khennouf, "Quasi-continuous stabilizing controllers for nonholonomic systems: design and robustness considerations," *Proc. 3rd European Control Conf.*, Roma, I, pp. 2630–2635, 1995.

[10] C. Canudas de Wit and O. J. Sørdalen, "Exponential stabilization of mobile robots with nonholonomic constraints," *IEEE Trans. on Automatic Control*, vol. 37, pp. 1791–1797, 1992.

[11] J.M. Coron, "Global asymptotic stabilization for controllable systems without drift," *Mathematics of Control, Signals, and Systems*, vol. 5, pp. 295–312, 1992.

[12] J.M. Coron, "On the stabilization of locally controllable systems by means of continuous time-varying feedback laws," *SIAM J. of Control and Optimization*, vol. 33, pp. 804–833, 1995.

[13] E.D. Dickmanns and A. Zapp, "Autonomous high speed road vehicle guidance by computer vision," *Prepr. 10th IFAC World Congr.*, München, D, vol. 4, pp. 232–237, 1987.

[14] L. Gurvits and Z.X. Li, "Smooth time-periodic feedback solutions for nonholonomic motion planning," in *Nonholonomic Motion Planning*, Z.X. Li and J. Canny (Eds.), Kluwer Academic Publishers, Boston, MA, pp. 53–108, 1993.

[15] H. Hermes, "Nilpotent and high-order approximations of vector field systems," *SIAM Review*, vol. 33, pp. 238–264, 1991.

[16] M. Kawski, "Homogeneous stabilizing feedback laws," *Control Theory and Advanced Techniques*, vol. 6, pp. 497–516, 1990.

[17] H. Khennouf and C. Canudas de Wit, "On the construction of stabilizing discontinuous controllers for nonholonomic systems," *Prepr. 4th IFAC Symp. on Nonlinear Control Systems Design*, Tahoe City, pp. 27–32, 1995.

[18] I. Kolmanovsky and N.H. McClamroch, "Developments in nonholonomic control problems," *IEEE Control Systems Mag.*, vol. 15, no. 6, pp. 20–36, 1995.

[19] I. Kolmanovsky and N.H. McClamroch, "Stabilization of nonholonomic Chaplygin systems with linear base space dynamics," *Proc. 34th IEEE Conf. on Decision and Control*, New Orleans, LA, pp. 4305–4310, 1995.

[20] P.S. Krishnaprasad and R. Yang, "Geometric phases, anholonomy, and optimal movement" *Proc. 1991 IEEE Int. Conf. on Robotics and Automation*, Sacramento, CA, pp. 2185–2189, 1991.

[21] A. Lafferriere and H.J. Sussmann, "A differential geometric approach to motion planning," in *Nonholonomic Motion Planning*, Z.X. Li and J. Canny (Eds.), Kluwer Academic Publishers, Boston, MA, pp. 235–270, 1993.

[22] J.-C. Latombe, *Robot Motion Planning*, Kluwer Academic Publishers, Boston, MA, 1991.

[23] Z. Li and J.F. Canny, *Nonholonomic Motion Planning*, Kluwer Academic Publishers, Boston, MA, 1993.

[24] R.T. M'Closkey and R.M. Murray, "Nonholonomic systems and exponential convergence: Some analysis tools," *Proc. 32nd IEEE Conf. on Decision and Control*, San Antonio, TX, pp. 943–948, 1993.

[25] R.T. M'Closkey and R.M. Murray, "Exponential stabilization of nonlinear driftless control systems via time-varying homogeneous feedback," *Proc. 34th IEEE Conf. on Decision and Control*, Lake Buena Vista, FL, pp. 1317–1322, 1994.

[26] R.M. Murray, Z. Li, and S.S. Sastry, *A Mathematical Introduction to Robotic Manipulation*, CRC Press, Boca Raton, FL, 1994.

[27] R.M. Murray and S.S. Sastry, "Steering nonholonomic systems in chained form," *Proc. 30th IEEE Conf. on Decision and Control*, Brighton, UK, pp. 1121–1126, 1991.

[28] R.M. Murray and S.S. Sastry, "Nonholonomic motion planning: Steering using sinusoids," *IEEE Trans. on Automatic Control*, vol. 38, pp. 700–713, 1993.

[29] R.M. Murray, G. Walsh, and S.S. Sastry, "Stabilization and tracking for nonholonomic systems using time-varying state feedbacks," *Prepr. 3rd IFAC Symp. on Nonlinear Control Systems Design*, Bordeaux, F, pp. 182–187, 1992.

[30] J.-B. Pomet, "Explicit design of time-varying stabilizing control laws for a class of controllable systems without drift," *Systems & Control Lett.*, vol. 18, pp. 147–158, 1992.

[31] J.-B. Pomet and C. Samson, "Time-varying exponential stabilization of nonholonomic systems in power form," *Prepr. IFAC Symp. on Robust Control Design*, Rio de Janeiro, BR, pp. 447–452, 1994.

[32] P. Rouchon, M. Fliess, J. Levine, and P. Martin, "Flatness, motion planning and trailer systems" *Proc. 32nd IEEE Conf. on Decision and Control*, Austin, TX, pp. 2700–2705, 1993.

[33] M. Sampei, T. Tamura, T. Itoh, and M. Nakamichi, "Path tracking control of trailer-like mobile robot," *Proc. IEEE/RSJ Int. Work. on Intelligent Robots and Systems*, Osaka, J, pp. 193–198, 1991.

[34] C. Samson, "Velocity and torque feedback control of a nonholonomic cart," in *Advanced Robot Control*, Lecture Notes in Control and Information Science, vol. 162, C. Canudas de Wit (Ed.), Springer-Verlag, Berlin, D, pp. 125–151, 1991.

[35] C. Samson, "Path following and time-varying feedback stabilization of a wheeled mobile robot," *Proc. Int. Conf. on Advanced Robotics and Computer Vision*, Singapore, vol. 13, pp. 1.1–1.5, 1992.

[36] C. Samson, *Mobile Robot Control, part 2: Control of chained systems with application to path following and time-varying point-stabilization of wheeled vehicles*, tech. rep. 1994, INRIA, Sophia-Antipolis, F, 1993.

[37] C. Samson, "Control of chained systems. Application to path following and time-varying point-stabilization of mobile robots," *IEEE Trans. on Automatic Control*, vol. 40, pp. 64–77, 1995.

[38] C. Samson and K. Ait-Abderrahim, "Feedback control of a nonholonomic wheeled cart in cartesian space," *Proc. 1991 IEEE Int. Conf. on Robotics and Automation*, Sacramento, CA, pp. 1136–1141, 1991.

[39] C. Samson and K. Ait-Abderrahim, "Feedback stabilization of a nonholonomic wheeled mobile robot," *Proc. IEEE/RSJ Int. Work. on Intelligent Robots and Systems*, Osaka, J, pp. 1242–1247, 1991.

[40] O.J. Sørdalen, "Conversion of the kinematics of a car with n trailers into a chained form," *Proc. 1993 IEEE Int. Conf. on Robotics and Automation*, Atlanta, GA, vol. 1, pp. 382–387, 1993.

[41] O.J. Sørdalen and O. Egeland, "Exponential stabilization of chained nonholonomic systems," *Proc. 2nd European Control Conf.*, Groningen, NL, pp. 1438–1443, 1993.

Appendix A

Control background

In this Appendix we present the background for the main control theory tools used throughout the book; namely, Lyapunov theory, singular perturbation theory, differential geometry theory, and input–output theory. We provide some motivation for the various concepts and also elaborate some aspects of interest for theory of robot control. No proofs of the various theorems and lemmas are given, and the reader is referred to the cited literature.

A.1 Lyapunov theory

We will use throughout the appendix a rather standard notation and terminology. R_+ will denote the set of nonnegative real numbers, and R^n will denote the usual n-dimensional vector space over R endowed with the Euclidean norm

$$\|x\| = \left(\sum_{j=1}^{n} |x_j|^2\right)^{1/2}.$$

Let us consider a nonlinear dynamic system represented as

$$\dot{x} = f(x, t), \tag{A.1}$$

where f is a nonlinear vector function and $x \in R^n$ is the state vector.

A.1.1 Autonomous systems

The nonlinear system (A.1) is said to be *autonomous* (or *time-invariant*) if f does not depend explicitly on time, i.e.,

$$\dot{x} = f(x); \tag{A.2}$$

otherwise the system is called *nonautonomous* (or *time-varying*). In this section, we briefly review the *Lyapunov theory* results for autonomous systems while nonautonomous systems will be reviewed in the next section. Lyapunov theory is the fundamental tool for stability analysis of dynamic systems, such as the robot manipulators and mobile robots treated in the book.

The basic stability concepts are summarized in the following definitions.

Definition A.1 (Equilibrium) A state $x^\star$ is an *equilibrium* point of (A.2) if $f(x^\star) = 0$.

□

Definition A.2 (Stability) The equilibrium point $x = 0$ is said to be *stable* if, for any $\rho > 0$, there exists $r > 0$ such that if $\|x(0)\| < r$, then $\|x(t)\| < \rho \; \forall t \geq 0$. Otherwise the equilibrium point is unstable.

□

Definition A.3 (Asymptotic stability) An equilibrium point $x = 0$ is *asymptotically stable* if it is stable, and if in addition there exists some $r > 0$ such that $\|x(0)\| < r$ implies that $x(t) \to 0$ as $t \to \infty$.

□

Definition A.4 (Marginal stability) An equilibrium point that is Lyapunov stable but not asymptotically stable is called *marginally stable*.

□

Definition A.5 (Exponential stability) An equilibrium point is *exponentially stable* if there exist two strictly positive numbers α and λ independent of time and initial conditions such that

$$\|x(t)\| \leq \alpha \exp(-\lambda t)\|x(0)\| \qquad \forall t > 0 \tag{A.3}$$

in some ball around the origin.

□

The above definitions correspond to *local* properties of the system around the equilibrium point. The above stability concepts become *global* when their corresponding conditions are satisfied for *any initial state*.

Lyapunov linearization method

Assume that $f(x)$ in (A.2) is continuously differentiable and that $x = 0$ is an equilibrium point. Then, using Taylor expansion, the system dynamics can be written as

$$\dot{x} = \left.\frac{\partial f}{\partial x}\right|_{x=0} + o(x) \tag{A.4}$$

A.1. LYAPUNOV THEORY

where o stands for higher-order terms in x. Linearization of the original nonlinear system at the equilibrium point is given by

$$\dot{x} = Ax \qquad (A.5)$$

where A denotes the Jacobian matrix of f with respect to x at $x = 0$, i.e.,

$$A = \left.\frac{\partial f}{\partial x}\right|_{x=0}.$$

A linear time-invariant system of the form (A.5) is (asymptotically) stable if A is a (strictly) stable matrix, i.e., if all the eigenvalues of A have (negative) nonpositive real parts. The stability of linear time-invariant systems can be determined according to the following theorem.

Theorem A.1 *The equilibrium point $x = 0$ of system (A.5) is asymptotically stable if and only if, given any matrix $Q > 0$, the solution P to the Lyapunov equation*

$$A^T P + PA = -Q \qquad (A.6)$$

is positive definite. If Q is only positive semi-definite ($Q \geq 0$), then only stability is concluded.

◇ ◇ ◇

In the above theorem, notice that if $Q = L^T L$ with (P, L) being an observable pair, then asymptotic stability is obtained again.

Local stability of the original nonlinear system can be inferred from stability of the linearized system as stated in the following theorem.

Theorem A.2 *If the linearized system is strictly stable (unstable), then the equilibrium point of the nonlinear system is locally asymptotically stable (unstable).*

◇ ◇ ◇

The above theorem does not allow us to conclude anything when the linearized system is marginally stable.

Lyapunov direct method

Let us consider the following definitions.

Definition A.6 ((Semi-)definiteness) A scalar continuous function $V(x)$ is said to be locally *positive (semi-)definite* if $V(0) = 0$ and $V(x) > 0$ ($V(x) \geq 0$) for $x \neq 0$. Similarly, $V(x)$ is said to be *negative (semi-)definite* if $-V(x)$ is positive (semi-)definite.

□

Definition A.7 (Lyapunov function) $V(x)$ is called a *Lyapunov function* for the system (A.2) if, in a ball B, $V(x)$ is positive definite and has continuous partial derivatives, and if its time derivative along the solutions to (A.2) is negative semi-definite, i.e., $\dot{V}(x) = (\partial V/\partial x)f(x) \leq 0$.

□

The following theorems can be used for local and global analysis of stability, respectively.

Theorem A.3 (Local stability) *The equilibrium point 0 of system (A.2) is (asymptotically) stable in a ball B if there exists a scalar function $V(x)$ with continuous derivatives such that $V(x)$ is positive definite and $\dot{V}(x)$ is negative semi-definite (negative definite) in the ball B.*

◊ ◊ ◊

Theorem A.4 (Global stability) *The equilibrium point of system (A.2) is globally asymptotically stable if there exists a scalar function $V(x)$ with continuous first-order derivatives such that $V(x)$ is positive definite, $\dot{V}(x)$ is negative definite and $V(x)$ is radially unbounded, i.e., $V(x) \to \infty$ as $\|x\| \to \infty$.*

◊ ◊ ◊

La Salle's invariant set theorem

La Salle's results extend the stability analysis of the previous theorems when $\dot{V}$ is only negative semi-definite. They are stated as follows.

Definition A.8 (Invariant set) A set S is an *invariant set* for a dynamic system if every trajectory starting in S remains in S.

□

Invariant sets include equilibrium points, limit cycles, as well as any trajectory of an autonomous system.

Theorem A.5 (La Salle) *Consider the system (A.2) with f continuous, and let $V(x)$ be a scalar function with continuous first partial derivatives. Consider a region Γ defined by $V(x) < \gamma$ for some $\gamma > 0$. Assume that the region Γ is bounded and $\dot{V}(x) \leq 0 \ \forall x \in \Gamma$. Let Ω be the set of all points in Γ where $\dot{V}(x) = 0$, and M be the largest invariant set in Ω. Then, every solution $x(t)$ originating in Γ tends to M as $t \to \infty$. On the other hand, if $\dot{V}(x) \leq 0 \ \forall x$ and $V(x) \to \infty$ as $\|x\| \to \infty$, then all solutions globally asymptotically converge to M as $t \to \infty$.*

◊ ◊ ◊

A.1.2 Nonautonomous systems

In this section we consider nonautonomous nonlinear systems represented by (A.1). The stability concepts are characterized by the following definitions.

Definition A.9 (Equilibrium) A state $x^\star$ is an *equilibrium* point of (A.1) if $f(x^\star, t) = 0 \ \forall t \geq t_0$.

□

Definition A.10 (Stability) The equilibrium point $x = 0$ is *stable* at $t = t_0$ if for any $\rho > 0$ there exists an $r(\rho, t_0) > 0$ such that $\|x(t_0)\| < \rho$ $\forall t \geq t_0$. Otherwise the equilibrium point $x = 0$ is unstable.

□

Definition A.11 (Asymptotic stability) The equilibrium point $x = 0$ is *asymptotically stable* at $t = t_0$ if it is stable and if it exists $r(t_0) > 0$ such that $\|x(t_0)\| < r(t_0) \Rightarrow x(t) \to 0$ as $t \to \infty$.

□

Definition A.12 (Exponential stability) The equilibrium point $x = 0$ is *exponentially stable* if there exist two positive numbers α and λ such that $\|x(t)\| \leq \alpha \exp(-\lambda(t - t_0))\|x(t_0)\|$ $\forall t \geq t_0$, for $x(t_0)$ sufficiently small.

□

Definition A.13 (Global asymptotic stability) The equilibrium point $x = 0$ is *globally asymptotically stable* if it is stable and $x(t) \to 0$ as $t \to \infty$ $\forall x(t_0)$.

□

The stability properties are called *uniform* when they hold independently of the initial time t_0 as in the following definitions.

Definition A.14 (Uniform stability) The equilibrium point $x = 0$ is *uniformly stable* if it is stable with $r = r(\rho)$ that can be chosen independently of t_0.

□

Definition A.15 (Uniform asymptotic stability) The equilibrium point $x = 0$ is *uniformly asymptotically stable* if it is uniformly stable and there exists a ball of attraction B, independent of t_0, such that $x(t_0) \in B \Rightarrow x(t) \to 0$ as $t \to \infty$.

□

Lyapunov linearization method

Using Taylor expansion, the system (A.1) can be rewritten as

$$\dot{x} = A(t)x + o(x,t) \qquad (A.7)$$

where

$$A(t) = \left.\frac{\partial f}{\partial x}\right|_{x=0}.$$

A linear approximation of (A.1) is given by

$$\dot{x} = A(t)x. \qquad (A.8)$$

The result of Theorem A.1 can be extended to linear time-varying systems of the form (A.8) as follows.

Theorem A.6 *A necessary and sufficient condition for the uniform asymptotic stability of the origin of system (A.8) is that a matrix $P(t)$ exists such that*

$$V = x^T P(t)x > 0$$

and

$$\dot{V} = x^T(A^T P + PA + \dot{P})x \leq k(t)V,$$

where $\lim_{t \to \infty} \int_{t_0}^{t} k(\tau)d\tau = -\infty$ uniformly with respect to t_0.

◇ ◇ ◇

We can now state the following result.

Theorem A.7 *If the linearized system (A.8) is uniformly asymptotically stable, then the equilibrium point $x = 0$ of the original system (A.1) is also uniformly asymptotically stable.*

◇ ◇ ◇

Lyapunov direct method

We present now the Lyapunov stability theorems for nonautonomous systems. The following definitions are required.

Definition A.16 (Function of class $\mathcal{K}$) *A continuous function $\kappa : [0,k) \to R_+$ is said to be of class $\mathcal{K}$ if*

(i) $\kappa(0) = 0$,

(ii) $\kappa(\chi) > 0 \quad \forall \chi > 0$,

A.1. LYAPUNOV THEORY

(iii) κ is nondecreasing.

Statements (ii) and (iii) can also be replaced with

(ii') κ is strictly increasing,

so that the inverse function κ^{-1} is defined. The function is said to be of class $\mathcal{K}_\infty$ if $k = \infty$ and $\kappa(\chi) \to \infty$ as $\chi \to \infty$.

$\square$

Based on the definition of function of class $\mathcal{K}$, a modified definition of exponential stability can be given.

Definition A.17 ($\mathcal{K}$-exponential stability) The equilibrium point $x = 0$ is $\mathcal{K}$-*exponentially stable* if there exist a function $\kappa(\cdot)$ of class $\mathcal{K}$ and a positive number λ such that $\|x(t)\| \leq \exp(-\lambda(t - t_0))\kappa(\|x(t_0)\|) \; \forall t \geq t_0$, for $x(t_0)$ sufficiently small.

$\square$

Definition A.18 (Positive definite function) A *function* $V(x, t)$ is said to be locally (globally) *positive definite* if and only if there exists a function α of class $\mathcal{K}$ such that $V(0, t) = 0$ and $V(x, t) \geq \alpha(\|x\|) \; \forall t \geq 0$ and $\forall x$ in a ball B.

$\square$

Definition A.19 (Decrescent function) A *function* $V(x, t)$ is locally (globally) *decrescent* if and only if there exists a function β of class $\mathcal{K}$ such that $V(0, t) = 0$ and $V(x, t) \leq \beta(\|x\|) \; \forall t > 0$ and $\forall x$ in a ball B.

$\square$

The main Lyapunov stability theorem can now be stated as follows.

Theorem A.8 *Assume that $V(x, t)$ has continuous first derivatives around the equilibrium point $x = 0$. Consider the following conditions on V and $\dot{V}$ where α, β and γ denote functions of class $\mathcal{K}$:*

$(i) \quad V(x,t) \geq \alpha(\|x\|) > 0$
$(ii) \quad \dot{V}(x,t) \leq 0$
$(iii) \quad V(x,t) \leq \beta(\|x\|)$
$(iv) \quad \dot{V} \leq -\gamma(\|x\|) < 0$
$(v) \quad \lim_{x \to \infty} \alpha(\|x\|) = \infty.$

Then the equilibrium point $x = 0$ is:

- *stable if conditions (i) and (ii) hold;*
- *uniformly stable if conditions (i)-(iii) hold;*
- *uniformly asymptotically stable if conditions (i)-(iv) hold;*
- *globally uniformly asymptotically stable if conditions (i)-(v) hold.*

⋄ ⋄ ⋄

Converse Lyapunov theorems

There exists a converse theorem for each of the Lyapunov stability theorems. Let us present in particular the following results.

Theorem A.9 *If the equilibrium point $x = 0$ of system (A.1) is stable (uniformly asymptotically stable), there exists a positive definite (decrescent) function $V(x,t)$ with a nonpositive (negative) definite derivative.*

⋄ ⋄ ⋄

Theorem A.10 *Consider system (A.1) with $\partial f/\partial x$ and $\partial f/\partial t$ bounded in a certain ball B $\forall t > 0$. Then, the equilibrium point $x = 0$ is exponentially stable if and only if there exists a function $V(x,t)$ and some positive constants α_i such that $\forall x \in B$ and $\forall t > 0$*

$$\alpha_1 \|x\|^2 \leq V(x,t) \leq \alpha_2 \|x\|^2$$

$$\dot{V} \leq -\alpha_3 \|x\|^2$$

$$\left\|\frac{\partial V}{\partial x}\right\| \leq \alpha_4 \|x\|.$$

⋄ ⋄ ⋄

Barbalat's lemma

La Salle's results are only applicable to autonomous systems. On the other hand, Barbalat's lemma can be used to obtain stability results when the Lyapunov function derivative is negative semi-definite.

Lemma A.1 (Barbalat) *If the differentiable function f has a finite limit as $t \to \infty$, and if $\dot{f}$ is uniformly continuous, then $\dot{f} \to 0$ as $t \to \infty$.*

⋄ ⋄ ⋄

This lemma can be applied for studying stability of nonautonomous systems with Lyapunov theory, as stated by the following result.

Lemma A.2 *If a scalar function $V(x,t)$ is lower bounded and $\dot{V}(x,t)$ is negative semi-definite, then $\dot{V}(x,t) \to 0$ as $t \to \infty$ if $\dot{V}(x,t)$ is uniformly continuous in time.*

⋄ ⋄ ⋄

A.1.3 Practical stability

In the previous section, we were interested in the stability of the equilibrium point(s) of ordinary differential equations in the Lyapunov sense. We have recalled those analytical tools at our disposal to study the asymptotic properties of the solutions, such as Lyapunov functions. We are generally interested in proving asymptotic stability, i.e., convergence of the state towards the equilibrium. Nevertheless, it may happen in certain circumstances that asymptotic convergence is difficult to obtain via a feedback control law, e.g., when uncertainties are taken into account in the model of the process. Note that it may even be difficult in this case to find the equilibrium point(s) of the complete differential equations.

The problem in this case is thus mainly to prove boundedness, i.e., at least no signal grows unbounded in the closed-loop system, and if possible to evaluate the domain within which the state asymptotically lies. This is called *practical stability* —or *(uniform) ultimate boundedness*— because although Lyapunov stability may not be guaranteed, the trajectories reach a neighbourhood S of the origin, and remain in it. If S is reached in finite time, the state is then said to be ultimately bounded with respect to the set S. We give now the definition of ultimate boundedness.

Definition A.20 (Ultimate boundedness) A solution $x(\cdot) : [t_0, \infty) \to R^n, x(t_0) = x_0$, is said to be *ultimately bounded* with respect to a compact set $S \subset R^n$ if there is a nonnegative constant time $t'(x_0, S, t_0) < \infty$ such that $x(t) \in S \ \forall t \geq t_0 + t'(x_0, S, t_0)$. If $t'(\cdot)$ does not depend on t_0, then the state is said to be *uniformly* ultimately bounded with respect to S.

□

A.2 Singular perturbation theory

In this section we briefly review the fundamental concepts of *singular perturbation theory*. This is very useful for the problem of controlling a robot manipulator with joint or link flexibility.

A system is said to be in singular perturbation form when the time derivatives of some components of its state are multiplied by a small positive parameter ϵ, i.e.,

$$\dot{x} = f(t, x, z, \epsilon) \qquad x(t_0) = \xi(\epsilon) \qquad x \in R^n \qquad (A.9)$$
$$\epsilon \dot{z} = g(t, x, z, \epsilon) \qquad z(t_0) = \eta(\epsilon) \qquad z \in R^m \qquad (A.10)$$

where f, g and their first partial derivatives with respect to x, z and t are continuous, and $\xi(\epsilon)$ and $\eta(\epsilon)$ are smooth functions of ϵ. This model is in standard form if and only if, by setting $\epsilon = 0$ in (A.10), the equation

$$0 = g(t, x, z, 0)$$

has $k \geq 1$ isolated real roots $z = h_i(t, x)$ for $i = 1, \ldots, k$.

Let us use the following change of variables

$$y = z - h(t, x)$$
$$\tau = \frac{t - t_0}{\epsilon}.$$

Then, the system (A.9) and (A.10) becomes

$$\dot{x} = f(t, x, y + h(t, x), \epsilon)$$
$$\frac{dy}{d\tau} = \epsilon \dot{y} = g(t, x, y + h(t, x), \epsilon) - \epsilon \frac{\partial h}{\partial t} - \epsilon \frac{\partial h}{\partial x} \dot{x},$$

and $y = 0$ is called the quasi-steady state.

For $\epsilon = 0$ and $y = 0$ we obtain the reduced (*slow*) model

$$\dot{x} = f(t, x, h(t, x), 0).$$

On the other hand, for $\epsilon = 0$ we obtain the boundary-layer (*fast*) model

$$\frac{dy}{d\tau} = g(t, x, y + h(t, x), 0)$$

which has an equilibrium point at $y = 0$.

Stability of a singularly perturbed system can be ascertained on the basis of Tikhonov's theorem, which is based on the following definition.

Definition A.21 (Exponential stability) The equilibrium point $y = 0$ of the boundary-layer system is *exponentially stable* uniformly in $(t, x) \in [0, t_1] \times B_r$ if there exist k, γ and ρ such that the solution to the boundary-layer model satisfy

$$\|y(\tau)\| \leq k \exp(-\gamma \tau) \|y(0)\| \qquad \forall \|y(0)\| < \rho \qquad \forall (t, x) \in [0, t_1] \times B_r.$$

$\square$

A.2. SINGULAR PERTURBATION THEORY

Theorem A.11 (Tikhonov) *Assume that for* $t \in [0, t_1]$, $x \in B_r$, $(z-h) \in B_\rho$ *and* $\epsilon \in [0, \epsilon_0]$ *the following conditions hold:*

(i) $h(t,x)$ *and* $\partial g(t,x,z,0)/\partial z$ *have continuous first partial derivatives;*

(ii) *the reduced model has a unique solution* $\bar{x}(t)$ *and* $\|\bar{x}(t)\| \leq r_1$ *for* $t \in [t_0, t_1]$;

(iii) *the origin of the boundary-layer model is exponentially stable uniformly in* (t,x) *and has a solution* $\hat{y}(t/\epsilon)$.

Then there exist μ and ϵ^* such that, $\forall \eta(0), \xi(0)$ *with* $\|\eta(0) - h(t_0, \xi(0))\| < \mu$ *and* $0 < \epsilon < \epsilon^*$, *it is*

$$x(t,\epsilon) - \bar{x}(t) = O(\epsilon)$$
$$z(t,\epsilon) - h(t, \bar{x}(t) - \hat{y}(t/\epsilon)) = O(\epsilon).$$

Moreover, $\forall t_b > t_0$, *there exists* $\epsilon_1^* \leq \epsilon^*$ *so that*

$$z(t,\epsilon) - h(t, \bar{x}) = O(\epsilon)$$

for $\epsilon < \epsilon_1^*$ *and* $t \in [t_b, t_1]$.

◇ ◇ ◇

Lyapunov theory can be used to prove a stability result for a singularly perturbed system.

Theorem A.12 *Assume that the following assumptions are satisfied for all* $t \in [0, \infty)$, $x \in B_r$, $\epsilon \in [0, \epsilon_0]$:

(i) $f(t,0,0,\epsilon) = 0$, $g(t,0,0,\epsilon) = 0$ *and* $h(t,0) = 0$;

(ii) f, g, h *and their partial derivatives up to order two are bounded in* $(z-h) \in B_\rho$;

(iii) *the origin of the reduced model is exponentially stable;*

(iv) *the origin of the boundary-layer system is exponentially stable uniformly in* (t,x).

Then, there exists an $\epsilon^* > 0$ *such that* $\forall \epsilon < \epsilon^*$ *the origin of system (A.9) and (A.10) is exponentially stable.*

◇ ◇ ◇

A.3 Differential geometry theory

In this section we briefly review the fundamental results of *differential geometry theory*. Consider a nonlinear affine single-input/single-output system of the form

$$\dot{x} = f(x) + g(x)u \qquad (A.11)$$
$$y = h(x) \qquad (A.12)$$

where $h(x) : R^n \to R$ and $f(x), g(x) : R^n \to R^n$ are smooth functions. For ease of presentation we assume that system (A.11) and (A.12) has an equilibrium point at $x = 0$.

Definition A.22 (Lie derivative) The *Lie derivative* of h with respect to f is the scalar

$$L_f h = \frac{\partial h}{\partial x} f,$$

and the higher derivatives satisfy the recursion

$$L_f^i h = L_f(L_f^{i-1} h)$$

with $L_f^0 h = h$.

□

Definition A.23 (Lie bracket) The *Lie bracket* of f and g is the vector

$$[f, g] = \frac{\partial g}{\partial x} f - \frac{\partial f}{\partial x} g,$$

and the recursive operation is established by

$$ad_f^i g = [f, ad_f^{i-1} g].$$

□

Some properties of Lie brackets are:

$$[\alpha_1 f_1 + \alpha_2 f_2, g] = \alpha_1 [f_1, g] + \alpha_2 [f_2, g]$$
$$[f, g] = -[g, f],$$

and the Jacobi identity

$$L_{ad_f g} h = L_f(L_g h) - L_g(L_f h).$$

To define nonlinear changes of coordinates we need the following concept.

A.3. DIFFERENTIAL GEOMETRY THEORY

Definition A.24 (Diffeomorphism) A function $\phi(x) : R^n \to R^n$ is said to be a *diffeomorphism* in a region $\Omega \in R^n$ if it is smooth, and $\phi^{-1}(x)$ exists and is also smooth.

□

A sufficient condition for a smooth function $\phi(x)$ to be a diffeomorphism in a neighbourhood of the origin is that the Jacobian $\partial \phi / \partial x$ be nonsingular at zero.

The conditions for feedback linearizability of a nonlinear system are strongly related with the following theorem.

Theorem A.13 (Frobenius) *Consider a set of linearly independent vectors* $\{f_1(x), \ldots, f_m(x)\}$ *with* $f_i(x) : R^n \to R^n$. *Then, the following statements are equivalent:*

(i) (complete integrability) there exist $n - m$ *scalar functions* $h_i(x) : R^n \to R$ *such that*

$$L_{f_j} h_i = 0 \quad 1 \leq i \quad j \leq n - m$$

where $\partial h_i / \partial x$ *are linearly independent;*

(ii) (involutivity) there exist scalar functions $\alpha_{ijk}(x) : R^n \to R$ *such that*

$$[f_i, f_j] = \sum_{k=1}^{m} \alpha_{ijk}(x) f_k(x).$$

◇ ◇ ◇

A.3.1 Normal form

In this section we present the normal form of a nonlinear system which has been instrumental for the development of the feedback linearization technique. For this, it is convenient to define the notion of relative degree of a nonlinear system.

Definition A.25 (Relative degree) The system (A.11) and (A.12) has *relative degree* r at $x = 0$ if

(i) $L_g L_f^k h(x) = 0$, $\forall x$ in a neighbourhood of the origin and $\forall k < r - 1$;

(ii) $L_g L_f^{r-1} h(0) \neq 0$.

□

It is worth noticing that in the case of linear systems, e.g., $f(x) = Ax$, $g(x) = Bx$, $h(x) = Cx$, the integer r is characterized by the conditions $CA^k B = 0 \ \forall k < r - 1$ and $CA^{r-1}B \neq 0$. It is well known that these are exactly the conditions that define the relative degree of a linear system.

Another interesting interpretation of the relative degree is that r is exactly the number of times we have to differentiate the output to obtain the input explicitly appearing.

The functions $L_f^i h$ for $i = 0, 1, \ldots, r - 1$ have a special significance as demonstrated in the following theorem.

Theorem A.14 (Normal form) *If system (A.11) and (A.12) has relative degree $r \leq n$, then it is possible to find $n - r$ functions $\phi_{r+1}(x), \ldots, \phi_n(x)$ so that*

$$\phi(x) = \begin{pmatrix} h(x) \\ L_f h(x) \\ \vdots \\ L_f^{r-1} h(x) \\ \phi_{r+1}(x) \\ \vdots \\ \phi_n(x) \end{pmatrix}$$

is a diffeomorphism $z = \phi(x)$ that transforms the system into the following normal form

$$\dot{z}_1 = z_2$$
$$\dot{z}_2 = z_3$$
$$\vdots$$
$$\dot{z}_{r-1} = z_r$$
$$\dot{z}_r = b(z) + a(z)u$$
$$\dot{z}_{r+1} = q_{r+1}(z)$$
$$\vdots$$
$$\dot{z}_n = q_n(z).$$

Moreover, $a(z) \neq 0$ in a neighbourhood of $z_0 = \phi(0)$.

◇ ◇ ◇

A.3.2 Feedback linearization

From the above theorem we see that the state feedback control law

$$u = \frac{1}{a(z)}(-b(z) + v) \qquad (A.13)$$

A.3. DIFFERENTIAL GEOMETRY THEORY

yields a closed-loop system consisting of a chain of r integrators and an $(n-r)$-dimensional autonomous system. In the particular case of $r = n$ we *fully linearize* the system. The first set of conditions for the triple $\{f(x), g(x), h(x)\}$ to have relative degree n is given by the partial differential equation

$$\frac{\partial h}{\partial x}\left(g(x) \quad ad_f g(x) \quad \ldots \quad ad_f^{n-2} g(x)\right) = 0.$$

Frobenius theorem shows that the existence of solutions to this equation is equivalent to the involutivity of $\{g(x), ad_f g(x), \ldots, ad_f^{n-2} g(x)\}$. It can be shown that the second condition, i.e., $L_g L_f^{n-1} h(x) \neq 0$ is ensured by the linear independence of $\{g(x), ad_f g(x), \ldots, ad_f^{n-1} g(x)\}$.

The preceding discussion is summarized by the following key theorem.

Theorem A.15 *For the system (A.11) there exists an output function $h(x)$ such that the triple $\{f(x), g(x), h(x)\}$ has relative degree n at $x = 0$ if and only if:*

(i) the matrix $(g(0) \quad ad_f g(0) \quad \ldots \quad ad_f^{n-1} g(0))$ is full-rank;

(ii) the set $\{g(x), ad_f g(x), \ldots, ad_f^{n-2} g(x)\}$ is involutive around the origin.

◊ ◊ ◊

The importance of the preceding theorem can hardly be overestimated. It gives (a priori verifiable) necessary and sufficient conditions for full linearization of a nonlinear affine system. However, it should be pointed out that this control design approach requires on one hand the solution to a set of partial differential equations. On the other hand, it is intrinsically nonrobust since it relies on exact cancellation of nonlinearities; in the linear case this is tantamount to pole-zero cancellation.

A.3.3 Stabilization of feedback linearizable systems

If the relative degree of the system $r < n$ then, under the action of the feedback linearizing controller (A.13), there remains an $(n - r)$-dimensional subsystem. The importance of this subsystem is underscored in the proposition below.

Theorem A.16 *Consider the system (A.11) and (A.12) assumed to have relative degree r. Further, assume that the trivial equilibrium point of the following $(n - r)$-dimensional dynamical system is locally asymptotically* *stable:*

$$\dot{z}_{r+1} = q_{r+1}(0, \ldots, 0, z_{r+1}, \ldots, z_n)$$
$$\vdots \qquad\qquad\qquad\qquad (A.14)$$
$$\dot{z}_n = q_n(0, \ldots, 0, z_{r+1}, \ldots, z_n)$$

where $q_{r+1}, \ldots, q_n$ are given by the normal form. Under these conditions, the control law (A.13) yields a locally *asymptotically stable closed-loop system.*

⋄ ⋄ ⋄

The $(n-r)$-dimensional system (A.14) is known as the *zero dynamics*. It represents the dynamics of the unobservable part of the system when the input is set equal to zero and the output is constrained to be identically zero.

It is worth highlighting the qualifier *local* in the above theorem; in other words, it can be shown that the conditions above are not enough to ensure *global* asymptotic stability.

A.4 Input–output theory

In this section we present some basic notions and definitions of *input–output theory*, according to which, systems are viewed as operators that map signals living in some well-defined function spaces. The "action" of the system is evaluated by establishing bounds on the size of the output trajectories in terms of bounds on the size of the inputs. These bounds are estimated using techniques from functional analysis. The stability of feedback configurations of such operators is then, roughly speaking, established by simply comparing the bounds. This in contrast to Lyapunov-based techniques, which concentrate on stability of the equilibria (or attractors) of state space systems. Input–output analysis hence provides an alternative paradigm for studying stability of feedback systems. It can be recognized that the input–output approach is most useful in control applications, since it can naturally incorporate unmodelled dynamics and disturbances, it automatically provides performance bounds, as well as it allows for easy cascade and feedback interconnection.

The main feature of the input–output approach we wish to emphasize here is that it provides a natural generalization to the nonlinear time-varying case of the fact that stability of a linear time-invariant feedback system depends on the amounts of gain and phase shift introduced in the loop. Furthermore, and perhaps more importantly, the measures of signal amplification and signal shift —which are suitably captured by the notion of *passivity* of the operator associated with the nonlinear time-varying system respectively— are physically motivated properties that are related with system *energy dissipation*. This is a special appealing feature of the input–output approach since it guides us in the search of a (Lyapunov-like) energy function via incorporation of system physical insight.

A.4.1 Function spaces and operators

Consider the set of functions $f : R_+ \to R^n$. The standard (Lebesgue) spaces are considered.

Definition A.26 (Space of integrable functions) $\mathcal{L}_p^n$ is the *space of integrable functions* f with norm

$$\|f\|_p = \left(\int_0^\infty \|f\|^p dt\right)^{1/p} < \infty \qquad p = 1, 2.$$

□

Definition A.27 (Space of uniformly bounded functions) $\mathcal{L}_\infty^n$ is the *space of uniformly bounded functions* f with norm

$$\|f\|_\infty = \sup_{t \geq 0} \|f(t)\| < \infty.$$

□

The dimensions of the above spaces are often omitted when clear from the context.

A key concept that allows us to address stability issues is that of *extended space*. Roughly speaking, the idea is to distinguish the signals that can grow unbounded (in norm) as $t \to \infty$ from those that remain bounded for all (finite) time. It should be remarked at this point that in the input–output formulation special care should be taken for systems that exhibit unbounded growth in finite time. Extended spaces can be suitably introduced as follows.

Definition A.28 (Truncation) Given any function $f : R_+ \to R^n$ and any $T \geq 0$, the function

$$f_T = \begin{cases} f(t) & t \in [0, T] \\ 0 & t \in (T, \infty). \end{cases}$$

is a *truncation* of f. Also, the truncated norm is

$$\|f\|_{pT} = \|f_T\|_p.$$

□

Definition A.29 (Extended space) Given any function $f : R_+ \to R^n$ and its truncation f_T, the *extended space* of integrable functions is

$$\mathcal{L}_{pe}^n = \{f \mid f_T \in \mathcal{L}_p^n, \forall T \geq 0\}.$$

□

Definition A.30 (Truncated inner product) Given any two functions $f_1, f_2 \in \mathcal{L}_{2e}^n$, the *truncated inner product* is

$$< f_1|f_2 >_t = \int_0^t f_1^T(\tau) f_2(\tau) d\tau.$$

□

A nonlinear time-varying system is represented in the input–output approach by a mapping (*operator*)

$$\mathcal{H} : u \mapsto y$$

where $u \in \mathcal{L}_{pe}^r$ and $y \in \mathcal{L}_{pe}^m$. We assume the system to be causal, i.e., the output $y(t)$ at any given time t depends only on past inputs $u(t')$ for $t' \leq t$. This is tantamount to requiring the following implication to hold true for the operator $\mathcal{H}$:

$$u_{1t} \equiv u_{2t} \implies y_{1t} \equiv y_{2t}$$

where y_1, y_2 are the outputs obtained from the inputs u_1, u_2 through $\mathcal{H}$. Here, we restrict ourselves to finite-dimensional systems that admit a state space representation of the form

$$\begin{aligned} \dot{x} &= f(x, u, t) \quad x(t_0) = x_0 \\ y &= h(x, u, t) \\ \mathcal{H} &: u \mapsto y \end{aligned} \quad (A.15)$$

where $u \in \mathcal{L}_{pe}^r$, $y \in \mathcal{L}_{pe}^m$, and $x \in R^n$. To avoid further technical discussions, we assume throughout this section that all functions, including the ones used in the various definitions, are "sufficiently" smooth.

A.4.2 Passivity

The concept of energy dissipation is fundamental in the study of dynamic systems in general, and mechanical systems in particular. In many practical cases this concept is well captured in the definition of *passivity*. This notion motivated by the energy storage (or dissipation) properties of RLC electric

A.4. INPUT–OUTPUT THEORY

circuits, has been widely used in order to analyze input–output stability of a general class of interconnected nonlinear systems. For systems with state representation, passivity allows for a more geometric interpretation of notions such as available, stored, and dissipated energy in terms of Lyapunov functions, which provides a clear link with the property of Lyapunov stability as studied above.

The property of passivity is defined only for operators that map signals belonging to spaces of the same dimension, i.e., $\mathcal{H} : \mathcal{L}_{2e}^m \to \mathcal{L}_{2e}^m$. For this class of systems we have the following characterization.

Definition A.31 (Passive operator) The *operator* $\mathcal{H}$ described by (A.15) is said to be *passive* relative to the function $V(x)$ if and only if $V(x) : R^n \to R$ is a positive function of the state and, $\forall u, x, y$ solutions to (A.15) and $\forall t_1, t_2$ with $t_1 < t_2$, we have

$$\int_{t_1}^{t_2} u^T(\tau) y(\tau) d\tau \geq V(x(t_2)) - V(x(t_1)).$$

The operator is said to be *strictly passive* relative to the functions $V(x)$ and $W(x)$ if $W(x) : R^n \to R$ is also a positive function of the state and

$$\int_{t_1}^{t_2} u^T(\tau) y(\tau) d\tau \geq \alpha \int_{t_1}^{t_2} W(x(\tau)) d\tau + V(x(t_2)) - V(x(t_1))$$

for some $\alpha > 0$.

□

Let us present a simple example. Consider the system $\dot{x} = f(x)$, and assume there exists a positive function $L(x) : R^n \to R$ such that

$$\frac{\partial L}{\partial x} f(x) \leq -\alpha L(x)$$

for some $\alpha > 0$. As shown above, if $L(x)$ is a positive definite function and $f(0) = 0$, this implies global asymptotic stability of the trivial equilibrium point. This however may not be true if $L(x)$ is not a function of the full state as is the case in some robotic systems. Now, let us define the system

$$\dot{x} = f(x) + u$$
$$y = \left(\frac{\partial L}{\partial x}\right)^T$$
$$\mathcal{H} : u \mapsto y;$$

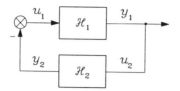

Figure A.1: Feedback interconnection of two passive systems.

we can show that $\mathcal{H}$ is strictly passive relative to the functions $V(x) = L(x)$ and $W(x) = L(x)$. To this end, we have that

$$\int_{t_1}^{t_2} y^T u\, d\tau = \int_{t_1}^{t_2} \frac{\partial L}{\partial x} \dot{x}\, d\tau - \int_{t_1}^{t_2} \frac{\partial L}{\partial x} f(x)\, d\tau$$

$$\geq L(x(t_2)) - L(x(t_1)) + \alpha \int_{t_1}^{t_2} L(x)\, d\tau.$$

Notice that our definitions of passivity and strict passivity differ from the standard definitions found in the literature in two respects. First, in the spirit of dissipative (state space) systems, we have included functions of the state $V(x)$ and $W(x)$. Second, no specific dependence of the function $W(x)$ on the system output is given. Our objective to consider this class of passive systems is twofold. First, to rigorously handle the effects of system initial conditions, which in our case of state space described systems are explicitly defined. Second, to dispose of an analysis framework that allows us to conclude, without additional arguments, not just input–output stability of the closed-loop system, but boundedness and internal stability as well. It is worth pointing out that the class of dissipative systems contains, as particular cases, passive, finite-gain and sector-bounded systems; each class of them resulting from a suitable choice of the so-called supply rate function.

Two key properties of passive systems are *stability* and *invariance* of feedback interconnection. These properties are summarized in the following propositions.

Theorem A.17 (Passivity) *Consider the systems $\mathcal{H}_1 : u_1 \mapsto y_1$ and $\mathcal{H}_2 : u_2 \mapsto y_2$ with states x_1 and x_2, respectively. Assume $\mathcal{H}_1$ is passive relative to $V_1(x_1)$ and $\mathcal{H}_2$ is strictly passive relative to $V_2(x_2)$ and $W_2(x_2)$ and they are interconnected in a negative feedback configuration (Fig. A.1), i.e.,*

$$u_2 = y_1$$
$$u_1 = -y_2.$$

A.4. INPUT–OUTPUT THEORY 383

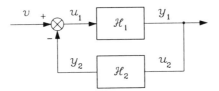

Figure A.2: Feedback interconnection of two passive systems with external input.

Under these conditions, all solutions defined on the interval $[0, t)$ satisfy

$$\int_0^t W_2(x_2) d\tau < \infty$$

and $V_1(x_1), V_2(x_2) \in \mathcal{L}_\infty$.

⋄ ⋄ ⋄

The above result implies that the feedback interconnection of two passive systems gives a passive closed-loop system. This is nicely complemented by the following result.

Theorem A.18 *For the systems of Theorem A.17, if an external input is added to the input of the first system (Fig. A.2), i.e.,*

$$u_1 = -y_2 + v,$$

then the map $v \mapsto y_1$ is strictly passive relative to $V(x_1, x_2) = V_1(x_1) + V_2(x_2)$ and $W(x_1, x_2) = W_2(x_2)$.

⋄ ⋄ ⋄

In the above theorem, if $W_2(x_2) = \|y_2\|^2$ then it can be shown that $\|y_1\|_2 \leq k\|v\|_2 + b$ for some $k > 0$ and $b \in R$. That is, if $\mathcal{H}_2$ is strictly passive relative to a function of the output, then the energy of the output of the closed-loop system is linearly bounded by the energy of the external input. In other words, the feedback interconnection of energy dissipating systems is also energy dissipating.

A.4.3 Robot manipulators as passive systems

The concept of passivity plays a central role in the analysis of control algorithms for *robot manipulators*. Manipulator dynamics defines a passive mapping from nonconservative joint torques τ —including friction torques and torques caused by external forces— to joint velocities $\dot{q}$. In the case of rigid manipulators this result may be obtained as follows (see Chapter 2 for a definition of all the terms). We know that the system kinetic energy is given by $T(q, \dot{q}) = \frac{1}{2}\dot{q}^T H(q)\dot{q}$ while potential energy is denoted by $U(q)$. Hence, we can introduce the system total energy

$$V(q,p) = T(q,p) + U(q) = \frac{1}{2}p^T H^{-1}(q)p + U(q) \geq 0$$

where we have defined the generalized momentum $p = H(q)\dot{q}$. Now, it is easy to verify that, in this set of coordinates, the robot dynamics is described by

$$\dot{q} = \frac{\partial T(q,p)}{\partial p} \qquad (A.16)$$

$$\dot{p} = -\frac{\partial T(q,p)}{\partial q} - \frac{\partial U(q)}{\partial q} + u.$$

Finally, taking the time derivative of the total energy along the solutions to (A.16), and then integrating from 0 to T we get the following restatement of the energy balance (Hamilton) principle

$$\int_0^T \tau^T(t)\dot{q}(t)dt = V(T) - V(0),$$

where the left-hand side is the supplied energy and the right-hand side is the energy at time T minus the initial energy. Passivity of the operator $\tau \mapsto \dot{q}$ relative to the function $V(q,p)$ follows from the equation above and positivity of the total energy.

A.4.4 Kalman-Yakubovich-Popov lemma

For linear time-invariant systems we have the following important proposition.

Lemma A.3 (Kalman-Yakubovich-Popov) *Consider a stable linear time-invariant system with minimal state space representation*

$$\dot{x} = Ax + Bu \qquad x(0) = x_0$$
$$y = Cx$$

where $x \in R^n$ and $u, y \in R^m$, and the corresponding transfer matrix $\mathcal{H}(s) = C(sI - A)^{-1}B$. The following statements are equivalent:

(i) $\mathcal{H}(s)$ is positive real, that is, all poles are in the open left-hand plane, those on the $j\omega$ axis are simple with Hermitian positive semi-definite residues, and $\mathcal{H}(j\omega) + \mathcal{H}^T(-j\omega) \geq 0$, $\forall \omega \in [0, \infty)$ which is not a pole of $\mathcal{H}(s)$;

(ii) there exist $P = P^T > 0$ and $Q = Q^T \geq 0$ such that
$$A^T P + PA = -Q \qquad PB = C^T; \qquad (A.17)$$

(iii) the map $\mathcal{H} : u \mapsto y$ is passive relative to $V(x) = x^T P x$.

Also, the statements below are equivalent:

(i') $\mathcal{H}(s)$ is strictly positive real, meaning that $\mathcal{H}(s - \epsilon)$ is positive real for some $\epsilon > 0$, that is, all poles are in the closed left-hand plane, $\mathcal{H}(j\omega) + \mathcal{H}^T(-j\omega) > 0$, $\forall \omega \in [0, \infty)$ and $\lim_{\omega \to \infty} \omega^2 \left(\mathcal{H}(j\omega) + \mathcal{H}^T(-j\omega) \right) > 0$;

(ii') there exist $P = P^T > 0$ and $Q = Q^T > 0$ such that (A.17) holds;

(iii') the map $\mathcal{H} : u \mapsto y$ is strictly passive relative to $V(x) = x^T P x$ and $W(x) = x^T Q x$.

◇ ◇ ◇

It is important to highlight the differences between the two parts of the proposition above. First, for positive realness we require only stability of the system, while for strict positive realness the system must be exponentially stable. As a consequence, the frequency domain condition (i) is strictly weaker than (i'). Second, while in the first part only passivity of the operator is ensured (Q is only positive *semi-definite*), in the second equivalence this is strengthened to strict passivity.

A basic lemma concerning stability of linear time-invariant systems is the following.

Lemma A.4 *Let the transfer function $\mathcal{H}(s) \in R^{n \times n}$ be strictly proper and exponentially stable, and let us denote by $u \in R^n$ the input and $y \in R^n$ the output of the system with transfer function $\mathcal{H}(s)$; then the following statements hold:*

(i) *if $u \in \mathcal{L}_1$, then $y \in \mathcal{L}_1 \cap \mathcal{L}_\infty$, $\dot{y} \in \mathcal{L}_1$, y is absolutely continuous and $y(t) \to 0$ as $t \to \infty$;*

(ii) *if $u \in \mathcal{L}_2$, then $y \in \mathcal{L}_2 \cap \mathcal{L}_\infty$, $\dot{y} \in \mathcal{L}_2$, y is continuous and $y(t) \to 0$ as $t \to \infty$;*

(iii) *if $u \in \mathcal{L}_\infty$, then $y \in \mathcal{L}_\infty$, $\dot{y} \in \mathcal{L}_\infty$ and y is uniformly continuous;*

(iv) *if $u \in \mathcal{L}_p$ and $1 < p < \infty$, then $y \in \mathcal{L}_p$ and $\dot{y} \in \mathcal{L}_p$.*

◇ ◇ ◇

A.5 Further reading

The original Lyapunov theory is contained in [10], while stability of nonlinear dynamic systems is widely covered in [8, 9]. The proofs of the theorems concerning Lyapunov stability theory can be found in [17, 6, 2]. The reader is referred to [7, 6] for a detailed discussion on singular perturbations. An extensive presentation of differential geometry methods can be found in [4] and the references therein. For the extension to the multivariable case and further details we refer the reader again to [4] as well as to [11]. Reference [15] also gives a nice presentation of stability theory and feedback linearization. This technique, which aims at linearizing the system dynamics via full state feedback, has been successfully used in robot manipulator applications under the name of inverse dynamics. The input/output approach for the analysis of dynamic systems was first introduced in [20, 21]. For an extensive treatment of this topic see [1, 18]. Dissipative state space systems were introduced in [19]. A review of some recent developments with an extensive reference list may be found in [12], while [16] contains a readable survey on the interconnection between input–output and state space theory. The importance of the input–output approach as a building block in robot control theory was first underscored in [13] (see also [5]). The proof of the Kalman-Yakubovich-Popov lemma can be found in [17, 6] (see also [14] for a recent elegant proof that relies only on elementary linear algebra). The extension of this result to the case of nonlinear systems can be found in [3].

References

[1] C.A. Desoer and M. Vidyasagar, *Feedback Systems: Input–Output Properties*, Academic Press, New York, NY, 1975.

[2] W. Hahn, *Stability of Motion*, Springer-Verlag, New York, NY, 1967.

[3] D. Hill and P. Moylan, "The stability of nonlinear dissipative systems," *IEEE Trans. on Automatic Control*, vol. 21, pp. 708–711, 1976.

[4] A. Isidori, *Nonlinear Control Systems*, (3rd ed.), Springer-Verlag, London, UK, 1995.

[5] R. Kelly, R. Carelli, and R. Ortega, "Adaptive motion control design of robot manipulators: An input–output approach," *Int. J. of Control*, vol. 50, pp. 2563–2581, 1989.

[6] H.K. Khalil, *Nonlinear Systems*, (2nd ed.), Prentice-Hall, Englewood Cliffs, NJ, 1996.

REFERENCES

[7] P.V. Kokotović, H.K. Khalil, and J. O'Reilly, *Singular Perturbation Methods in Control: Analysis and Design*, Academic Press, New York, NY, 1986.

[8] J. La Salle, S. Lefschetz, *Stability by Liapunov's Direct Method*, Academic Press, New York, NY, 1961.

[9] S. Lefschetz, *Stability of Nonlinear Control Systems*, Academic Press, New York, NY, 1962.

[10] A.M. Lyapunov, *The General Problem of Motion Stability*, in Russian, 1892; translated in French, Ann. Faculté des Sciences de Toulouse, pp. 203–474, 1907.

[11] H. Nijmeijer and A.J. van der Schaft, *Nonlinear Dynamical Control Systems*, Springer-Verlag, New York, 1990.

[12] R. Ortega, "Applications of input–output techniques to control problems," *Proc. 1st European Control Conf.*, Grenoble, F, pp. 1307–1315, 1991.

[13] R. Ortega and M.W. Spong, "Adaptive motion control of rigid robots: a tutorial," *Automatica*, vol. 25, pp. 877–888, 1989.

[14] A. Rantzer, "A note on the Kalman-Yacubovich-Popov lemma," *Proc. 3rd European Control Conf.*, Rome, I, pp. 1792–1795, 1995.

[15] J.-J.E. Slotine and W. Li, *Applied Nonlinear Control*, Prentice-Hall, Englewood Cliffs, NJ, 1991.

[16] E. Sontag, "State-space and I/O stability for nonlinear systems," in *Feedback Control, Nonlinear Systems, and Complexity*, Lecture Notes in Control and Information Science, vol. 202, B. Francis and A. Tannenbaum (Eds.), Springer-Verlag, Berlin, D, pp. 215–235, 1995.

[17] M. Vidyasagar, *Nonlinear Systems Analysis*, (2nd ed.), Prentice-Hall, Englewood Cliffs, NJ, 1993.

[18] J.C. Willems, *The Analysis of Feedback Systems*, MIT Press, Cambridge, MA, 1971.

[19] J.C. Willems, "Dissipative dynamical systems, part I: general theory," *Arch. Rational Mechanics Analysis*, vol. 45, pp. 321–351, 1972.

[20] G. Zames, "On the input–output stability of nonlinear time-varying feedback systems, part I," *IEEE Trans. on Automatic Control*, vol. 11, pp. 228–238, 1966.

[21] G. Zames, "On the input–output stability of nonlinear time-varying feedback systems, part II," *IEEE Trans. on Automatic Control*, vol. 11, pp. 465–477, 1966.

Index

actuator configuration, 296
adaptive control, 95
adaptive gravity compensation, 95
adaptive inverse dynamics control, 98
adaptive passivity-based control, 101
analytical Jacobian, 19, 128, 131, 143
anthropomorphic manipulator, 9, 14, 18, 19, 25, 35
assumed modes, 230
asymptotic stability, 364, 367
augmented task space, 126
autonomous system, 363

Barbalat's lemma, 370
base dynamic parameters, 30, 34, 45, 47
base frame, 6
base kinematic parameters, 43, 44

car-like robot, 275, 302
castor wheel, 268
centrifugal forces, 22
chained system, 333
Christoffel symbols, 22
clamped-free model, 230
composite control, 199, 248
configuration coordinates, 270

configuration dynamic model, 294, 295
configuration kinematic model, 290
constrained dynamics, 157
constrained mode, 224
contact with environment, 141, 167
controllability, 289
conventional wheel, 267
converse Lyapunov theorems, 370
coordinate projection, 349
Coriolis forces, 22

damped least-squares inverse, 120, 123
damping matrix, 145, 229
decrescent function, 369
degree of freedom, 8, 12, 13
degree of manoeuvrability, 286
degree of mobility, 272, 286
degree of nonholonomy, 291
degree of steerability, 273
Denavit-Hartenberg notation, 6
diffeomorphism, 375
differential geometry theory, 374
differential kinematics, 16, 19, 124, 133
differential kinematics inversion, 116
differentially flat system, 321
direct dynamics, 29

direct kinematics, 5, 231
direct task space control, 131
disturbance, 68, 81, 82
dynamic extension algorithm, 319
dynamic model, 21, 25, 26, 45, 143, 233
dynamic model properties, 61, 184, 235
dynamic parameters, 24, 29, 44
dynamic state feedback, 202, 308, 318

elastic joint, 179
end effector, 5, 11
end-effector force, 24
end-effector frame, 6
end-effector position and orientation, 12, 19
end-effector velocity, 16
energy model, 47
equations of motion, 22
equilibrium, 364, 367
Euler-Bernoulli beam equations, 221
exponential stability, 364, 367, 372
extended space, 380

feedback linearization, 376
finite-dimensional model, 228
first moment of inertia, 23
fixed wheel, 267
flexible link, 220, 221
flexible manipulators, 177
force control, 141
frequency domain inversion, 250
Frobenius theorem, 375
front-wheel drive robot, 316
function of class $\mathcal{K}$, 368
function spaces, 379

generalized coordinates, 21

geometric Jacobian, 16, 19, 23, 41, 116, 129, 143
global asymptotic stability, 367
global stability, 366
gravity compensation, 71, 131

Hamiltonian, 30
hybrid force/motion control, 156
hybrid piecewise control, 351
hybrid task specification, 166

identification of dynamic parameters, 44
identification of kinematic parameters, 38
impedance control, 142
inertia matrix, 22, 61, 145
inertia tensor, 23
input–output theory, 378
instantaneous center of rotation, 272
integral action, 68, 81, 83, 84, 151
invariant set, 366
inverse dynamics, 29
inverse dynamics control, 73, 81, 133, 145, 161, 311
inverse kinematics, 12
inverse kinematics algorithm, 124
inversion control, 242
irreducibility, 287

Jacobian, 16
Jacobian pseudoinverse, 125, 129
Jacobian transpose, 125, 127
joint flexibility, 179
joint space, 11, 12
joint space control, 59
joint variable, 6, 11, 12, 21, 40, 42
joints, 5

INDEX

𝒦-exponential stability, 369
Kalman-Yakubovich-Popov lemma, 384
kinematic calibration, 42
kinematic control, 116
kinematic model, 4
kinematic parameters, 6, 38, 40, 42
kinetic energy, 21, 183, 233

La Salle's invariant set theorem, 366
Lagrange formulation, 21, 233, 294
Lie bracket, 374
Lie derivative, 374
linear feedback control, 336, 341
linearity in the parameters, 24, 63
link flexibility, 219
link variable, 190
links, 5
local stability, 366
Lyapunov-based control, 74, 82, 133
Lyapunov direct method, 365, 368
Lyapunov equation, 365
Lyapunov function, 366
Lyapunov linearization method, 364, 368
Lyapunov stability, 369
Lyapunov theory, 363

marginal stability, 364
mass, 23
mobile robots, 263
modal analysis, 224
model parameter uncertainty, 84
model transformation, 332
motor variable, 190

negative definiteness, 365

negative semi-definiteness, 365
Newton-Euler formulation, 26
nonautonomous system, 367
nonlinear feedback control, 337, 341
nonlinear regulation, 209, 252
normal form, 375

omnidirectional robot, 274, 277, 289, 292, 296, 297, 310
operator, 380
orientation coordinates, 270

parallel control, 150
parameter drift, 107
parameter identifiability, 42
passivity, 380
passivity-based control, 78, 84
path following, 339, 348
PD control, 63, 95, 131, 147, 191, 237
persistent excitation, 46
PID control, 68, 153
piecewise continuous control, 349, 354
point tracking, 308
positive definite function, 369
positive definiteness, 365
positive semi-definiteness, 365
posture coordinates, 270
posture dynamic model, 300, 320
posture kinematic model, 282, 283, 320
posture stabilization, 343
posture tracking, 308, 325, 335, 344
potential energy, 21, 183, 233
practical stability, 371
prismatic joint, 5
pseudoinverse, 42, 45, 117

reduced model, 186

redundancy, 12, 118, 125, 126, 130
regressor, 25, 40
regulation, 63, 131, 153, 189, 237
relative degree, 197, 243, 375
restricted mobility robot, 289, 312
revolute joint, 5
rigid manipulators, 1
robust control, 80, 84
robust inverse dynamics control, 84
robust passivity-based control, 91
rotation coordinates, 270

saturating control, 80
singular perturbation theory, 371
singular value decomposition, 20, 117, 118, 120, 122, 123, 126
singularity, 19, 118, 119, 121, 124, 126, 127, 322
singularly perturbed model, 187, 247
skew-symmetry of matrix $\dot{H} - 2C$, 22, 61, 184
sliding mode control, 80
smooth state feedback, 334
smooth time-varying control, 344
spherical wrist, 13, 14
stability, 364, 367
stabilizability, 289
stabilization of feedback linearizable systems, 377
static state feedback, 196, 290, 308, 310
steering wheel, 268
stiffness control, 149
stiffness matrix, 145, 148, 183, 229

Swedish wheel, 267, 269

task frame, 167
task priority strategy, 127
task space, 11, 12, 19, 40
task space control, 115
three-wheel robot, 276
Tikhonov's theorem, 373
time-invariant system, 363
time-varying system, 364
torque control, 309
tracking control, 72, 133, 195, 242, 248
truncated inner product, 380
truncation, 379
two-time scale control, 199, 246
Type (1,1) robot, 275, 280, 285, 299, 316
Type (1,2) robot, 275, 281, 285, 299, 317
Type (2,0) robot, 274, 278, 284, 293, 297, 316, 320
Type (2,1) robot, 275, 279, 284, 298, 316
Type (3,0) robot, 274, 277, 284, 292, 296, 297

ultimate boundedness, 91, 371
unconstrained mode, 224
unicycle robot, 278, 331
uniform asymptotic stability, 367
uniform stability, 367
user-defined accuracy, 123

velocity control, 310
velocity scaling, 337
vibration damping control, 240

wheeled mobile robot, 265
wheels, 266

zero dynamics, 130, 196, 204, 244, 249, 378

DATE DE RETOUR L.-Brault

15 MAR. 2007